Generalized Linear Models and Extensions

Second Edition

Generalized Linear Models and Extensions

Second Edition

James W. Hardin
Department of Epidemiology and Biostatistics
University of South Carolina

Joseph M. Hilbe
Department of Sociology and Statistics
Arizona State University

A Stata Press Publication
StataCorp LP
College Station, Texas

Stata Press, 4905 Lakeway Drive, College Station, Texas 77845

 First edition 2001
Second edition 2007
Typeset in $\LaTeX 2_\varepsilon$
Printed in the United States of America
10 9 8 7 6 5 4 3 2 1

ISBN-10: 1-59718-014-9
ISBN-13: 978-1-59718-014-6

For our

wives

Mariaelena Castro-Hardin
Cheryl Hilbe

and

children

Taylor Hardin and Conner Hardin
Michael Hilbe, Mitchell Hilbe,
and Heather Hilbe O'Meara

who were affected by our time away preparing this book (again).

Contents

Tables

Figures

Listings

Preface

This second edition of *Generalized Linear Models and Extensions* is written for the active researcher as well as for the theoretical statistician. Our goal has been to clarify the nature and scope of generalized linear models (GLMs) and to demonstrate how all the families, links, and variations of GLMs fit together in an understandable whole.

In a step-by-step manner, we detail the foundations and provide working algorithms that readers can use to construct and better understand models that they wish to develop. In a sense, we offer readers a workbook or handbook of how to deal with data using GLM and GLM extensions.

Many people have contributed to the ideas presented in the new edition of this book. John Nelder has been the foremost influence. Other important and influential people include Peter Bruce, David Collett, David Hosmer, Stanley Lemeshow, James Lindsey, J. Scott Long, Roger Newson, Scott Zeger, Kung-Yee Liang, Raymond J. Carroll, H. Joseph Newton, Henrik Schmiediche, Norman Breslow, Berwin Turlach, Gordon Johnston, Thomas Lumley, Bill Sribney, Vince Wiggins, Mario Cleves, William Greene, and many others. We also thank William Gould, president of StataCorp, for his encouragement in this project. His statistical computing expertise and his contributions to statistical modeling have had a deep impact on this book.

We also thank StataCorp's editorial staff for their equanimity in reading and editing our manuscript, especially Roberto Gutierrez, Patricia Branton, and Lisa Gilmore for their insightful and patient contributions in this area.

Stata Press allowed us to dictate some of the style of this text. In writing this material in other forms for short courses, we have always included equation numbers for all equations rather than only for those equations mentioned in text. Although this is not the standard editorial style for textbooks, we enjoy the benefits of students' being able to more easily (and efficiently) communicate questions and comments on all parts of the material. We hope that readers will find this practice as beneficial as our short-course participants have found it.

James Hardin
Joseph Hilbe

December 2006

1 Introduction

Contents

In updating this text, our primary goal is to convey the practice of analyzing data via generalized linear models to researchers across a broad spectrum of scientific fields. To accomplish this goal, we lay out the framework used for describing various aspects of data and for communicating tools for data analysis. This part of the text contains no examples. Rather, we focus on the lexicon that we will use in later chapters. These later chapters use examples from such fields of interest as biostatistics, economics, and survival analysis.

In developing analysis tools, we illustrate techniques via their genesis in estimation algorithms. We believe that motivating the discussion through the estimation algorithms clarifies the origin and usefulness of the techniques. Instead of detailed theoretical exposition, we refer to texts and papers that present this type of material so that we may focus our detailed presentations on the algorithms and their justification. Our detailed presentations are mostly algebraic; we have minimized matrix notation whenever possible. We also provide a list of the statistics and tests associated with this area along with their formulas and utility.

1.1 Origins and motivation

We wrote this text for researchers who want to understand the theory, scope, and application of generalized linear models. For brevity's sake, we use the acronym GLM to refer to the *generalized linear model*, but we acknowledge that GLM has been used as an acronym for the *general linear model*. The latter usage, of course, refers to the area of statistical modeling that is based solely on the normal or Gaussian probability distribution. GLZ is another term referring to generalized linear models found in some software packages.

We take GLM to be the generalization of the general, for that is precisely what GLMs are; they are the result of extending ordinary least squares (OLS) regression, or the normal model, to a model that is appropriate for a variety of response distributions. We examine exactly how this extension is accomplished. We also aim to provide the reader with a firm understanding of how GLMs are to be evaluated and when their use is appropriate. We even advance a bit beyond the traditional GLM and give the reader a look at how GLMs can be extended to model certain types of data that do not fit exactly within the GLM framework.

Nearly every text that addresses a statistical topic uses one or more statistical computing packages to calculate and display results. We use Stata exclusively, though we do refer to other software packages.

The individual statistical models that make up GLMs are often found as standalone software modules, typically fitted using maximum likelihood methods based on quantities from model-specific derivations. Stata has several such commands for specific GLMs such as `poisson`, `logistic`, and `regress`. Some of these procedures were included in the Stata package from its first version. More models have been addressed through commands written by users employing Stata's programming language that has led to the creation of highly complex statistical models. Most of these user-written commands have since been incorporated into the official Stata package. Stata's `glm` command was itself created as a user program, first in 1993, and then 3 years later as part of Stata 6.0. The `glm` command, which we wrote to augment the first version of this text, was adopted and is now the official Stata `glm` command distributed with the commercial software; the software company maintains and supports the command. Examples in this edition reflect StataCorp's updates to the command.

Stata already offers many of the statistical models that are discussed in this text. Moreover, Stata allows the user to write complex statistical algorithms for themselves. You will see many examples throughout this text, including the `glm` command.

Readers of technical books often need to know about prerequisites, especially how much math and statistics background is required. To gain full advantage from the text and follow its every statement and algorithm, you should have an understanding equal to a two-semester calculus-based course on statistical theory. Without a background in statistical theory, the reader can accept the presentation of the theoretical underpinnings and follow the (mostly) algebraic derivations that do not require more than a mastery of simple derivatives. We assume prior knowledge of multiple regression but no other specialized knowledge is required.

We believe that you can best understand GLMs if their computational basis has been made clear. Hence, we begin our exposition with an explanation of the foundations and computation of GLMs; there are two major methodologies for developing algorithms. We then show how simple changes to the base algorithms lead to different GLM families, links, and even further extensions. In short, we attempt to lay the GLM open to inspection and to make every part of it as clear as possible. In this fashion, the reader can understand exactly how and why GLM algorithms can be used, as well as altered, to better model a desired dataset.

Perhaps more than any other text in this area, we examine alternatively two major computational GLM algorithms:

1. Iteratively reweighted least squares (and modifications)
2. Newton–Raphson (and modifications)

Interestingly, some of the models we present can be calculated only by using one of the above methods. Iteratively reweighted least squares (IRLS) is the more specialized technique and is applied less often. Yet it is typically the algorithm of choice for quasi-likelihood models such as generalized estimating equations (GEEs). On the other hand, truncated models that do not fit neatly into the exponential family of distributions are modeled using Newton–Raphson methods—and for this, too, we show why. Once again, focusing on the details of calculation should help the reader understand both the scope and the limits of a particular model.

Whenever possible, we present the log likelihood for the model under discussion. In writing the log likelihood, we include offsets so that interested programmers can see how those elements enter estimation. In fact, we attempt to offer programmers the ability to understand and write their own working GLMs, plus many useful extensions. As programmers ourselves, we believe that there is value in such a presentation; we would have much enjoyed having it at our fingertips when we first entered this statistical domain.

1.2 Notational conventions

We use L to denote the likelihood and the script $\mathcal{L}$ to denote the log likelihood. We use X to denote the design matrix of independent (explanatory) variables. When appropriate, we use boldface type $\mathbf{X}$ to emphasize that we are referring to a matrix; a lowercase letter with a subscript $\mathbf{x}_i$ will refer to the ith row from this matrix.

We use Y to denote the dependent (response) variable and refer to the vector $\boldsymbol{\beta}$ as the coefficients of the design matrix. We will use $\widehat{\boldsymbol{\beta}}$ when we wish to discuss or emphasize the fitted coefficients. Throughout the text, we will discuss the role of the (vector) linear predictor $\eta = \mathbf{X}\boldsymbol{\beta}$. In generalizing this concept, we will also refer to the augmented (by an offset) version of the linear predictor $\xi = \eta + \text{offset}$.

Finally, we will use the $\mathrm{E}(\cdot)$ notation to refer to the expectation of a random variable and the $\mathrm{V}(\cdot)$ notation to refer to the variance of a random variable. We will describe other notational conventions at the time of their first use.

1.3 Applied or theoretical?

A common question regarding texts concerns their focus. Is the text applied or theoretical? Our text is both. However, we would argue that it is basically applied. We show enough technical details for the theoretician to understand the underlying basis of GLMs. However, we believe that understanding the use and limitations of a GLM includes understanding its estimation algorithm. For some, dealing with formulas and algorithms appears thoroughly theoretical. We believe that it aids understanding the scope and limits of proper application. Perhaps we can call the text a bit of both and not worry about classification. In any case, for those who fear formulas, each formula and algorithm is thoroughly explained and that by book's end the formulas and algorithms will seem simple and meaningful. For completeness, we give the reader references to texts that discuss more advanced topics and theory.

1.4 Road map

Part I of the text deals with the basic foundations of GLM. We detail the various components of GLM, including various family, link, variance, deviance, and log-likelihood functions. We also provide a thorough background and detailed particulars of both the Newton–Raphson and IRLS algorithms. The chapters that follow highlight this discussion, which describes the framework through which the models of interest arise.

We also give the reader an overview of GLM residuals, introducing some that are not widely known, but that nevertheless can be extremely useful for analyzing a given model's worth. Finally, in part I we discuss the general notion of goodness of fit and provide a framework through which you can derive more extensions to GLM.

Part II concerns the continuous family of distributions, including the Gaussian, gamma, inverse Gaussian, and power families. We derive the related formulas and relevant algorithms for each family and then discuss the ancillary or scale parameters appropriate to each model. We also examine noncanonical links and generalizations to the basic model. Finally, we give examples, showing how a given dataset may be analyzed using each family and link. We give examples dealing with model application, including discussion of the appropriate criteria for the analysis of fit. We have expanded the number of examples in this new edition to highlight both model fitting and assessment.

Part III deals with binomial response models. It includes exposition of the general binomial model and of the various links. Major links described include the canonical logit, as well as the noncanonical links probit, log-log, and complementary log-log. We also cover other links. We present examples and criteria for analysis of fit throughout. This new edition includes extensions to generalized binomial regression resulting from a special case of building a regression model from the generalized negative binomial probability function.

We also give considerable space to overdispersion. We discuss the problem's nature, how it is identified, and how it can be dealt with in the context of discovery and analysis.

Pursuant to the latter, we explain how to adjust the binomial model to accommodate overdispersion. You can accomplish this task by internal adjustment to the base model, or you may need to reformulate the base model itself. We also introduce methods of adjusting of the variance–covariance matrix of the model to produce robust standard errors. The problem of dealing with overdispersion continues in the chapters on data.

Part IV concerns count response data. We include examinations of the Poisson, the geometric, and the negative binomial models. With respect to the negative binomial, we show how the standard models can be further extended to derive a class called heterogeneous negative binomial models. There are several "brands" of negative binomial, and it is wise for the researcher to know how each is best used. The distinction of these models is typically denoted NB-1 and NB-2 and relates to the variance-to-mean ratio of the resulting derivation of the model. We have updated this discussion to include the generalized Poisson regression model, which is similar to NB-1.

Part V deals with the categorical response regression models. Typically considered extensions to the basic GLM algorithm, categorical response models are divided into two general varieties: unordered responses, also known as multinomial models, and ordered responses models. We begin by considering ordered-response models. In such models, the discrete number of outcomes are ordered, but the integer labels applied to the ordered levels of outcome are not necessarily equally spaced. A simple example is the set of outcomes "bad", "average", and "good". We also cover unordered multinomial responses, whose outcomes are given no order. For an example of an unordered outcome, consider choosing the type of entertainment that is available for an evening. The following choices may be given as "movie", "restaurant", "dancing", or "reading". Ordered response models are themselves divisible into two varieties: 1) ordered binomial including ordered logit, ordered probit, ordered complementary log-log, or ordered log-log and 2) the generalized ordered binomial model with the same links as the nongeneralized parameterization. We have expanded our discussion to include more ordered-outcome models, generalized ordered-outcome models, and continuation ratio models.

Finally, part VI is about extensions to the GLM family of models. In particular, we examine the following models:

1. Fixed-effects models
2. Random-effects models
3. Quasilikelihood models
4. GEEs
5. Generalized additive models

We attempt throughout to give the reader a thorough outline or overview of GLMs. We have attempted to cover nearly every major development in the area. We have also tried to show the direction in which statistical modeling is moving, hence laying a foundation for future research and for ever-more-appropriate GLMs. Moreover, we have expanded each section of the original version of this text to bring new and expanded regression models into focus. Our attempt, as always, is to illustrate these new models within the context of the GLM.

1.5 Installing the support materials

All the data used in this book are freely available for you to download from the Stata Press web site, http://www.stata-press.com. In fact, when we introduce new datasets, we merely load them into Stata the same way that you would. For example,

```
. use http://www.stata-press.com/data/hh2/medpar
```

To download the datasets, do-files, and user-written commands for this book, type

```
. net from http://www.stata-press.com/data/hh2/
. net install glme2-ado1
. net install glme2-ado2
. net get glme2-data
```

The user-written commands will be automatically installed for your copy of Stata. The datasets and do-files will be downloaded to your current working directory. We suggest that you create a new directory into which the materials will be downloaded.

Part I

Foundations of Generalized Linear Models

2 GLMs

Contents

Nelder and Wedderburn (1972) introduced the theory of GLMs. They discovered an underlying unity for an entire class of regression models. This class consisted of models whose single-response variable, the variable that the model is to explain, is hypothesized to follow a member of the single-parameter exponential family of probability distributions. This family of distributions includes the Gaussian or normal, binomial, Poisson, gamma, inverse Gaussian, geometric, and negative binomial.

To establish a basis, we begin discussion of GLMs by initially recalling important results on linear models, specifically those results for linear regression. The standard linear regression model relies on several assumptions, among which are the following:

1. Each observation of the response variable is characterized by the normal or Gaussian distribution; $y_i \sim \text{N}(\mu_i, \sigma_i^2)$.
2. The distributions for all observations have a common variance; $\sigma_i^2 = \sigma^2$ for all i.
3. There is a direct or "identical" relationship between the linear predictor (linear combination of covariate values and associated parameters) and the expected values of the model; $\mathbf{x}_i\boldsymbol{\beta} = \mu_i$.

The purpose of GLMs, and the linear models that they generalize, is to specify the relationship between the observed response variable and some number of covariates. The outcome variable is viewed as a realization from a random variable.

Nelder and Wedderburn discovered that general models could be developed by relaxing the assumptions of the linear model. By restructuring the relationship between the linear predictor and the fit, we can "linearize" relationships that initially seem to be nonlinear. Nelder and Wedderburn accordingly dubbed the models that allowed this type of restructuring "generalized linear models".

Most models that were placed under the GLM framework were well established and popular—some more than others. However, these models had historically been fitted using individually developed maximum likelihood (ML) algorithms. ML algorithms, as we will call them, can be hard to implement. Starting or initial estimates for parameters must be provided, and considerable work is required to derive model-specific quantities to ultimately obtain estimates and their standard errors. We will later show exactly how much effort is involved.

Ordinary least squares (OLS) extends maximum-likelihood linear regression such that the properties of OLS estimates depend only on the assumptions of constant variance and independence. Maximum-likelihood linear regression carries the more restrictive distributional assumption of normality. Similarly, although we may derive likelihoods from specific distributions in the exponential family, the second-order properties of our estimates are shown to depend only on the assumed mean–variance relationship and on the independence of the observations rather than on a more restrictive assumption that observations follow a particular distribution.

The classical linear model assumes that the observations that our dependent variable y represents are independent normal variates with constant variance σ^2. Also covariates are related to the expected value of the independent variable such that

$$\mathrm{E}(y) = \mu \tag{2.1}$$

$$\mu = \mathbf{X}\boldsymbol{\beta} \tag{2.2}$$

This last equation shows the "identical" or identity relationship between the linear predictor $\mathbf{X}\boldsymbol{\beta}$ and the mean μ.

Whereas the linear model conceptualizes the outcome y as the sum of its mean μ and a random variable ϵ, Nelder and Wedderburn linearized each GLM family member by means of a link function. They then altered a previously used algorithm called *iterative weighted least squares*, which was used in estimating weighted-least squares regression models. Aside from introducing the link function relating the linear predictor to the fitted values, they also introduced the variance function as an element in the weighting of the regression. The iterations of the algorithm updated parameter estimates to produce appropriate linear predictors, fitted values, and standard errors. We will clarify exactly how all this falls together in the section on the IRLS algorithm.

The estimation algorithm allowed researchers to more easily estimate many models previously considered to be nonlinear by restructuring them into GLMs. Later, it was discovered that an even more general class of linear models results from more relaxations of assumptions for GLMs.

However, even though the historical roots of GLMs are based on IRLS methodology, many generalizations to the linear model still require Newton–Raphson techniques common to maximum likelihood methodologies. We take the position here that GLMs should not be constrained to those models first discussed by Nelder and Wedderburn but rather that they encompass all such linear generalizations to the standard model.

Many other books and journal articles followed the cornerstone article by Nelder and Wedderburn (1972) as well as the text by McCullagh and Nelder (1989) (the original text was published in 1983). Lindsey (1997) illustrates the application of GLMs to biostatistics, most notably focusing on survival models. Hilbe (1994) gives an overview of the GLM and its support from various software packages. Software was developed early on. In fact, Nelder was instrumental in developing the first statistical program based entirely on GLM principles—generalized linear interactive modeling (GLIM). Published by the Numerical Algorithms Group (NAG), the software package has been widely used since the mid-1970s. Other vendors began offering GLM capabilities in the 1980s, including GENSTAT and S-Plus. Stata and SAS included it in their software offerings in 1993 and 1994, respectively.

This text covers much of the same foundation material as other books. What distinguishes our presentation of the material is twofold. First, we focus on the estimation of various models via the estimation technique. Second, we present our derivation of the methods of estimation at a more widely accessible level than that found in other sources. In fact, where possible, we present complete algebraic derivations that include nearly every step in the illustrations. Pedagogically, we have found that this manner of exposition imparts a more solid understanding and "feel" of the area than do other approaches. The idea is this: if you can write your own GLM, then you are probably more able to know how it works, when and why it does not work, and how it is to be evaluated. Of course, we also discuss methods of fit assessment and testing. To model data without subjecting them to evaluation is like taking a test without checking the answers. Hence, we will spend considerable time dealing with model evaluation as well as algorithm construction.

2.1 Components

Cited in various places such as Hilbe (1993a) and Francis, Green, and Payne (1993), GLMs are characterized by an expanded itemized list given by the following:

1. A random component for the response, $\mathbf{y}$, which has a distribution following the exponential family.
2. A linear systematic component relating the linear predictor, $\eta = \mathbf{X}\boldsymbol{\beta}$, to the product of the design matrix $\mathbf{X}$ and the parameters $\boldsymbol{\beta}$.
3. A known monotonic, one-to-one, differentiable link function $g(\cdot)$ relating the linear predictor to the fitted values. Since the function is one-to-one, there is an inverse function relating the mean expected response, $\mathrm{E}(y) = \mu$, to the linear predictor such that $\mu = g^{-1}(\eta) = \mathrm{E}(y)$.
4. The variance may change with the covariates only as a function of the mean.
5. There is one IRLS algorithm that suffices to fit all members of the class.

Item 5 is of special interest. The traditional formulation of the theory certainly supposed that there was one algorithm that could fit all GLMs. We will see later how this was implemented. However, there have been extensions to this traditional viewpoint. Adjustments to the weight function have been added to more closely match usual Newton–Raphson algorithms and so that more appropriate standard errors may be calculated for noncanonical link models. Such features as scaling and robust variance estimators have also been added to the basic algorithm. More importantly, sometimes a traditional GLM must be restructured and fitted using a model-specific Newton–Raphson algorithm. Of course, one may simply define a GLM as a model requiring only the standard approach, but doing so would severely limit the range of possible models. We prefer to think of a GLM as a model that is ultimately based on the probability function belonging to the exponential family of distributions, but with the proviso that this criterion may be relaxed to include quasilikelihood models as well as certain types of multinomial, truncated, censored, and inflated models. Most of the latter type require a Newton–Raphson approach rather than the traditional IRLS algorithm.

Early GLM software development constrained GLMs to those models that could be fitted using the originally described estimation algorithm. As we will illustrate, the traditional algorithm is relatively simple to implement and requires little computing power. In the days when RAM was scarce and expensive, this was an optimal production strategy for software development. Because this is no longer the case, a wider range of GLMs can more easily be fitted using a variety of algorithms. We will discuss these implementation details at length.

In the classical linear model, the observations of the dependent variable $\mathbf{y}$ are independent normal variates with constant variance σ^2. We assume that the mean value of $\mathbf{y}$ may depend on other quantities (predictors) denoted by the column vectors $\mathbf{X}_1, \mathbf{X}_2, \ldots, \mathbf{X}_{p-1}$. In the simplest situation, we assume that this dependency is linear and write

$$\mathrm{E}(\mathbf{y}) = \beta_0 + \beta_1\mathbf{X}_1 + \cdots + \beta_{p-1}\mathbf{X}_{p-1} \tag{2.3}$$

and attempt to estimate the vector $\boldsymbol{\beta}$.

GLMs specify a relationship between the mean of the random variable $\mathbf{y}$ and a function of the linear combination of the predictors. This generalization admits a model specification allowing for continuous or discrete outcomes and allows a description of the variance as a function of the mean.

2.2 Assumptions

The *link function* relates the mean $\mu = \mathrm{E}(y)$ to the linear predictor $\mathbf{X}\boldsymbol{\beta}$ and the *variance function* relates the variance as a function of the mean $\mathrm{V}(y) = a(\phi)\mathrm{V}(\mu)$, where $a(\phi)$ is the scale factor. For the Poisson, binomial, and negative binomial variance models, $a(\phi) = 1$.

Breslow (1996) points out that the critical assumptions in the GLM framework may be stated as follows:

1. Statistical independence of the n observations.
2. The variance function $\mathrm{V}(\mu)$ is correctly specified.
3. $a(\phi)$ is correctly specified (1 for Poisson, binomial, and negative binomial).
4. The link function is correctly specified.
5. Explanatory variables are of the correct form.
6. There is no undue influence of the individual observations on the fit.

As a simple illustration, below we demonstrate the effect of the assumed variance function on the model and fitted values of a simple GLM.

Table 2.1: Predicted values for various choices of variance function

Observed (y)	1.00	2.00	9.00	
Predicted (Normal: $\mathrm{V}(\mu) = \phi$)	0.00	4.00	8.00	$\widehat{y} = -4.00 + 4.00x$
Predicted (Poisson: $\mathrm{V}(\mu) = \mu$)	0.80	4.00	7.20	$\widehat{y} = -2.40 + 3.20x$
Predicted (Gamma: $\mathrm{V}(\mu) = \phi\mu^2$)	0.94	3.69	6.43	$\widehat{y} = -1.80 + 2.74x$
Predicted (Inverse Gaussian: $\mathrm{V}(\mu) = \phi\mu^3$)	0.98	3.33	5.69	$\widehat{y} = -1.37 + 2.35x$

NOTE: The models are all fitted using the identity link and the data consist of 3 observations $(y, x) = \{(1, 1), (2, 2), (9, 3)\}$. The fitted models are included in the last column.

2.3 Exponential family

GLMs are traditionally formulated within the framework of the exponential family of distributions. In this representation, we can derive a general model that may be fitted using the scoring process (IRLS) detailed in section 3.3. Many people confuse the estimation method with the class of GLMs. This is a mistake because there are many estimation methods. Some software implementations allow specification of more diverse models than others. We will point this out throughout the text.

The exponential family is usually (there are other algebraically equivalent forms in the literature) written

$$f_y(y; \theta, \phi) = \exp\left\{\frac{y\theta - b(\theta)}{a(\phi)} + c(y, \phi)\right\} \tag{2.4}$$

where θ is the canonical (natural) parameter and ϕ is the scale required to produce standard errors following a distribution in the exponential family of distributions including Gaussian, gamma, inverse Gaussian, and others. In fact, the exponential family provides a notation that allows us to model continuous, discrete, proportional, count, and binary outcomes.

In the exponential family presentation, we construe each of the y_i observations as being defined in terms of the parameters θ. Because the observations are independent, the joint density of the sample of observations y_i, given parameters θ and ϕ, is defined by the product of the density over the individual observations (review sec. 2.2). Interested readers can review Barndorff-Nielsen (1976) for the theoretical justification that allows this factorization:

$$f_{y_1,y_2,\ldots,y_n}(y_1, y_2, \ldots, y_n; \theta, \phi) = \prod_{i=1}^{n} \exp\left\{\frac{y_i\theta_i - b(\theta_i)}{a(\phi)} + c(y_i, \phi)\right\} \tag{2.5}$$

Conveniently, the joint probability density function may be expressed as a function of θ and ϕ given the observations y_i. This function is called the likelihood, L, and is written as

$$L(\theta, \phi; y_1, y_2, \ldots, y_n) = \prod_{i=1}^{n} \exp\left\{\frac{y_i\theta_i - b(\theta_i)}{a(\phi)} + c(y_i, \phi)\right\} \tag{2.6}$$

We wish to obtain estimates of (θ, ϕ) that maximize the likelihood function. Given the product in the likelihood, it is more convenient to work with the log likelihood,

$$\mathcal{L}(\theta, \phi; y_1, y_2, \ldots, y_n) = \sum_{i=1}^{n} \left\{\frac{y_i\theta_i - b(\theta_i)}{a(\phi)} + c(y_i, \phi)\right\} \tag{2.7}$$

since the values that maximize the likelihood are the same values that maximize the log likelihood.

Throughout the text, we will derive each distributional family member from the exponential family notation so that the components are clearly illustrated. The log likelihood for the exponential family is in a relatively basic form, admitting simple calculations of first and second derivatives for maximum likelihood estimation. The IRLS algorithm takes advantage of this form of the log likelihood.

First, we generalize the log likelihood to include an offset to the linear predictor. This generalization will allow us to investigate simple equality constraints on the parameters.

The idea of an offset is simple. To fit models with covariates, we specify that θ is a function of specified covariates, $\mathbf{X}$, and their associated coefficients, $\boldsymbol{\beta}$. Within the linear combination of the covariates and their coefficients $\mathbf{X}\boldsymbol{\beta}$, we may further wish to constrain a particular subset of the coefficients β_i to particular values. For example, we may know or wish to test that $\beta_3 = 2$ in a model with a constant, X_0, and three covariates X_1, X_2, and X_3. If we wish to enforce the $\beta_3 = 2$ restriction on the estimation, then we will want the optimization process to calculate the linear predictor as

$$\eta = \widehat{\beta}_0 + \widehat{\beta}_1 X_1 + \widehat{\beta}_2 X_2 + 2X_3 \tag{2.8}$$

at each step. We know (or wish to enforce) that the linear predictor is composed of a linear combination of the unrestricted parameters plus two times the X_3 covariate. If we consider that the linear predictor is generally written as

$$\eta = \mathbf{X}\boldsymbol{\beta} + \text{offset} \tag{2.9}$$

then we can appreciate the implementation of a program that allows an offset. We could generate a new variable equal to two times the variable containing the X_3 observations and specify that generated variable as the offset. By considering this issue from the outset, we can include an offset in our derivations, which will allow us to write programs that include this functionality.

The offset is a given (nonstochastic) component in the estimation problem. By including the offset we gain the ability to fit (equality) restricted models without adding unnecessary complexity to the model: the offset plays no role in derivative calculations. If we do not include an offset in our derivations and subsequent programs, we can still fit restricted models, but the justification is not as clear; see the arguments of Nyquist (1991) for obtaining restricted estimates in a GLM.

2.4 Example: Using an offset in a GLM

Here we show the use of an offset with Stata's `glm` command. From an analysis presented in chapter 12, consider the output of the following model:

```
. use http://www.stata-press.com/data/hh2/medpar
. glm los hmo white type2 type3, family(poisson) link(log) nolog
Generalized linear models                          No. of obs      =       1495
Optimization     : ML                              Residual df     =       1490
                                                   Scale parameter =          1
Deviance         =  8142.666001                    (1/df) Deviance =   5.464877
Pearson          =  9327.983215                    (1/df) Pearson  =   6.260391
Variance function: V(u) = u                        [Poisson]
Link function    : g(u) = ln(u)                    [Log]
                                                   AIC             =   9.276131
Log likelihood   = -6928.907786                    BIC             =  -2749.057

------------------------------------------------------------------------------
             |                 OIM
         los |      Coef.   Std. Err.      z    P>|z|     [95% Conf. Interval]
-------------+----------------------------------------------------------------
         hmo |  -.0715493    .023944    -2.99   0.003    -.1184786    -.02462
       white |   -.153871   .0274128    -5.61   0.000    -.2075991   -.100143
       type2 |   .2216518   .0210519    10.53   0.000     .1803908   .2629127
       type3 |   .7094767    .026136    27.15   0.000     .6582512   .7607022
       _cons |   2.332933   .0272082    85.74   0.000     2.279606    2.38626
------------------------------------------------------------------------------
```

We would like to test whether the coefficient on `white` is equal to $-.20$. We could use Stata's `test` command to obtain a Wald test

```
. test white=-.20
 ( 1)  [los]white = -.2
           chi2(  1) =    2.83
         Prob > chi2 =    0.0924
```

which provides that $-.15$ (coefficient on `white`) is not significantly different, at a 5% level, from $-.20$. However, we want to use a likelihood-ratio test, a usually more reliable test of parameter estimate significance. Stata provides a command that stores the likelihood from the unrestricted model (above) and then compares it with a restricted model. Having fitted the unrestricted model, our attention now turns to fitting a model satisfying our specific set of constraints. Our constraint is that the coefficient on `white` be restricted to the constant value $-.20$.

First, we store the log-likelihood value from the unrestricted model, and then we generate a variable indicative of our constraint. This new variable contains the restrictions that we will then supply to the software for fitting the restricted model. In short, the software will add our restriction any time that it calculates the linear predictor $\mathbf{x}_i\beta$. Since we envision a model for which the coefficient of `white` is equal to $-.20$, we need to generate a variable that is equal to $-.20$ times the variable `white`.

```
. estimates store Unconstrained

. gen offvar = -.20*white

. glm los hmo type2 type3, family(poisson) link(log) offset(offvar) nolog

Generalized linear models                          No. of obs      =       1495
Optimization     : ML                              Residual df     =       1491
                                                   Scale parameter =          1
Deviance         =  8145.531652                    (1/df) Deviance =   5.463133
Pearson          =  9334.640731                    (1/df) Pearson  =   6.260658

Variance function: V(u) = u                        [Poisson]
Link function    : g(u) = ln(u)                    [Log]

                                                   AIC             =    9.27671
Log likelihood   = -6930.340612                    BIC             =  -2753.502

------------------------------------------------------------------------------
             |                 OIM
         los |      Coef.   Std. Err.      z    P>|z|     [95% Conf. Interval]
-------------+----------------------------------------------------------------
         hmo |  -.0696133   .0239174    -2.91   0.004    -.1164906    -.022736
       type2 |    .218131    .020951    10.41   0.000     .1770677    .2591942
       type3 |   .7079687   .0261214    27.10   0.000     .6567717    .7591658
       _cons |   2.374881   .0107841   220.22   0.000     2.353744    2.396017
      offvar |   (offset)
------------------------------------------------------------------------------

. lrtest Unconstrained

Likelihood-ratio test                                  LR chi2(1)  =      2.87
(Assumption: . nested in Unconstrained)                Prob > chi2 =    0.0905
```

Since we restricted one coefficient from our full model, the likelihood-ratio statistic is distributed as a chi-squared random variable with 1 degree of freedom. We fail to reject the hypothesis that the coefficient on `white` is equal to $-.20$ at the 5% level.

Restricting coefficients for likelihood-ratio tests is just one use for offsets. Later, we will discuss how to use offsets to account for exposure in count data models.

2.5 Summary

The class of GLMs extends traditional linear models so that a linear predictor is mapped through a link function to model the mean of a response characterized by any member of the exponential family of distributions. Since we are able to develop one algorithm to estimate the entire class of models, we can support estimation of such useful statistical models as logit, probit, and Poisson.

The traditional linear model is not appropriate when assuming that data are normally distributed is unreasonable or if the response variable has a limited outcome set. Furthermore, in many instances in which homoskedasticity is an untenable requirement, the linear model is again inappropriate. The GLM allows these extensions to the linear model.

A GLM is constructed by first selecting explanatory variables for the response variable of interest. A probability distribution that is a member of the exponential family is selected, and an appropriate link function is specified, for which the mapped range of values supports the implied variance function of the distribution.

3 GLM estimation algorithms

Contents

This chapter covers the theory behind GLMs. We present the material in a general fashion showing all the results in terms of the exponential family of distributions. We illustrate two computational approaches for obtaining the parameter estimates of interest and discuss the assumptions that each method inherits. Parts II through VI of this text will then illustrate the application of this general theory to specific members of the exponential family.

The goal of our presentation is a thorough understanding of the underpinnings of the GLM method. We also wish to highlight the assumptions and limitations that algorithms inherit from their associated framework.

Traditionally, GLMs are estimated by applying Fisher scoring within the Newton–Raphson method applied to the entire single-parameter exponential family of distributions. After making simplifications (which we will detail), the estimation algorithm is then referred to as IRLS. Before the publication of this algorithm, models based on a member distribution of the exponential family were fitted using distribution-specific Newton–Raphson algorithms.

GLM theory showed how these models could be unified and fitted using one IRLS algorithm that does not require starting values for the coefficients $\widehat{\beta}$; rather, it substitutes easy-to-compute fitted values $\widehat{y}_i$. This is the beauty and attraction of GLM. Estimation and theoretical presentation are simplified by addressing the entire family of distributions. Results are valid regardless of the inclusion of Fisher scoring in the Newton–Raphson computations.

In what follows, we highlight the Newton–Raphson method for finding the zeros (roots) of a real-valued function. In the simplest case, we view this problem as changing the point of view from maximizing the log likelihood to that of determining the root of the derivative of the log likelihood. Since the values of interest are obtained by setting the derivative of the log likelihood to zero and solving, that equation is referred to as the *estimating equation*.

We begin our investigation by assuming that we have a measured response variable that follows a member distribution of the exponential family. We also assume that the responses are independent and identically distributed (i.i.d.). Focusing on the exponential family of distributions admits two powerful generalizations. First, we are allowed a rich collection of models that includes members for binary, discrete, and continuous outcomes. Second, by focusing on the form of the exponential family, we derive general algorithms that are suitable for all members of the entire distribution family; historically, each member-distribution was addressed in separate distribution-specific model specifications.

A probability function is a means to describe data on the basis of given parameters. Using the same formula for the probability function, we reverse the emphasis to the description of parameters on the basis of given data. Essentially, given a dataset, we determine those parameters that would most likely admit the given data within the scope of the specific probability function. This reversal of emphasis on the probability function of what is given is then called the likelihood function.

Although the joint i.i.d. probability function is given as

$$f(\mathbf{y};\theta,\phi) = \prod_{i=1}^{n} f(y_i;\theta,\phi) \tag{3.1}$$

the likelihood is understood as

$$L(\theta,\phi;\mathbf{y}) = \prod_{i=1}^{n} f(\theta,\phi;y_i) \tag{3.2}$$

where the difference appears in that the probability function is a function of the unknown data $\mathbf{y}$ for the given parameters θ and ϕ, whereas the likelihood is a function of the unknown θ and ϕ for the given data $\mathbf{y}$.

Moreover, since the joint likelihood of the function is multiplicative, we log-transform the likelihood function, L, to obtain a function on an additive scale. This function is known as the log likelihood, $\mathcal{L}$, and it is the focus of maximum likelihood theory. Thus maximum likelihood is a reasonable starting point in the examination of GLMs because the estimating equation of interest (equivalence of the derivative of the log likelihood function and zero) provides the same results as those that maximize the log-likelihood function.

We begin with the specification of the log likelihood under standard likelihood theory, cf. Kendall and Stuart (1979), for the exponential family.

$$\mathcal{L} = \sum_{i=1}^{n} \left\{ \frac{y_i\theta_i - b(\theta_i)}{a(\phi)} + c(y_i, \phi) \right\} \tag{3.3}$$

The individual elements of the log likelihood are identified through descriptive terms; θ is the canonical parameter (it is the parameter vector of interest), $b(\theta)$ is the cumulant (it describes moments), ϕ is the dispersion parameter (it is a scale or ancillary parameter), and $c()$ is a normalization term. The normalization term is not a function of θ and simply scales the range of the underlying density to integrate (or sum) to one.

To obtain the maximum likelihood estimate $\widehat{\theta}$ of θ, we first rely on the facts

$$\mathrm{E}\left(\frac{\partial \mathcal{L}}{\partial \theta}\right) = 0 \tag{3.4}$$

$$\mathrm{E}\left\{\frac{\partial^2 \mathcal{L}}{\partial \theta^2} + \left(\frac{\partial \mathcal{L}}{\partial \theta}\right)^2\right\} = 0 \tag{3.5}$$

See Gould, Pitblado, and Sribney (2006) for examples wherein the two previous equations are presented as a lemma.

Using (3.4) for our sample of i.i.d. observations, we can write

$$0 = \frac{\mathrm{E}(y_i) - b'(\theta_i)}{a(\phi)} \tag{3.6}$$

which implies that

$$b'(\theta_i) = \mathrm{E}(y_i) = \mu_i \tag{3.7}$$

where μ_i is the mean value parameter, and from (3.5), we may write

$$0 = -\frac{b''(\theta_i)}{a(\phi)} + \frac{1}{a(\phi)^2}\mathrm{E}\left\{y_i - b'(\theta_i)\right\}^2 \tag{3.8}$$

$$= -\frac{b''(\theta_i)}{a(\phi)} + \frac{1}{a(\phi)^2}\mathrm{E}\left(y_i - \mu_i\right)^2 \tag{3.9}$$

$$= -\frac{b''(\theta_i)}{a(\phi)} + \frac{1}{a(\phi)^2}\mathrm{V}(y_i) \tag{3.10}$$

Since the observations are i.i.d., we can drop the subscript and see that

$$b''(\theta) = \frac{1}{a(\phi)} \mathrm{V}(y) \tag{3.11}$$

Solving for $\mathrm{V}(y)$, we see that

$$\begin{aligned} \mathrm{V}(y) &= b''(\theta)a(\phi) & (3.12) \\ &= \mathrm{V}(\mu)a(\phi) & (3.13) \end{aligned}$$

since the variance is a function of the mean (expected value of y) via $b''(\theta)$. As such, we refer to $b''(\theta)$ as the variance function. From (3.7) and (3.11)–(3.13), we know that

$$\frac{\partial \mu}{\partial \theta} = \mathrm{V}(\mu) \tag{3.14}$$

Using only the properties (3.4) and (3.5) of the maximum likelihood estimator for θ, we then specify models that allow a parameterization of the mean μ in terms of covariates $\mathbf{X}$ with associated coefficients $\boldsymbol{\beta}$. Covariates are introduced via a known (invertible) function that links the mean μ to the linear predictor $\eta = \mathbf{X}\boldsymbol{\beta} + \text{offset}$ (the sum of the products of the covariates and coefficients).

$$\begin{aligned} g(\mu) &= \eta + \text{offset} & (3.15) \\ g^{-1}(\eta + \text{offset}) &= \mu & (3.16) \end{aligned}$$

Finally, we also know that since $\eta = \sum_{j=1}^{p} x_j \beta_j + \text{offset}$,

$$\frac{\partial \eta}{\partial \beta_j} = x_j \tag{3.17}$$

The linear predictor is not constrained to lie within a specific range. Rather, we have $\eta \in \Re$. Thus one purpose of the link function is to map the linear predictor to the range of the response variable. Mapping to the range of the response variable ensures nonnegative variances associated with a particular member distribution of the exponential family. The unconstrained linear model in terms of the form of the coefficients and associated covariates is constrained through the range-restricting properties of the link function.

The natural or canonical link is the result of equating the canonical parameter to the linear predictor, $\eta = \theta$; it is the link function obtained relating η to μ when $\eta = \theta$. In practice, we are not limited to this link function. We can choose any monotonic link function that maps the unrestricted linear predictor to the range implied by the variance function (see table A.1). In fact, in section 10.5 we investigate the use of link functions that could theoretically map the linear predictor to values outside the range implied by the variance function. This concept is somewhat confusing at first glance. After all, why would one want to consider different parameterizations? We demonstrate various interpretations and advantages to various parameterizations throughout the rest of the book.

Summarizing the current points: we can express the scalar derivatives of the vector $\boldsymbol{\beta}$ (thus we may find the estimating equation) by using the chain rule for our sample of n observations,

$$\frac{\partial \mathcal{L}}{\partial \beta_j} = \sum_{i=1}^{n} \left(\frac{\partial \mathcal{L}_i}{\partial \theta_i}\right)\left(\frac{\partial \theta_i}{\partial \mu_i}\right)\left(\frac{\partial \mu_i}{\partial \eta_i}\right)\left(\frac{\partial \eta_i}{\partial \beta_j}\right) \tag{3.18}$$

$$= \sum_{i=1}^{n} \left(\frac{y_i - b'(\theta_i)}{a(\phi)}\right)\left(\frac{1}{\mathrm{V}(\mu_i)}\right)\left(\frac{\partial \mu}{\partial \eta}\right)_i (x_{ji}) \tag{3.19}$$

$$= \sum_{i=1}^{n} \frac{y_i - \mu_i}{a(\phi)\mathrm{V}(\mu_i)}\left(\frac{\partial \mu}{\partial \eta}\right)_i x_{ji} \tag{3.20}$$

where $i = 1, \ldots, n$ indexes the observations and x_{ji} is the ith observation for the jth covariate $\mathbf{X}_j$, $j = 1, \ldots, p$.

To find estimates $\widehat{\boldsymbol{\beta}}$, we can use Newton's method. This method is a linear Taylor series approximation where we expand the derivative of the log likelihood (the gradient or the estimating equation) in a Taylor series. In the following, we use $\mathcal{L}' = \partial\mathcal{L}/\partial\boldsymbol{\beta}$ for the gradient and $\mathcal{L}'' = \partial^2\mathcal{L}/(\partial\boldsymbol{\beta}\partial\boldsymbol{\beta}^{\mathrm{T}})$. We wish to solve

$$\mathcal{L}'(\boldsymbol{\beta}) = 0 \tag{3.21}$$

Solving this equation provides estimates of $\boldsymbol{\beta}$. Thus this is called the estimating equation. The Taylor series expansion is

$$0 = \mathcal{L}'(\boldsymbol{\beta}^{(0)}) + (\boldsymbol{\beta} - \boldsymbol{\beta}^{(0)})\mathcal{L}''(\boldsymbol{\beta}^{(0)}) + \frac{(\boldsymbol{\beta} - \boldsymbol{\beta}^{(0)})^2}{2!}\mathcal{L}'''(\boldsymbol{\beta}^{(0)}) + \cdots \tag{3.22}$$

Keeping only the first two terms reduces the estimating equation to the linear equation given by

$$0 \approx \mathcal{L}'(\boldsymbol{\beta}^{(0)}) + (\boldsymbol{\beta} - \boldsymbol{\beta}^{(0)})\mathcal{L}''(\boldsymbol{\beta}^{(0)}) \tag{3.23}$$

such that we may write (solving for $\boldsymbol{\beta}$)

$$\boldsymbol{\beta} \approx \boldsymbol{\beta}^{(0)} - \left\{\mathcal{L}''(\boldsymbol{\beta}^{(0)})\right\}^{-1} \mathcal{L}'(\boldsymbol{\beta}^{(0)}) \tag{3.24}$$

We may iterate this estimation using

$$\boldsymbol{\beta}^{(r)} = \boldsymbol{\beta}^{(r-1)} - \left\{\mathcal{L}''(\boldsymbol{\beta}^{(r-1)})\right\}^{-1} \mathcal{L}'(\boldsymbol{\beta}^{(r-1)}) \tag{3.25}$$

for $r = 1, 2, \ldots$ and a reasonable vector of starting values $\boldsymbol{\beta}^{(0)}$.

This linearized Taylor series approximation is exact if the function is truly quadratic. This is the case for the linear regression model (Gaussian variance with the identity link) as illustrated in an example in section 5.8. As such, only one iteration is needed. Other models require iteration from a reasonable starting point. The benefit of the derivation so far is that we have a general solution applicable to the entire exponential family of distributions. Without the generality of considering the entire family of distributions, separate derivations would be required for linear regression, Poisson regression, logistic regression, etc.

In the next two sections, we present derivations of two estimation approaches. Both are Newton–Raphson algorithms. In fact, there are many so-called Newton–Raphson algorithms. In (3.25), we see that the second derivative matrix of the log likelihood (first derivative of the estimating equation) is required to obtain an updated estimate of the parameter vector $\boldsymbol{\beta}$. This matrix is, numerically speaking, difficult. Clearly, one approach is to form the second derivatives from analytic specification or from numeric approximation using the estimating equation. Instead, we can use the outer matrix product of the estimating equation, see (3.5), as is done in IRLS.

Also there are many other techniques that Stata has conveniently made available through the `ml` command. Users of this program can specify an estimating equation and investigate the performance of several numeric algorithms for a given command. Larger investigations are also possible through simulation, which is also conveniently packaged for users in the `simul` command. We recommend that Stata users take time to study these commands, which enable quick development and deployment of software to support estimation for new models. The best source of information for investigating built-in support for optimization within Stata is Gould, Pitblado, and Sribney (2006).

The obvious choice for the required matrix term in (3.25) is to use the second-derivative matrix, the so-called observed Hessian matrix. When we refer to the Newton–Raphson algorithm, we mean the algorithm using the observed Hessian matrix. The second obvious choice for the required matrix term is (Fisher scoring) the outer product of the estimating equation, the so-called expected Hessian matrix. When we refer to IRLS, we mean the Newton–Raphson algorithm using the expected Hessian matrix. We discuss these choices in section 3.6.

Although there are conditions under which the two estimation methods (Newton–Raphson and IRLS) are equivalent, finding small numeric differences in software (computer output) is common, even for problems with unique solutions. These small discrepancies are due to differences in starting values and convergence paths. Such differences from the two estimation algorithms should not concern you. When the two algorithms are not numerically equivalent, they are still equal in the limit (they have the same expected value).

If we choose the canonical link in a model specification, then there are certain simplifications. For example, since the canonical link has the property that $\eta = \theta$, the estimating equation may be written

$$\frac{\partial \mathcal{L}}{\partial \beta_j} = \sum_{i=1}^{n} \left(\frac{\partial \mathcal{L}_i}{\partial \theta_i}\right)\left(\frac{\partial \theta_i}{\partial \mu_i}\right)\left(\frac{\partial \mu_i}{\partial \eta_i}\right)\left(\frac{\partial \eta_i}{\partial \beta_j}\right) \tag{3.26}$$

$$= \sum_{i=1}^{n} \left(\frac{\partial \mathcal{L}_i}{\partial \eta_i}\right)\left(\frac{\partial \eta_i}{\partial \mu_i}\right)\left(\frac{\partial \mu_i}{\partial \eta_i}\right)\left(\frac{\partial \eta_i}{\partial \beta_j}\right) \tag{3.27}$$

$$= \sum_{i=1}^{n} \left(\frac{\partial \mathcal{L}_i}{\partial \eta_i}\right)\left(\frac{\partial \eta_i}{\partial \beta_j}\right) \tag{3.28}$$

$$= \sum_{i=1}^{n} \frac{y_i - \mu_i}{a(\phi)} x_{ji} \tag{3.29}$$

Thus using the canonical link leads to setting (3.29) to zero such that $\widehat{\mu}_i$ is seen to play the role of $\widehat{y}_i$. Interpretation of predicted values (transformed via the inverse canonical link) are the predicted outcomes. Interpretation of predicted values under other (noncanonical) links is more difficult, as we will see in the discussion of negative binomial regression in chapter 13.

3.1 Newton–Raphson (using the observed Hessian)

Maximum likelihood begins with specifying a likelihood. The likelihood is a restatement of the joint probability density or joint probability mass function in which the data are taken as given and the parameters are estimated.

To find maximum likelihood estimates, we proceed with the derivation originating with the log likelihood given in (3.3) and use the first derivatives given in (3.20). We now derive the (observed) matrix of second derivatives, which will then give us the necessary ingredients for the Newton–Raphson algorithm to find the maximum likelihood estimates.

This algorithm solves or optimizes (3.21), using variations on (3.25). The Newton–Raphson algorithm implements (3.25) without change. Research in optimization has found that this algorithm can be improved. If there are numeric problems in the optimization, these problems are usually manifested as singular (estimated) second-derivative matrices. Various algorithms address this situation. These other modifications to the Newton–Raphson algorithm include the Marquardt (1963) modification used in the heart of Stata's `ml` program. This software includes many other approaches. For quick assistance see, for example, the documentation on the `bfgs` and `bhhh` options; for more in depth explanation see Gould, Pitblado, and Sribney (2006).

(Continued on next page)

The observed matrix of second derivatives (Hessian matrix) is given by

$$\left(\frac{\partial^2 \mathcal{L}}{\partial \beta_j \partial \beta_k}\right) = \sum_{i=1}^{n} \frac{1}{a(\phi)} \left(\frac{\partial}{\partial \beta_k}\right) \left\{ \frac{y_i - \mu_i}{\mathrm{V}(\mu_i)} \left(\frac{\partial \mu}{\partial \eta}\right)_i x_{ji} \right\} \tag{3.30}$$

$$\begin{aligned}
&= \sum_{i=1}^{n} \frac{1}{a(\phi)} \left[\left(\frac{\partial \mu}{\partial \eta}\right)_i \left\{ \left(\frac{\partial}{\partial \mu}\right)_i \left(\frac{\partial \mu}{\partial \eta}\right)_i \left(\frac{\partial \eta}{\partial \beta_k}\right)_i \right\} \frac{y_i - \mu_i}{\mathrm{V}(\mu_i)} \right. \\
&\quad \left. + \frac{y_i - \mu_i}{\mathrm{V}(\mu_i)} \left\{ \left(\frac{\partial}{\partial \eta}\right)_i \left(\frac{\partial \eta}{\partial \beta_k}\right)_i \right\} \left(\frac{\partial \mu}{\partial \eta}\right)_i \right] x_{ji}
\end{aligned} \tag{3.31}$$

$$\begin{aligned}
&= -\sum_{i=1}^{n} \frac{1}{a(\phi)} \left[\frac{1}{\mathrm{V}(\mu_i)} \left(\frac{\partial \mu}{\partial \eta}\right)_i^2 \right. \\
&\quad \left. - (\mu_i - y_i) \left\{ \frac{1}{\mathrm{V}(\mu_i)^2} \left(\frac{\partial \mu}{\partial \eta}\right)_i^2 \frac{\partial \mathrm{V}(\mu_i)}{\partial \mu} - \frac{1}{\mathrm{V}(\mu_i)} \left(\frac{\partial^2 \mu}{\partial \eta^2}\right)_i \right\} \right] x_{ji} x_{ki}
\end{aligned} \tag{3.32}$$

Armed with the first two derivatives, one can easily implement a Newton–Raphson algorithm to obtain the maximum likelihood estimates. Without the derivatives written out analytically, one could still implement a Newton–Raphson algorithm by programming numeric derivatives (calculated using difference equations).

After optimization is achieved, one must estimate a variance matrix for $\boldsymbol{\beta}$. An obvious and suitable choice is based on the estimated observed Hessian matrix. This is the most common default choice in software implementations because the observed Hessian matrix is a component of the estimation algorithm. As such, it is already calculated and available. We discuss in section 3.6 that there are other choices for estimated variance matrices.

Thus the Newton–Raphson algorithm provides

1. An algorithm for estimating the coefficients for all single-parameter exponential family GLM members

2. Estimated standard errors of estimated coefficients: square roots of the diagonal elements of the inverse of the estimated observed negative Hessian matrix

Our illustration of the Newton–Raphson algorithm could be extended in several ways. The presentation did not show the estimation of the scale parameter, ϕ. If we estimate ϕ as an ancillary parameter, the estimates of $\boldsymbol{\beta}$ are properly called restricted maximum likelihood estimates instead of ML estimates.

Other implementations could include the scale parameter in the derivatives and cross derivatives to obtain ML estimates—sometimes called full information maximum likelihood (FIML).

3.2 Starting values for Newton–Raphson

To implement an algorithm for obtaining estimates of $\boldsymbol{\beta}$, we must have an initial guess for the parameters. There is no global mechanism for good starting values, but there is a reasonable solution for obtaining starting values when there is a constant in the model.

If the model includes a constant, then a common practice is to find the estimates for a constant only model. For maximum likelihood, this is a part of the model of interest and knowing the likelihood for the constant-only model then admits a likelihood-ratio test for the parameters of the model of interest.

Often the maximum likelihood estimate for the constant-only model may be found analytically. For example, in chapter 12 we introduce the Poisson model. That model has a log likelihood given by

$$\mathcal{L} = \sum_{i=1}^{n} \{y_i(\mathbf{x}_i\boldsymbol{\beta}) - \exp(\mathbf{x}_i\boldsymbol{\beta}) - \ln\Gamma(y_i+1)\} \tag{3.33}$$

If we assume that there is only a constant term in the model, then the log likelihood may be written

$$\mathcal{L} = \sum_{i=1}^{n} \{y_i\beta_0 - \exp(\beta_0) - \ln\Gamma(y_i+1)\} \tag{3.34}$$

The maximum likelihood estimate of β_0 is found by setting the derivative

$$\frac{\partial\mathcal{L}}{\partial\beta_0} = \sum_{i=1}^{n} \{y_i - \exp(\beta_0)\} \tag{3.35}$$

to zero and solving

$$\begin{aligned} 0 &= \sum_{i=1}^{n} \left\{y_i - \exp(\widehat{\beta}_0)\right\} && (3.36)\\ n\exp(\widehat{\beta}_0) &= \sum_{i=1}^{n} y_i && (3.37)\\ \widehat{\beta}_0 &= \ln(\overline{y}) && (3.38) \end{aligned}$$

We can now use $\boldsymbol{\beta} = (\widehat{\beta}_0, 0, \ldots, 0)^{\mathrm{T}}$ as the starting values (initial vector) in the Newton–Raphson algorithm.

Using this approach affords us two advantages. First, we start our iterations from a reasonable subset of the parameter space. Second, for maximum likelihood, since we know the solution for the constant-only model, we may compare the log likelihood obtained for our model of interest with the log likelihood obtained at the initial step in our algorithm. This comparison is a likelihood-ratio test that all of the covariates (except the constant) are zero.

If there is not a constant in the model of interest, or if we cannot analytically solve for the constant-only model, then we must use more sophisticated methods. We can either iterate to a solution for the constant-only model, or we can use a search method to look for reasonable points at which to start our Newton–Raphson algorithm.

3.3 IRLS (using the expected Hessian)

Here we discuss the estimation algorithm known as IRLS. We begin by rewriting the (usual) updating formula from the Taylor series expansion presented in (3.25) as

$$\Delta\beta^{(r-1)} = -\left\{\frac{\partial^2\mathcal{L}}{\partial(\boldsymbol{\beta}^{(r-1)})^{\mathrm{T}}\partial\boldsymbol{\beta}^{(r-1)}}\right\}^{-1}\frac{\partial\mathcal{L}}{\partial\boldsymbol{\beta}^{(r-1)}} \tag{3.39}$$

and we replace the calculation of the observed Hessian (second derivatives) with its expectation. This substitution is known as the method of Fisher scoring. Since we know that $\mathrm{E}\{(y_i-\mu_i)^2\} = \mathrm{V}(\mu_i)a(\phi)$, we may write

$$-\mathrm{E}\left(\frac{\partial^2\mathcal{L}}{\partial\beta_j\partial\beta_k}\right) = \mathrm{E}\left(\frac{\partial\mathcal{L}}{\partial\beta_j}\frac{\partial\mathcal{L}}{\partial\beta_k}\right) \tag{3.40}$$

$$= \sum_{i=1}^{n}\left(\frac{\partial\mu}{\partial\eta}\right)_i^2\frac{1}{\mathrm{V}(\mu_i)a(\phi)}x_{ji}x_{ki} \tag{3.41}$$

Substituting (3.41) and (3.20) into (3.39) and rearranging, we see that $\delta\boldsymbol{\beta}^{(r-1)}$ is the solution to

$$\left\{\sum_{i=1}^{n}\frac{1}{\mathrm{V}(\mu_i)a(\phi)}\left(\frac{\partial\mu}{\partial\eta}\right)_i^2 x_{ji}x_{ki}\right\}\Delta\boldsymbol{\beta}^{(r-1)} = \sum_{i=1}^{n}\frac{y_i-\mu_i}{\mathrm{V}(\mu_i)a(\phi)}\left(\frac{\partial\mu}{\partial\eta}\right)_i\mathbf{x}_i^{\mathrm{T}} \tag{3.42}$$

Using the $(r-1)$ superscript to emphasize calculation with $\boldsymbol{\beta}^{(r-1)}$, we may refer to the linear predictor as

$$\eta_i^{(r-1)} - \mathrm{offset}_i = \sum_{k=1}^{p}x_{ki}\beta_k^{(r-1)} \tag{3.43}$$

which may be rewritten as

$$\left\{\sum_{i=1}^{n}\frac{1}{\mathrm{V}(\mu_i)a(\phi)}\left(\frac{\partial\mu}{\partial\eta}\right)_i^2 x_{ji}x_{ki}\right\}\boldsymbol{\beta}^{(r-1)} =$$

$$\sum_{i=1}^{n}\frac{1}{\mathrm{V}(\mu_i)a(\phi)}\left(\frac{\partial\mu}{\partial\eta}\right)_i^2\left(\eta_i^{(r-1)} - \mathrm{offset}_i\right)\mathbf{x}_i^{\mathrm{T}} \tag{3.44}$$

Summing (3.42) and (3.44) and then substituting (3.39), we obtain

$$\left\{\sum_{i=1}^{n} \frac{1}{\mathrm{V}(\mu_i)a(\phi)}\left(\frac{\partial\mu}{\partial\eta}\right)_i^2 x_{ji}x_{ki}\right\}\left(\Delta\boldsymbol{\beta}^{(r-1)}+\boldsymbol{\beta}^{(r-1)}\right) = \sum_{i=1}^{n} \frac{1}{\mathrm{V}(\mu_i)a(\phi)}\left(\frac{\partial\mu}{\partial\eta}\right)_i^2 \left\{(y_i-\mu_i)\left(\frac{\partial\eta}{\partial\mu}\right)_i + \left(\eta_i^{(r-1)} - \text{offset}_i\right)\right\}\mathbf{x}_i^{\mathrm{T}} \tag{3.45}$$

$$\left\{\sum_{i=1}^{n} \frac{1}{\mathrm{V}(\mu_i)a(\phi)}\left(\frac{\partial\mu}{\partial\eta}\right)_i^2 x_{ji}x_{ki}\right\}\boldsymbol{\beta}^{r} = \sum_{i=1}^{n} \frac{1}{\mathrm{V}(\mu_i)a(\phi)}\left(\frac{\partial\mu}{\partial\eta}\right)_i^2 \left\{(y_i-\mu_i)\left(\frac{\partial\eta}{\partial\mu}\right)_i + \left(\eta_i^{(r-1)} - \text{offset}_i\right)\right\}\mathbf{x}_i^{\mathrm{T}} \tag{3.46}$$

Now again using the $(r-1)$ superscript to emphasize calculation with $\boldsymbol{\beta}^{(r-1)}$, we let

$$\mathbf{W}^{(r-1)} = \mathrm{diag}\left\{\frac{1}{\mathrm{V}(\mu)a(\phi)}\left(\frac{\partial\mu}{\partial\eta}\right)^2\right\}_{(n\times n)} \tag{3.47}$$

$$\mathbf{z}^{(r-1)} = \left\{(y-\mu)\left(\frac{\partial\eta}{\partial\mu}\right) + \left(\eta^{(r-1)} - \text{offset}\right)\right\}_{(n\times 1)} \tag{3.48}$$

so that we may rewrite (3.46) in matrix notation as

$$(\mathbf{X}^{\mathrm{T}}\mathbf{W}^{(r-1)}\mathbf{X})\boldsymbol{\beta}^{(r)} = \mathbf{X}^{\mathrm{T}}\mathbf{W}^{(r-1)}\mathbf{z}^{(r-1)} \tag{3.49}$$

This algorithm then shows that a new estimate of the coefficient vector is obtained via weighted OLS (weighted linear regression).

After optimization is achieved, one must estimate a suitable variance matrix for $\boldsymbol{\beta}$. An obvious and suitable choice is the estimated expected Hessian matrix. This is the most common default choice in software implementations since the expected Hessian matrix is a component of the estimation algorithm. As such, it is already calculated and available. We discuss in section 3.6 that there are other choices for estimated variance matrices.

Thus the IRLS algorithm provides

1. An algorithm for estimating the coefficients for all single-parameter exponential family GLM members
2. An algorithm that uses weighted OLS (linear regression) and thus may be easily incorporated into almost any statistical analysis software
3. Estimated standard errors of estimated coefficients: square roots of the diagonal elements of $(\mathbf{X}^{\mathrm{T}}\mathbf{W}\mathbf{X})^{-1}$

The usual estimates of standard error are constructed using the expected Hessian. These are equal to the observed Hessian when the canonical link is used. Otherwise, the calculated standard errors will differ from usual Newton–Raphson standard error estimates.

Because the expected Hessian is commonly used by default for calculating standard errors, some refer to these as naive standard errors. They are naive in that the estimation assumes that the conditional mean is specified correctly.

In the end, our algorithm needs specification of the variance as a function of the mean $V(\mu)$, the derivative of the inverse link function, the dispersion parameter $a(\phi)$, the linear predictor (the systematic component), and the random variable y. These are the components listed in section 2.1.

This algorithm may be generalized in several important ways. First, the link function may be different for each observation (may be subscripted by i). The algorithm may include an updating step for estimating unknown ancillary parameters and include provisions for ensuring that the range of the fitted linear predictor does not fall outside specified bounds. A program that implements these extensions may do so automatically or via user-specified options. As such, you may see in the literature somewhat confusing characterizations of the power of GLM, when in fact the characterizations are of the IRLS algorithm, which supports estimation of GLMs. This substitution of the class of models for the estimation algorithm is common.

John Nelder, in his response to Green (1984), included an "etymological quibble" in which he argued that the use of "*re*weighted" over "weighted" placed unnecessary emphasis on the updated weights. That we update both the weights and the (synthetic) dependent variable at each iteration is the heart of Nelder's quibble in naming the algorithm IRLS. As such, although we refer to this estimation algorithm as IRLS, we will also see reference in various places to iterative weighted least squares. This is the term originally used in Nelder and Wedderburn (1972).

Wedderburn's extension shows that the specification of variance as a function of the mean implies a quasilikelihood. We will use this result throughout the text in developing various models, but see chapter 17 for motivating details.

The important fact from Wedderburn's paper is that the applicability of the IRLS algorithm is extended to specifying the ingredients of the algorithm and that we do not have to specify a likelihood. Rather, a quasilikelihood is implied by the assumed (specified) moments. Estimates from the IRLS algorithm are maximum likelihood (ML) when the implied quasilikelihood is a true likelihood. Otherwise the estimates are maximum quasilikelihood (MQL). Thus we can think of IRLS as an algorithm encompassing the assumptions and basis of maximum likelihood (Newton–Raphson method) or as an algorithm encompassing the assumptions and basis of the methodology of maximum quasilikelihood.

In reference information in the appendix (tables A.1, A.2, and A.4), we present a large collection of tabular material showing typical variance and link functions. Most of these are derived from likelihoods. These derivations will be highlighted in later chapters of the book. The power family of the variance functions listed in the tables is not derived from a likelihood, and its use is a direct result of the extension given by Wedderburn (1974).

3.4 Starting values for IRLS

Unlike Newton–Raphson, the IRLS algorithm does not require an initial guess for the parameter vector of interest $\boldsymbol{\beta}$. Instead, we need initial guesses for the fitted values $\widehat{\mu}_i$. This is much easier to implement.

A reasonable approach is to initialize the fitted values to the inverse link of the mean of the response variable. We must ensure that the initial starting values are within the implied range of the variance function specified in the GLM that we are fitting. For example, this prevents the situation where the algorithm may try to compute the logarithm (or square root) of a nonpositive value.

Another approach is to set the initial fitted values to $(y_i + \overline{y})/2$ for a nonbinomial model and $k_i(y_i + .5)/(k_i + 1)$ for a binomial(k_i, p_i) model. In fact, this is how initial values are assigned in the Stata `glm` command.

3.5 Goodness of fit

In developing a model, we hope to generate fitted values $\widehat{\mu}$ that are close to the data y. For a dataset with n observations, we may consider candidate models with one to n parameters. The simplest model would include only one parameter. The best one-parameter model would result in $\widehat{\mu}_i = \mu$ (for all i). Although the model is parsimonious, it does not model the variability in the data. The saturated model (with n parameters) would include one parameter for each observation and result in $\widehat{\mu}_i = y_i$. This model exactly reproduces the data but is uninformative since there is no summarization of the data.

We define a measure of the fit of the model to the data as twice the difference between the log likelihoods of the model of interest and the saturated model. Since this difference is a measure of the deviation of the model of interest from a perfectly fitting model, the measure is called the deviance. Our competing goals in modeling are to find the simplest model (fewest parameters) that has the smallest deviance (reproduces the data).

The deviance, D, is given by

$$D = \sum_{i=1}^{n} 2\left[y\{\theta(y_i) - \theta(\mu_i)\} - b\{\theta(y_i)\} + b\{\theta(\mu_i)\}\right] \tag{3.50}$$

where the equation is given in terms of the mean parameter μ instead of the canonical parameter θ. In fitting a particular model, we seek the values of the parameters that minimize the deviance. Thus optimization in the IRLS algorithm is achieved when the difference in deviance calculations between successive iterations is small (less than some tolerance). The values of the parameters that minimize the deviance are the same as the values of the parameters that maximize the likelihood.

The quasideviance is an extension of this concept for models that do not have an underlying likelihood. We discuss this idea more fully in chapter 17.

3.6 Estimated variance matrices

It is a natural question to ask how the Newton–Raphson (based on the observed Hessian) variance estimates compare with the usual (based on the expected Hessian) variance estimates obtained using the IRLS algorithm outlined in the preceding section. The matrix of second derivatives in the IRLS algorithm is equal to using the first term in (3.32). As Newson (1999) points out, the calculation of the expected Hessian is simplified from that of the observed Hessian since we assume that $\mathrm{E}(\mu - y) = 0$ or, equivalently, if the conditional mean of y given $\mathbf{X}$ is correct. As such, the IRLS algorithm assumes that the conditional mean is specified correctly. Both approaches result in parameter estimates that differ only because of numeric roundoff or because of differences in optimization criteria.

This distinction is especially important in the calculation of sandwich estimates of variance. The Hessian may be calculated as given above in (3.32) or may be calculated using the more restrictive (naive) assumptions of the IRLS algorithm as

$$\mathrm{E}\left(\frac{\partial^2 \mathcal{L}}{\partial\beta_j \partial\beta_k}\right) = -\sum_{i=1}^{n} \frac{1}{a(\phi)} \frac{1}{\mathrm{V}(\mu_i)} \left(\frac{\partial\mu}{\partial\eta}\right)_i^2 x_{ji} x_{ki} \tag{3.51}$$

Equations (3.32) and (3.51) are equal when the canonical link is used. This occurs because for the canonical link, we can make the substitution that $\theta = \eta$, and thus that $\mathrm{V}(\mu) = \partial\mu/\partial\theta = \partial\mu/\partial\eta$. The second term in (3.32) then collapses to zero since

$$\begin{aligned}
(\mu_i - y_i)&\left\{\frac{1}{\mathrm{V}(\mu_i)^2}\left(\frac{\partial\mu}{\partial\eta}\right)_i^2 \frac{\partial \mathrm{V}(\mu_i)}{\partial\mu} - \frac{1}{\mathrm{V}(\mu_i)}\left(\frac{\partial^2\mu}{\partial\eta^2}\right)_i\right\} \\
&= (\mu_i - y_i)\left\{\frac{1}{(\partial\mu/\partial\eta)_i^2}\left(\frac{\partial\mu}{\partial\eta}\right)_i^2 \frac{\partial}{\partial\mu_i}\left(\frac{\partial\mu}{\partial\eta}\right)_i - \frac{1}{(\partial\mu/\partial\eta)_i}\left(\frac{\partial^2\mu}{\partial\eta^2}\right)_i\right\} && (3.52) \\
&= (\mu_i - y_i)\left\{\frac{\partial}{\partial\mu_i}\left(\frac{\partial\mu}{\partial\eta}\right)_i - \left(\frac{\partial\eta}{\partial\mu}\right)_i\left(\frac{\partial^2\mu}{\partial\eta^2}\right)_i\right\} && (3.53) \\
&= (\mu_i - y_i)\left\{\left(\frac{\partial^2\mu}{\partial\mu\partial\eta}\right)_i - \left(\frac{\partial^2\mu}{\partial\mu\partial\eta}\right)_i\right\} && (3.54) \\
&= 0 && (3.55)
\end{aligned}$$

An estimation algorithm is not limited to using one version of the Hessian or the other. Users of `PROC GENMOD` in SAS, for example, should be familiar with the `SCORING` option. This option allows the user to specify how many of the initial iterations are performed using the Fisher scoring method, expected Hessian calculated using (3.32), before all subsequent iterations are performed using the observed Hessian calculated with (3.51). The overall optimization is still the Newton–Raphson method (optimizes the log likelihood). The Stata software described in association with this text also includes this capability (see the documentation on the `fisher()` option in the *Stata Base Reference Manual* for the `glm` entry.

Although there is an obvious choice for the estimated variance matrix for a given implementation of Newton–Raphson, one is not limited to that specific estimator used in the optimization (estimation algorithm); the IRLS algorithm could, in fact, calculate the observed Hessian after convergence and report standard errors based on the observed Hessian.

Later, we will want to alter IRLS algorithms to substitute this observed information matrix (OIM) for the IRLS expected information matrix (EIM). To do so, we merely have to alter the **W** matrix by the factor listed in (3.52). Here we identify the ingredients of this factor in general form. For specific applications, the individual terms are specified in the tables in appendix A.

$$-(\mu_i - y_i)\left\{\frac{1}{\mathrm{V}(\mu_i)^2}\left(\frac{\partial\mu}{\partial\eta}\right)_i^2\frac{\partial\mathrm{V}(\mu_i)}{\partial\mu} - \frac{1}{\mathrm{V}(\mu_i)}\left(\frac{\partial^2\mu}{\partial\eta^2}\right)_i\right\}$$

$$= (y_i - \mu_i)\left\{\frac{1}{\mathrm{V}(\mu_i)^2}\left(\frac{\partial\mu}{\partial\eta}\right)_i^2\frac{\partial\mathrm{V}(\mu_i)}{\partial\mu} - \frac{1}{\mathrm{V}(\mu_i)}\left(\frac{\partial^2\mu}{\partial\eta^2}\right)_i\right\} \tag{3.56}$$

$$= (y_i - \mu_i)\left(\frac{\mathrm{V}'(\mu_i)}{\mathrm{V}(\mu_i)^2}\left[\{g^{-1}(\eta_i)\}'\right]^2 - \frac{1}{\mathrm{V}(\mu_i)}\{g^{-1}(\eta_i)\}''\right) \tag{3.57}$$

$$= (y_i - \mu_i)\left[\frac{\mathrm{V}'(\mu_i)}{\mathrm{V}(\mu_i)^2}\frac{1}{\{g'(\mu_i)\}^2} - \frac{1}{\mathrm{V}(\mu_i)}\frac{-g''(\mu_i)}{\{g'(\mu_i)\}^3}\right] \tag{3.58}$$

$$= (y_i - \mu_i)\left[\frac{\mathrm{V}'(\mu_i)g'(\mu_i) + \mathrm{V}(\mu_i)g''(\mu_i)}{\mathrm{V}^2(\mu_i)\{g'(\mu_i)\}^3}\right] \tag{3.59}$$

Tables A.4 and A.6 provide the information for using (3.57). We have also provided the derivatives in tables A.3 and A.5 for using (3.59). Which set of derivatives you choose in a given application is motivated by how easily they might be calculated. For this discussion on the derivation of the GLM theory, we focus on the derivatives of the inverse link function, but for developing algorithms, we often use the derivatives of the link function. The derivation from (3.57) to (3.59) is justified (leaving off the subscripts) by

(*Continued on next page*)

$$g = g(\mu) = \eta \tag{3.60}$$

$$g' = g'(\mu) = \frac{\partial \eta}{\partial \mu} \tag{3.61}$$

$$g'' = g''(\mu) = \frac{\partial}{\partial \mu} g' = \frac{\partial^2 \eta}{\partial \mu^2} \tag{3.62}$$

$$g^{-1} = g^{-1}(\eta) = \mu \tag{3.63}$$

$$(g^{-1})' = \{g^{-1}(\eta)\}' = \frac{\partial \mu}{\partial \eta} = \frac{1}{g'} \tag{3.64}$$

$$\begin{aligned}(g^{-1})'' &= \{g^{-1}(\eta)\}'' = \frac{\partial^2 \mu}{\partial \eta^2} = \frac{\partial}{\partial \eta}\frac{1}{g'} = -\frac{1}{(g')^2}\frac{\partial}{\partial \eta} g' \\ &= -\frac{1}{(g')^2}\frac{\partial}{\partial \mu}\frac{\partial \mu}{\partial \eta} g' = -\frac{1}{(g')^2}\frac{\partial \mu}{\partial \eta}\frac{\partial}{\partial \mu} g' = -\frac{1}{(g')^2}\frac{1}{g'} g'' = -\frac{g''}{(g')^3}\end{aligned} \tag{3.65}$$

Davidian and Carroll (1987) provide details for estimating variance functions in general, and Oakes (1999) shows how to calculate the information matrix (see sec. 3.6.1) by using the EM algorithm. In the following subsections, we present details on calculating various variance matrix estimates. Throughout, $\partial\mu/\partial\eta$ is calculated at $\mu = \widehat{\mu}$, and $\widehat{\phi}$ is an estimator of the dispersion parameter $a(\phi)$. We will also use the hat diagonals defined later in (4.5).

3.6.1 Hessian

The usual variance estimate in statistical packages is calculated (numerically or analytically) as the inverse matrix of (negative) second derivatives. For generalized linear models, we may calculate the Hessian in two ways.

Assuming that the conditional mean is specified correctly, we may calculate the expected Hessian as

$$\widehat{\mathrm{V}}_{\mathrm{EH}} = \left\{ \mathrm{E}\left(-\frac{\partial^2 \mathcal{L}}{\partial \boldsymbol{\beta} \partial \boldsymbol{\beta}^{\mathrm{T}}} \right) \right\}^{-1} \tag{3.66}$$

where the matrix elements $-\partial^2\mathcal{L}/(\partial\beta_j\partial\beta_k)$ are calculated using (3.51). Sandwich estimates of variance formed using this calculation of the Hessian are called semirobust or semi-Huber since they are not robust to the conditional mean's being misspecified.

More generally, we may calculate the observed Hessian without the additional assumption of correct conditional mean specification using

$$\widehat{\mathrm{V}}_{\mathrm{OH}} = \left\{ \left(-\frac{\partial^2 \mathcal{L}}{\partial \boldsymbol{\beta} \partial \boldsymbol{\beta}^{\mathrm{T}}} \right) \right\}^{-1} \tag{3.67}$$

where the matrix elements $-\partial^2\mathcal{L}/(\partial\beta_j\partial\beta_k)$ are calculated using (3.32). Sandwich estimates of variance formed using this calculation of the Hessian are called robust or full Huber.

Since the two calculations are equal when using the canonical link, the result in those cases is always robust or full Huber. Throughout the remaining subsections, we refer generally to $\widehat{\mathrm{V}}_{\mathrm{H}}$ and make clear in context whether the inference is affected by using the observed versus expected Hessian.

3.6.2 Outer product of the gradient

Berndt et al. (1974) describe the conditions under which we may use the outer product of the gradient (OPG) vector in optimization. Known in the economics literature as the "B-H-cubed" method (from the initials of the last names of the authors of the previously cited paper), we may estimate the variance matrix without calculating second derivatives. Instead, we use the OPG. The gradient is the estimating equation or the vector of first derivatives when the derivation is the result of a likelihood approach. This is similar to the expected Hessian matrix except that we are not taking the expected values of the outer product. In the equation, $\mathbf{x}_i$ refers to the ith row of the matrix $\mathbf{X}$.

$$\widehat{\mathrm{V}}_{\mathrm{OPG}} = \sum_{i=1}^{n} \mathbf{x}_i^{\mathrm{T}} \left\{ \frac{y_i - \widehat{\mu}_i}{\mathrm{V}(\widehat{\mu}_i)} \left(\frac{\partial \mu}{\partial \eta} \right)_i \widehat{\phi} \right\}^2 \mathbf{x}_i \tag{3.68}$$

3.6.3 Sandwich

The validity of the covariance matrix based on the Hessian depends on the correct specification of the variance function (3.32) or the correct specification of both the variance and the link functions (3.51). The sandwich estimate of variance provides a consistent estimate of the covariance matrix of parameter estimates even when the specification of the variance function is incorrect.

The sandwich estimate of variance is a popular estimate of variance that is known by many different names; see Hardin (2003). The formula for calculating this variance estimate (either in algebraic or matrix form) has a score factor "sandwiched" between two copies of the Hessian. Huber (1967) is generally credited with first describing this variance estimate. Since his discussion addressed maximum likelihood estimates under misspecification, the variance estimate is sometimes called the robust variance estimate or Huber's estimate of variance.

White (1980) independently showed that this variance estimate is consistent under a model including heteroskedasticity, and so it is sometimes labeled the heteroskedasticity covariance matrix or heteroskedasticity consistent covariance matrix. Because of this same reference, the sandwich estimate of variance is also called White's estimator.

From the derivation of the variance estimate and the discussion of its properties in survey statistics literature papers such as Binder (1983), Kish and Frankel (1974), or Gail, Tan, and Piantadosi (1988), the sandwich estimate of variance is sometimes called the survey variance estimate or the design-based variance estimate. The statistics literature has also referred to the sandwich estimate of variance as the empirical estimate of variance. This name derives from the point of view that the variance estimate is not relying on the model being correct. In fact, we have heard in private conversation the descriptive "model-agnostic variance estimate", though we have not yet seen this one in the literature.

The estimated score is calculated using properties of the variance function (family) and the link function:

$$\text{score}(\beta)_i = \left(\frac{\partial \mathcal{L}}{\partial \eta}\right)_i = \left(\frac{\partial \mathcal{L}}{\partial \mu}\right)_i \left(\frac{\partial \mu}{\partial \eta}\right)_i = \frac{y_i - \widehat{\mu}_i}{\text{V}(\widehat{\mu})_i} \left(\frac{\partial \mu}{\partial \eta}\right)_i \tag{3.69}$$

The contribution from the estimated scores is then calculated, where $\mathbf{x}_i$ refers to the ith row of the matrix $\mathbf{X}$ using

$$\widehat{B}_{\text{S}} = \sum_{i=1}^{n} \mathbf{x}_i^{\text{T}} \left\{ \frac{y_i - \widehat{\mu}_i}{\text{V}(\widehat{\mu}_i)} \left(\frac{\partial \mu}{\partial \eta}\right)_i \widehat{\phi} \right\}^2 \mathbf{x}_i \tag{3.70}$$

and the (usual) sandwich estimate of variance is then

$$\widehat{\text{V}}_{\text{S}} = \widehat{\text{V}}_{\text{H}}^{-1} \widehat{B}_{\text{S}} \widehat{\text{V}}_{\text{H}}^{-1} \tag{3.71}$$

The OPG estimate of variance is the "middle" of the sandwich estimate of variance

$$\widehat{\text{V}}_{\text{OPG}} = \widehat{B}_{\text{S}} \tag{3.72}$$

Royall (1986) discusses the calculation of robust confidence intervals for maximum likelihood estimates using the sandwich estimate of variance.

3.6.4 Modified sandwich

If observations may be grouped because of some correlation structure (perhaps because the data are really panel data), then the sandwich estimate is calculated using the setup where we have n_i observations for each panel i and $\mathbf{x}_{ij}$ refers to the row of the matrix $\mathbf{X}$ associated with the jth observation for subject i.

The contribution from the estimated scores is then calculated, where $\mathbf{x}_{ij}$ refers to the (i, j)th row of the matrix $\mathbf{X}$ using

$$\widehat{B}_{\text{MS}} = \sum_{i=1}^{n} \left\{ \sum_{j=1}^{n_i} \mathbf{x}_{ij}^{\text{T}} \frac{y_{ij} - \widehat{\mu}_{ij}}{\text{V}(\widehat{\mu}_{ij})} \left(\frac{\partial \mu}{\partial \eta}\right)_{ij} \widehat{\phi} \right\} \left\{ \sum_{j=1}^{n_i} \frac{y_{ij} - \widehat{\mu}_{ij}}{\text{V}(\widehat{\mu}_{ij})} \left(\frac{\partial \mu}{\partial \eta}\right)_{ij} \widehat{\phi}\, \mathbf{x}_{ij} \right\} \tag{3.73}$$

as the modified (summed or partial) scores.

In either case, the calculation of $\widehat{\text{V}}_{\text{H}}$ is the same and the modified sandwich estimate of variance is then

$$\widehat{\text{V}}_{\text{MS}} = \widehat{\text{V}}_{\text{H}}^{-1} \widehat{B}_{\text{MS}} \widehat{\text{V}}_{\text{H}}^{-1} \tag{3.74}$$

The problem with referring to the sandwich estimate of variance as "robust" and the modified sandwich estimate of variance as "robust cluster" is that it implies to the unenlightened that robust standard errors are bigger than usual (Hessian) standard errors and robust cluster standard errors are bigger still. This variability is a false conclusion. See Carroll et al. (1998) for a lucid comparison of usual and robust standard errors. A comparison of the sandwich estimate of variance and the modified sandwich estimate of variance depends on the within-panel correlation of the score terms. If the within-panel correlation is negative, then the panel score sums of residuals will be small and the panel score sums will have less variability than the variability of the individual scores. This will lead to the modified sandwich standard errors being smaller than the sandwich standard errors.

Strictly speaking, the calculation of the modified sandwich estimate of variance is a statement that the fitted model is not represented by a true likelihood, even though the chosen family and link might otherwise imply a true likelihood. The fitted model is optimized assuming that the distribution of the sample may be calculated as the product of the distribution for each observation. However, when we calculate a modified sandwich estimate of variance, we acknowledge that the observations within a panel are not independent but that the sums over individual panels are independent. Our assumed or implied likelihood does not take into account this intrapanel correlation and therefore is not the true likelihood.

3.6.5 Unbiased sandwich

The sandwich estimate of variance has been applied in many cases and is becoming more common in statistical software. One area of active research concerns the small-sample properties of this variance estimate. There are two main modifications to the sandwich estimate of variance in constructing confidence intervals. The first is a degrees-of-freedom correction (scale factor) and the second is the use of a more conservative distribution ("fatter-tailed" than the normal).

Acknowledging that the usual sandwich estimate is biased, we may calculate an unbiased sandwich estimate of variance that has improved small-sample performance in coverage probability. This modification is a scale factor adjustment motivated by the knowledge that the variance of the estimated residuals are biased

$$\mathrm{V}[\widehat{\epsilon}_i] = \sigma^2(1 - h_i) \tag{3.75}$$

We can adjust for the bias of the contribution from the scores, where $\mathbf{x}_i$ refers to the ith row of the matrix $\mathbf{X}$ using

$$\widehat{B}_{\mathrm{US}} = \sum_{i=1}^{n} \mathbf{x}_i^{\mathrm{T}} \left\{ \frac{y_i - \widehat{\mu}_i}{\mathrm{V}(\widehat{\mu}_i)} \left(\frac{\partial \mu}{\partial \eta} \right)_i \widehat{\phi} \right\}^2 \mathbf{x}_i / (1 - \widehat{h}_i) \tag{3.76}$$

where the (unbiased) sandwich estimate of variance is then

$$\widehat{\mathrm{V}}_{\mathrm{US}} = \widehat{\mathrm{V}}_{\mathrm{H}}^{-1} \widehat{B}_{\mathrm{US}} \widehat{\mathrm{V}}_{\mathrm{H}}^{-1} \tag{3.77}$$

The Stata program includes this variance matrix in linear regression using the `hc2` option and cites MacKinnon and White (1985). This reference includes three scale factor adjustments. MacKinnon and White label the first adjustment HC1. The HC1 adjustment applies the scale factor $n/(n-k)$. This is a degree-of-freedom correction from Hinkley (1977). The second adjustment is labeled HC2 and is a scale factor adjustment of $1/(1-h_i)$. For linear regression, this scale factor is discussed by Belsley, Kuh, and Welsch (1980) and by Wu (1986). The final adjustment is labeled HC3 and is an application of the scale factor $1/(1-h_i)^2$. This scale factor is a jackknife approximation discussed by Efron in the Wu paper. The unadjusted sandwich variance matrix is labeled HC0—which makes 11 different names if you are counting.

We use the descriptive name "unbiased sandwich estimate of variance" found in Carroll et al. (1998) rather than the HC2 designation. Another excellent survey of the sandwich estimator may be found in the working paper by Long and Ervin (1998).

3.6.6 Modified unbiased sandwich

If observations may be grouped because of some correlation structure (perhaps because the data are really panel data), then the unbiased sandwich estimate is calculated using the setup where we have n_i observations for each panel i and $\mathbf{x}_{ij}$ refers to the row of the matrix $\mathbf{X}$ associated with the jth observation for subject i. This is the same adjustment as was made for the modified sandwich estimate of variance.

$$\begin{aligned}\widehat{B}_{\text{MUS}} &= \sum_{i=1}^{n}\left[\left\{\sum_{j=1}^{n_i}\mathbf{x}_{ij}^{\text{T}}\frac{y_{ij}-\widehat{\mu}_{ij}}{\text{V}(\widehat{\mu}_{ij})}\left(\frac{\partial\mu}{\partial\eta}\right)_{ij}\frac{\widehat{\phi}}{\sqrt{1-\widehat{h}_{ij}}}\right\}\right. \\ &\qquad\left.\times\left\{\sum_{j=1}^{n_i}\frac{y_{ij}-\widehat{\mu}_{ij}}{\text{V}(\widehat{\mu}_{ij})}\left(\frac{\partial\mu}{\partial\eta}\right)_{ij}\frac{\widehat{\phi}}{\sqrt{1-\widehat{h}_{ij}}}\mathbf{x}_{ij}\right\}\right] \end{aligned} \tag{3.78}$$

are the modified (summed or partial) scores. Here h_{ij} is the ijth diagonal of the hat matrix described in (4.5). The modified unbiased sandwich estimate of variance is then

$$\widehat{\text{V}}_{\text{MUS}} = \widehat{\text{V}}_{\text{H}}^{-1}\widehat{B}_{\text{MUS}}\widehat{\text{V}}_{\text{H}}^{-1} \tag{3.79}$$

3.6.7 Weighted sandwich: Newey–West

These variance estimators are referred to as HAC variance estimators since they are heteroskedasticity and autocorrelation consistent estimates of the variances (of parameter estimators). A weighted sandwich estimate of variance calculates a (possibly) different middle of the sandwich. Instead of using only the usual score contributions, a weighted sandwich estimate of variance calculates a weighted mean of score contributions and lagged score contributions.

Newey and West (1987) discuss a general method for combining the contributions for each considered lag. The specific implementation then assigns a weight to each lagged score contribution. In related sections, we present various weight functions for use with this general approach.

In the following, let n be the number of observations, p be the number of predictors, G be the maximum lag, C be an overall scale factor, and q be the prespecified bandwidth. The overall scale factor is usually defined as one but could be defined as $n/(n-p)$ to serve as a small sample scale factor adjustment.

$$\widehat{\mathrm{V}}_{\mathrm{NW}} = \widehat{\mathrm{V}}_{\mathrm{H}}^{-1}\widehat{B}_{\mathrm{NW}}\widehat{\mathrm{V}}_{\mathrm{H}}^{-1} \tag{3.80}$$

$$\widehat{B}_{\mathrm{NW}} = C\left\{\widehat{\Omega}_0 + \sum_{j=1}^{G}\omega\left(\frac{j}{q+1}\right)\left(\widehat{\Omega}_j + \widehat{\Omega}_j'\right)\right\} \tag{3.81}$$

$$\widehat{\Omega}_j = \sum_{i=j+1}^{n}\mathbf{x}_i\widehat{r}_i^{\,S}\,\widehat{r}_{i-j}^{\,S}\mathbf{x}_i^{\mathrm{T}} \tag{3.82}$$

$$\omega(z) = \text{sandwich weights} \tag{3.83}$$

$$\widehat{r}_i^{\,S} = \text{score residuals (see sec. 4.5.9)} = \nabla_i(y_i-\mu_i)/v_i \tag{3.84}$$

To calculate the Newey–West variance–covariance matrix for a given model, we need the link function, the derivative of the link function, the mean-variance function, and the sandwich weights

$$\{\mu_i, \nabla_i, v_i, \omega(z)\} = \{\mu_i, (\partial\mu/\partial\eta)_i, \mathrm{V}(\mu_i), w_i\} \tag{3.85}$$

The needed quantities are defined in tables A.1, A.2, A.4, and A.8.

Specifically for $G=0$ (no autocorrelations), $\widehat{B}_{\mathrm{NW}} = \widehat{\Omega}_0 = \widehat{B}_{\mathrm{S}}$ from (3.70) so that $\widehat{\mathrm{V}}_{\mathrm{NW}}$ is equal to a robust sandwich estimate of variance $\widehat{\mathrm{V}}_{\mathrm{S}}$ (see sec. 3.6.3) for all the above weight functions.

In table A.8, the Newey and West (1987) estimator uses weights derived from the Bartlett kernel for a user-specified bandwidth (number of autocorrelation lags). These are the usual kernel weights in software packages implementing this weighted sandwich estimate of variance. For $G=0$, the kernel is more appropriately called the truncated kernel, resulting in the sandwich variance estimator.

In table A.8, the Gallant (1987) estimator (see p. 533) uses weights derived from the Parzen (1957) kernel for a user-specified bandwidth (number of autocorrelation lags).

In table A.8, the Anderson estimator uses weights derived from the quadratic spectral kernel. These weights are based on a spectral decomposition of the variance matrix.

A general class of kernels is given in Andrews (1991), where he refers to using the Tukey–Hanning kernel

$$\omega(z) = \begin{cases} \{1+\cos(\pi z)\}/2 & \text{if } |z|\le 1 \\ 0 & \text{otherwise} \end{cases} \tag{3.86}$$

In fact, Andrews mentions an entire class of admissible kernels and his paper investigates bandwidth selection.

Automatic lag selection is also investigated in Newey and West (1994). In further research, Lumley and Heagerty (1999) provide details on weighted sandwich estimates of variance where the weights are data driven (rather than kernel driven). These are presented under the name *weighted empirical adaptive variance estimates* (WEAVEs), given that the weights are computed based on estimated correlations from the data instead of on fixed weights based on user-specified maximum lag value.

3.6.8 Jackknife

The jackknife estimate of variance estimates variability in fitted parameters by comparing results from leaving out one observation at a time in repeated estimations. Jackknifing is based on a data resampling procedure in which the variability of an estimator is investigated by repeating an estimation with a subsample of the data. Subsample estimates are collected and compared with the full sample estimate to assess variability. Introduced by Quenouille (1949), an excellent review of this technique and extensions is available in Miller (1974).

The sandwich estimate of variance is related to the jackknife. Asymptotically, it is equivalent to the one-step and iterated jackknife estimates, and as shown in Efron (1981), the sandwich estimate of variance is equal to the infinitesimal jackknife.

There are two general methods for calculating jackknife estimates of variance. One approach is to calculate the variability of the individual estimates from the full sample estimate. We supply formulas for this approach. A less conservative approach is to calculate the variability of the individual estimates from the average of the individual estimates. You may see references to this approach in other sources. Generally, we prefer the approach outlined here because of its more conservative nature.

3.6.8.1 Usual jackknife

A jackknife estimate of variance may be formed by estimating the coefficient vector for $n-1$ observations n different times. Each of the n estimates is obtained for a different subsample of the original dataset. The first estimate considers a subsample in which only the first observation is deleted from consideration. The second estimate considers a subsample where only the second observation is deleted from consideration. The process continues until n different subsamples have been processed and the n different estimated coefficient vectors are collected. The n estimates are then used as a sample from which the variability of the estimated coefficient vector may be estimated.

The jackknife estimate of variance is then calculated as

$$\widehat{\mathrm{V}}_{\mathrm{J}} = \frac{n-p}{n} \sum_{i=1}^{n} \left(\widehat{\boldsymbol{\beta}}_{(i)} - \widehat{\boldsymbol{\beta}}\right) \left(\widehat{\boldsymbol{\beta}}_{(i)} - \widehat{\boldsymbol{\beta}}\right)^{\mathrm{T}} \tag{3.87}$$

where $\widehat{\beta}_{(i)}$ is the estimated coefficient vector leaving out the ith observation and p is the number of predictors (possibly including a constant).

3.6.8.2 One-step jackknife

The updating equations for one-step jackknife type estimates come from

$$\widehat{\boldsymbol{\beta}}^{*}_{(i)} = \widehat{\boldsymbol{\beta}} - \frac{(X^{\mathrm{T}}\widehat{\mathrm{V}}X)^{-1}\mathbf{x}_i^{\mathrm{T}}\widehat{r}_i^R}{1-\widehat{h}_i} \tag{3.88}$$

where $\widehat{\mathrm{V}} = \mathrm{diag}\{\mathrm{V}(\widehat{\mu})\}$ is an $(n \times n)$ matrix, $\mathbf{X}$ is the $(n \times p)$ matrix of covariates, $\widehat{h}_i$ is the ith diagonal of the hat matrix described in (4.5), $\widehat{\beta}$ is the estimated coefficient vector using all observations, and $\widehat{r}_i^R$ is the estimated response residual (see sec. 4.5.1). In this approach, instead of fitting (to full convergence) each of the n candidate models, we instead use the full observation estimate as a starting point and take only one Newton–Raphson step (iteration).

The one-step jackknife estimate of variance is then calculated as

$$\widehat{\mathrm{V}}_{\mathrm{J1}} = \frac{n-p}{n}\sum_{i=1}^{n}\left(\widehat{\boldsymbol{\beta}}^{*}_{(i)} - \widehat{\boldsymbol{\beta}}\right)\left(\widehat{\boldsymbol{\beta}}^{*}_{(i)} - \widehat{\boldsymbol{\beta}}\right)^{\mathrm{T}} \tag{3.89}$$

where $\widehat{\boldsymbol{\beta}}^{*}_{(i)}$ is defined in (3.88) and p is the number of predictors (possibly including a constant).

3.6.8.3 Weighted jackknife

Just as the subsample estimates may be combined with various weighting schemes representing different underlying assumptions, we may also generalize the jackknife resampling estimate. Hinkley (1977) is credited with introducing this extension. Though an important extension in general, this particular estimate of variance is not included in the associated software for this text.

3.6.8.4 Variable jackknife

The variable jackknife is an extension that allows different subset sizes. For a clear review of this and other resampling methods, see Wu (1986) or Efron (1981). This flexible choice of subset size allows specifying the independent subsamples of the data.

A variable jackknife estimate of variance may be formed by generalizing the strategy for what is left out at each estimation step. We can leave out a group of observations at a time. The usual jackknife variance estimate is then a special case of the variable jackknife where each group is defined as 1 observation.

The variable jackknife estimate of variance is then calculated as

$$\widehat{\mathrm{V}}_{\mathrm{VJ}} = \frac{n-p}{ng} \sum_{i=1}^{g} \left(\widehat{\boldsymbol{\beta}}_{(i)}^{\mathrm{VJ}} - \widehat{\boldsymbol{\beta}} \right) \left(\widehat{\boldsymbol{\beta}}_{(i)}^{\mathrm{VJ}} - \widehat{\boldsymbol{\beta}} \right)^{\mathrm{T}} \tag{3.90}$$

where $\widehat{\boldsymbol{\beta}}_{(i)}^{\mathrm{VJ}}$ is the estimated coefficient vector leaving out the ith group of observations, g is the number of groups, and p is the number of predictors (possibly including a constant).

3.6.9 Bootstrap

The bootstrap estimate of variance is based on a data resampling procedure in which the variability of an estimator is investigated by repeating an estimation with a subsample of the data. Subsamples are drawn with replacement from the original sample and subset estimates are collected and compared with the full (original) sample estimate.

There are two major types of bootstrap variance estimates: parametric and nonparametric. Parametric bootstrap variance estimates are obtained by resampling residuals and nonparametric variance estimates are obtained by resampling cases (as described above). Because the provided software calculates nonparametric bootstrap variance estimates, that is what we describe in the following sections.

Within the area of nonparametric bootstrap variance estimates, one may calculate the variability about the mean of the bootstrap samples or about the original estimated coefficient vector. The former is more common, but we generally use (and have implemented) the latter because it is more conservative.

As a note of caution: applying this variance estimate usually calls for art as well as science because there are usually some bootstrap samples that are "bad". They are bad in the sense that estimation results in infinite coefficients, collinearity, or some other difficulty in obtaining estimates. The software provided does not allow the user to apply art to the calculation in the form of removing a subset of bootstrap results.

3.6.9.1 Usual bootstrap

A bootstrap estimate of variance may be formed by estimating the coefficient vector k different times. Each of the k estimates is obtained by fitting the model using a random sample of the n observations (sampled with replacement) drawn from the original sample dataset.

A bootstrap estimate of variance may then calculated as

$$\widehat{\mathrm{V}}_{\mathrm{BS}} = \frac{n-p}{nk} \sum_{i=1}^{k} \left(\widehat{\boldsymbol{\beta}}_{i}^{\mathrm{BS}} - \widehat{\boldsymbol{\beta}} \right) \left(\widehat{\boldsymbol{\beta}}_{i}^{\mathrm{BS}} - \widehat{\boldsymbol{\beta}} \right)^{\mathrm{T}} \tag{3.91}$$

where $\widehat{\boldsymbol{\beta}}_{i}^{\mathrm{BS}}$ is the estimated coefficient vector for the ith sample (drawn with replacement) and p is the number of predictors (possibly including a constant).

3.6.9.2 Grouped bootstrap

A grouped (or cluster) bootstrap estimate of variance may be formed by estimating the coefficient vector k different times. Each of the k estimates is obtained by fitting the model using a random sample of the g groups of observations (sampled with replacement) drawn from the original sample dataset of g groups. The sample drawn for the groups in this case is such that all members of a group are included in the sample if the group is selected.

A bootstrap estimate of variance may then calculated as

$$\widehat{\mathrm{V}}_{\mathrm{GBS}} = \frac{n-p}{nk} \sum_{i=1}^{k} \left(\widehat{\boldsymbol{\beta}}_i^{\mathrm{GBS}} - \widehat{\boldsymbol{\beta}}\right) \left(\widehat{\boldsymbol{\beta}}_i^{\mathrm{GBS}} - \widehat{\boldsymbol{\beta}}\right)^{\mathrm{T}} \tag{3.92}$$

where $\widehat{\boldsymbol{\beta}}_i^{\mathrm{GBS}}$ is the estimated coefficient vector for the ith sample (drawn with replacement) and p is the number of predictors (possibly including a constant).

3.7 Estimation algorithms

Here we present algorithms for fitting GLMs. The algorithms do not illustrate all the details such as choosing starting values or alternative criteria for convergence. Instead, they focus on the major steps of implementing the algorithms. Actual software implementations will have modifications of the above, but for those implementations limited to the list of variance functions and link functions that are outlined in the tables in the appendix, these algorithms are an accurate representation. The Newton–Raphson algorithm assumes that the ϕ parameter is ancillary so that the estimates of $\boldsymbol{\beta}$ are properly called restricted maximum likelihood. The algorithm may be generalized by including ϕ in the parameter list to obtain ML estimates of all parameters.

In addition to the issues raised previously, other modifications can be made to the algorithms. In particular, we raised the issue of whether we can use the OIM or EIM in the implementation of the IRLS algorithm. Also we mentioned that `PROC GENMOD` used the EIM in initial iterations of the Newton–Raphson maximum likelihood routines before switching to the use of the OIM in later iterations. These types of modifications are easily implemented in the following algorithms:

(*Continued on next page*)

Listing 3.1: IRLS algorithm

```
1  Initialize μ
2  Calculate η = g(μ)                                              Table A.2
3  Set dev_old = 0
4  Set dev_new = 1
5  Set Δdev = 1
6  while(abs(Δdev) > tolerance) {
7      Calculate W = 1/[(g')^2 V(μ)a(φ)]         (3.47), Table A.1, Table A.4
8      Calculate z = (y − μ)g' + η − offset                (3.48), Table A.4
9      Regress z on X with weights W
10     Calculate η = Xβ+offset
11     Calculate μ = g^{-1}(η)                                     Table A.2
12     Set dev_old = dev_new
13     Calculate dev_new                                   (3.50), Table A.11
14     Calculate Δdev = dev_new - dev_old
15 }
16 Calculate variance matrix                                     Section 3.6
```

Listing 3.2: Newton–Raphson algorithm

```
1  Initialize β
2  while(||β_new − β_old|| > tol and abs(L_new − L_old) > ltol) {
3      Calculate g = ∂L/∂β                               Table A.1, Table A.2
4      Calculate H = ∂^2L/∂β^2   Table A.1, Table A.2, Table A.4, Table A.6
5      Set β_old = β_new
6      Calculate β_new = β_old − H^{-1} g
7      Set L_old = L_new
8      Calculate L_new                                             Table A.7
9  }
10 Calculate variance matrix                                     Section 3.6
```

3.8 Summary

This chapter covers several topics. The IRLS method is the simplification that results when the Newton–Raphson method uses the expected Hessian matrix. Either way, these approaches are the result of theoretical arguments and assumptions that admit a function (an estimating equation) to be optimized.

The Newton–Raphson method provides an updating equation for optimizing the parameter vector of interest. The method may then be codified in algorithms in several ways. For example, algorithm development may implement various techniques to choose step sizes and directions for the updating equation. The IRLS method does not specifically rely on an assumed likelihood; it was further justified based only on moment assumptions.

Algorithms are intended to optimize a function. Algorithms derived from methods in ML and MQL will result in the same estimated coefficient vector except for roundoff and differences in convergence criteria (when there is a true underlying likelihood). This statement is true regardless of whether the model uses the canonical link.

Variance matrices are estimated for the parameter vector of interest. The variance matrix is calculated after the function has been optimized. Since IRLS usually uses V_{EH} as part of its algorithm, that most software implementations provide estimated standard errors based on this matrix is not surprising. Since Newton–Raphson algorithms usually use V_{OH}, that most software implementations provide standard errors based on that matrix is also no surprise. In both cases, the usual supplied standard errors, or variance matrix, are selected on the basis of an efficiency (in the programming sense rather than the statistical sense) argument; the matrix is already calculated and available from the optimization algorithm.

For either approach, we are, however, free to calculate variance matrices described in section 3.6. We do not equate the estimated variance matrix with the method or the algorithm. If we obtain estimates for a model using the expected Hessian for optimization, for example, that does not preclude us from calculating and using V_{OH} to estimate standard errors.

4 Analysis of fit

Contents

This chapter presents various statistics and techniques useful for assessing the fit of a GLM. Our presentation here is a catalog of formulas and reference material. Since illustrations of the concepts and techniques cataloged here are left to the individual chapters later in the book, first-time readers may find glancing ahead at those examples useful as they read through this material. This organization makes introducing the material slightly more inconvenient but makes introducing it as reference material much more convenient.

In assessing a model, we assume that the link and variance functions are correctly specified. Another assumption of our model is that the linear structure expresses the relationship of the explanatory variables to the response variable. We present information in this chapter useful for assessing the validity of these assumptions. One detail of GLMs not covered in this text is that of correcting for the bias in coefficient vector estimates. Interested readers can find such information in Cordeiro and McCullagh (1991) or Neuhaus and Jewell (1993).

4.1 Deviance

In a maximum likelihood approach, a standard assessment is to compare the fitted model with a fully specified model (a model with as many independent parameters as observations). The scaled deviance S is given in terms of the likelihoods of the (fitted) model m and the full model f by

$$S = -2\ln(L_m/L_f) \tag{4.1}$$

Thus it has the familiar form (and limiting χ^2 distribution) of a likelihood-ratio test statistic.

We may generalize this concept to GLMs such that

$$S = D/\phi \tag{4.2}$$

where D is the deviance.

A limitation of this statistic is that its distribution is exactly chi-squared for the Gaussian family with an identity link but is asymptotically chi-squared for other variance functions and links.

The deviance is calculated for each family by using the canonical parameter, which is the function $\theta(\mu)$ illustrated in the canonical links throughout the text. Our goal in assessing deviance is to determine the utility of the parameters added to the null model.

The deviance for GLMs is calculated as

$$D(u;\widehat{\mu}) = \sum_i 2\left[y\{\theta(y_i) - \theta(\widehat{\mu}_i)\} - b\{\theta(y_i)\} + b\{\theta(\widehat{\mu}_i)\}\right] \tag{4.3}$$

where $\theta()$ is the canonical parameter and $b()$ is the cumulant. Brown (1992) offers more advice on graphically assessing the deviance, and the deviance concept is generalized to quasilikelihoods in chapter 17. Throughout the text, we derive properties for a given family and then illustrate deriving the deviances (which have an underlying likelihood) in more succinct terms using

$$D = 2\phi\left\{\mathcal{L}(y;y) - \mathcal{L}(y;\mu)\right\} \tag{4.4}$$

where $\mathcal{L}(y;y) = \ln(L_f)$ and $\mathcal{L}(y;\mu) = \ln(L_m)$.

The notation $\mathcal{L}(y;y)$ for the full model is a reflection that the saturated model perfectly captures the outcome variable. Thus the model's predicted values $\widehat{\mu} = y$. So, the difference in deviance statistics between the saturated and fitted model captures the distance between the predicted values and the outcomes.

Models may be compared using deviance-based statistics. When competing models are nested, this comparison is a straightforward application of likelihood-ratio tests. Wahrendorf, Becher, and Brown (1987) discuss bootstrap comparison of nonnested GLM.

4.2 Diagnostics

We will use $\widehat{\mathbf{W}}$ from (3.47), and $\widehat{h}_{ii}$ is the ith diagonal of

$$\widehat{\mathbf{H}} = \widehat{\mathbf{W}}^{1/2}\mathbf{X}(\mathbf{X}^{\mathrm{T}}\widehat{\mathbf{W}}\mathbf{X})^{-1}\mathbf{X}^{\mathrm{T}}\widehat{\mathbf{W}}^{1/2} \tag{4.5}$$

and denotes the hat diagonals as a generalization of the same concept in linear regression. Pregibon (1981) presents this definition in assessing "leverage" for logistic distribution. The hat diagonals in regression have many useful properties. The property of leverage (or potential) in linear regression relates to the fact that as the hat diagonal for the ith observation approaches one, the corresponding residual approaches zero. Williams (1987) discusses diagnostics based on the deviance and single-case deletions.

4.2.1 Cook's distance

By using the one-step approximations to the change in the estimated coefficients when an observation is left out of the estimation procedure, we approximate Cook's distance with

$$C_i = \left(\widehat{\boldsymbol{\beta}}^*_{(i)} - \widehat{\boldsymbol{\beta}}\right)^{\mathrm{T}} I \left(\widehat{\boldsymbol{\beta}}^*_{(i)} - \widehat{\boldsymbol{\beta}}\right) \tag{4.6}$$

where I is the Fisher information (inverse Hessian; see sec. 3.6.1), and $\widehat{\boldsymbol{\beta}}^*_{(i)}$, defined in (3.88), is the one-step approximation to the jackknife-estimated coefficient vector. These measures can be used to detect observations with undue influence. Standard approaches include declaring those observations with measures greater than $4/(n-p-1)$ to be problematic and those measures greater than $4/n$ to be worth investigating.

4.2.2 Overdispersion

Overdispersion is a phenomenon that occurs with data fitted using the binomial, Poisson, or negative binomial distribution. If the estimated dispersion after fitting is not near the assumed value, then the data may be overdispersed if the value is greater than expected or underdispersed if the value is less than expected. Overdispersion is far more common. For overdispersion, the simplest remedy is to assume a multiplicative factor in the usual implied variance. As such, the resulting covariance matrix will be inflated through multiplication by the estimated scale parameter. Care should be exercised, because an inflated estimated dispersion parameter may result from model misspecification rather than overdispersion, indicating that the model should be assessed for appropriateness by the researcher. Smith and Heitjan (1993) discuss testing and adjusting for departures from the nominal dispersion, Breslow (1990) discusses Poisson regression as well as other quasilikelihood models, and Cox (1983) gives a general overview of overdispersion. A discussion of this topic is in chapter 11.

Ganio and Schafer (1992), Lambert and Roeder (1995), and Dean and Lawless (1989) discuss diagnostics for overdispersion in GLMs.

A score test effectively compares the residuals with their expectation under the model. A test for overdispersion of the Poisson model, which compares the variance with the variance of a negative binomial, is given by

$$T_1^2 = \frac{\left[\sum_{i=1}^n \left\{(y_i - \widehat{\mu}_i)^2 - (1 - \widehat{h}_i)\widehat{\mu}_i\right\}\right]^2}{2\sum_{i=1}^n \widehat{\mu}_i^2} \sim \chi_1^2 \tag{4.7}$$

This formula tests the null hypothesis H_0: $\phi = 0$, assuming that the variance is described by the function $v(\mu, \phi) = \mu + \phi\mu^2$ (negative binomial variance).

Another test for the Poisson model is given by

$$T_2^2 = \frac{1}{2n}\left\{\sum_{i=1}^n \frac{(y_i - \widehat{\mu}_i)^2 - (1 - \widehat{h}_i)\widehat{\mu}_i}{\widehat{\mu}_i}\right\}^2 \sim \chi_1^2 \tag{4.8}$$

This formula tests the null hypothesis H_0: $\phi = 1$, assuming that the variance is described by the function $v(\mu, \phi) = \phi\mu$ (scaled Poisson variance).

4.3 Assessing the link function

We may wish to investigate whether the link function is appropriate. For example, in Poisson regression we may wish to examine whether the usual log-link (multiplicative effects) is appropriate compared with an identity link (additive effects). For binomial regression, we may wish to compare the logit link (symmetric about one half) with the complementary log-log link (asymmetric about one half).

Pregibon (1980) advocates the comparison of two link functions by embedding them in a parametric family of link functions. The Box–Cox family of power transforms

$$g(\mu; \lambda) = \frac{\mu^\lambda - 1}{\lambda} \tag{4.9}$$

and yields the log-link at $\lim_{\lambda\to 0} g(\mu; \lambda)$ and the identity link at $\lambda = 1$. Likewise, the family

$$g(\mu; \lambda) = \ln\left\{\frac{(1 - \mu)^{-\lambda} - 1}{\lambda}\right\} \tag{4.10}$$

gives the logit link at $\lambda = 1$ and the complementary log-log link at $\lim_{\lambda\to 0} g(\mu; \lambda)$.

We may test the hypothesis H_0: $\lambda = \lambda_0$ by fitting the model with the standard link ($\lambda = \lambda_0$) and by obtaining the fitted values $\widehat{\mu}_i$. We then construct the variable

$$z = -\left.\frac{\partial g(\widehat{\mu};\lambda)}{\partial \lambda}\right|_{\lambda=\lambda_0} \tag{4.11}$$

and use the standard (chi-squared) score test for added variables.

For instance, to test the log link ($\lambda = 0$) embedded in the Box–Cox power family, we see that

$$-\left.\frac{\partial g(\widehat{\mu};\lambda)}{\partial \lambda}\right|_{\lambda=0} = -\left.\left\{\frac{(\ln\widehat{\mu})\exp(\lambda\ln\widehat{\mu})}{\lambda} - \frac{\exp(\lambda\ln\widehat{\mu})-1}{\lambda^2}\right\}\right|_{\lambda=0} \tag{4.12}$$

$$= -\lim_{\lambda\to 0}\left\{\frac{(\ln\widehat{\mu})\exp(\lambda\ln\widehat{\mu})}{\lambda} - \frac{\exp(\lambda\ln\widehat{\mu})-1}{\lambda^2}\right\} \tag{4.13}$$

$$= -\lim_{\lambda\to 0}\left\{(\ln\widehat{\mu})^2\exp(\lambda\ln\widehat{\mu}) - \frac{(\ln\widehat{\mu})^2}{2}\exp(\lambda\ln\widehat{\mu})\right\} \tag{4.14}$$

$$= -\frac{(\ln\widehat{\mu})^2}{2} \tag{4.15}$$

(using two steps of l'Hôpital's Rule). The test is then a standard (chi-squared) score test for an omitted variable. That is, the originally hypothesized link function is used and fitted values $\hat{\mu}$ are obtained. The example above shows how those fitted values are then used to generate a new predictor variable. Adding this predictor variable (which forms the systematic difference between the original link and the hypothesized alternative) is then assessed through inspecting the change in deviance.

4.4 Checks for systematic departure from the model

To check for systematic departure from the model, the standard diagnostic plots include a plot of the standardized deviance residuals versus the linear predictor or versus the linear predictor transformed to the constant-information scale of the error distribution. Also plot a normal quantile plot of the deviance or Anscombe residuals.

To check individual covariates, plot residuals versus one of the predictor variables. There should be no trend because the null pattern is the same as that for residuals versus fitted values.

Added variable plots check whether an omitted covariate, u, should be included in the linear predictor. First we obtain the unstandardized residuals for u as response, using the same linear predictor and quadratic weights as for y. The unstandardized residuals for y are then plotted against the residuals for u. There should be no trend if u is correctly omitted.

Plot the absolute residuals versus the fitted values as an informal check of the adequacy of the assumed variance function. The constant information scale for the fitted values is usually helpful in spreading out the points on the horizontal scale.

A standard approach to investigating the link function for a given model is to generate a covariate equal to the square of the linear predictor for that model. Refitting the model with the new predictor, one assesses the goodness of the specification by looking at the p-value associated with the additional covariate. If that covariate is significant, then there is evidence of lack of fit. This process is easily done in Stata via the `linktest` command; see the `linktest` manual entry in the *Stata Base Reference Manual.*

Exploratory data analysis plots of the outcome and various predictors are useful. Such plots allow the investigator to assess the linearity assumption as well as to diagnose outliers.

4.5 Residual analysis

In assessing the model, residuals measure the discrepancy between our observed and fitted values for each observation. The degree to which one observation affects the estimated coefficients is a measure of influence.

Pierce and Schafer (1986) and Cox and Snell (1968) provide excellent surveys of various definitions for residuals in GLMs. In the following sections, we present the definitions of several residuals that have been proposed for GLMs.

Discussions on residuals are hampered by a lack of uniform terminology throughout the literature, so we will expand our descriptions to facilitate comparison to other books and papers.

We will use the following conventions when naming residuals:

- A residual is named based on the statistic from which it was derived or the author that proposed its use.
- The adjective *modified* means that the denominator of a residual has been modified to be a reasonable estimate of the variance of y. The base residual has been multiplied by the factor $(k/w_i)^{-1/2}$, where k is a scale parameter (defaulting to one if not set).
- The adjective *standardized* means that the variance of the residual has been standardized to take into account the correlation between y and $\widehat{\mu}$. The base residual has been multiplied by the factor $(1-h)^{-1/2}$.
- The adjective *studentized* means that the residuals have been scaled by an estimate of the unknown scale parameter. The base residual has been multiplied by the factor $\widehat{\phi}^{-1/2}$. In many texts and papers, it is the standardized residuals that are then scaled (studentized). However, the result is often called *studentized residuals* instead of the more accurate *standardized studentized residuals*.

- The adjective *adjusted* means that the residual has been adjusted (per variance function family) from the original definition. The adjustment has to do with including an extra term to bring higher moments of the resulting distribution in line with expected values for a normal distribution. We have seen this adjustment only with deviance residuals, Pierce and Schafer (1986), so we include a specific section on it.

Using the above lexicon, we prefer the more descriptive "modified standardized Pearson residuals" over "Cox–Snell residuals", even though the latter is fairly common (because of the description mentioned at the beginning of this section).

Under these conventions, we present software at the end of the book that allows generating residuals using these adjectives as option names. This schema enables users of the software to exactly specify desired calculations.

A general aim for defining residuals is that their distribution is approximately normal. This goal is accomplished by considering linear equations, transformed linear equations, and deviance contributions. For transformed linear equations, our choice of transformation can be made to ensure first-order skewness of zero (Anscombe residuals) or to ensure that the transformed variance is constant (variance-stabilizing residuals) or at least comparable (Pearson residuals).

For achieving approximate normality, the Anscombe and adjusted deviance residuals are similar and perform well. If we are considering a model using the binomial family and if the denominator is less than 10, the Anscombe residuals are preferred. Otherwise, the adjusted deviance residuals may be used.

In the following sections, we catalog the formulas for residuals of interest. Throughout these formulas, the quantity $\partial\eta/\partial\mu$ is evaluated at $\widehat{\mu}$.

4.5.1 Response residuals

These residuals are simply the difference between the observed and fitted outcomes.

$$r_i^R = y_i - \widehat{\mu}_i \tag{4.16}$$

4.5.2 Working residuals

These residuals are the difference between the working (synthetic) response and the linear predictor at convergence

$$r_i^W = (y_i - \widehat{\mu}_i)\left(\frac{\partial\eta}{\partial\mu}\right)_i \tag{4.17}$$

and are associated with the working response and the linear predictor in the IRLS algorithm given in (3.48).

4.5.3 Pearson residuals

Pearson residuals are a rescaled version of the working residuals. The sum of the squared Pearson residuals is equal to the Pearson chi-squared statistic. The scaling puts the residuals on similar scales of variance so that standard plots of Pearson residuals versus an individual predictor or versus the outcome will reveal dependencies of the variance on these other factors (if such dependencies exist).

$$r_i^P = \frac{y_i - \widehat{\mu}_i}{\sqrt{\mathrm{V}(\widehat{\mu}_i)}} \tag{4.18}$$

Large absolute values of the residual indicate the model's failure to fit a particular observation. A common diagnostic for detecting outliers is to plot the standardized Pearson residuals versus the observation number.

4.5.4 Partial residuals

Partial residuals are used to assess the form of a predictor and are thus calculated for each predictor. Hines and Carter (1993) discuss the graphical use of these residuals in assessing model fit.

$$r_{ki}^T = (y_i - \widehat{\mu}_i)\left(\frac{\partial \eta}{\partial \mu}\right)_i + (x_{ik}\widehat{\beta}_k) \tag{4.19}$$

where $k = 1, \ldots, p$ and p is the number of predictors. In the above equation, $(x_{ki}\widehat{\beta}_k)$ refers to the ith observation of the kth predictor times the kth fitted coefficient.

4.5.5 Anscombe residuals

In response to a paper by Hotelling, Anscombe (1972) contributed the following definition of residuals. First, let

$$A(\cdot) = \int \frac{d\mu}{\mathrm{V}^{1/3}(\mu)} \tag{4.20}$$

where the residual is

$$r_i^A = \frac{A(y_i) - A(\widehat{\mu}_i)}{A'(\widehat{\mu}_i)\sqrt{\mathrm{V}(\widehat{\mu}_i)}} \tag{4.21}$$

The choice of the function $A(\cdot)$ was made so that the resulting residuals would be as normal as possible (have a distribution that was approximately normal).

These residuals may be difficult to compute since they involve ${}_2F_1()$, the hypergeometric ${}_2F_1()$ function (not common in statistical packages). Among the commonly supported variance functions, dependence on this function is limited to the binomial and negative binomial family. However, this function is available in the Stata package and so residuals are fully supported even for these variance functions. Formulas for the residuals are given for common families in table A.10.

4.5.6 Deviance residuals

The deviance plays a key role in our derivation of generalized linear modeling and in the inference of our results. The deviance residual is the increment to the overall deviance for each observation. These residuals are common and are often standardized, studentized, or both. The residuals are based on the χ^2 distribution.

$$r_i^D = \text{sign}(y_i - \widehat{\mu}_i)\sqrt{\widehat{d}_i^2} \tag{4.22}$$

Computation formulas of d_i^2 for individual families are given in table A.11.

In general, the deviance residual (standardized or not) is preferred to the Pearson residual for model checking since its distributional properties are closer to the residuals arising in linear regression models.

4.5.7 Adjusted deviance residuals

The deviance residual may be adjusted (corrected) to make the convergence to the limiting normal distribution faster. The adjustment removes an $O(n^{-1/2})$ term.

$$r_i^{D_a} = r_i^D + \frac{1}{6}\rho_3(\theta) \tag{4.23}$$

where $\rho_3(\theta)$ is defined for individual families in table A.11.

4.5.8 Likelihood residuals

Likelihood residuals are a combination of standardized Pearson residuals, $r_i^{P'}$, and standardized deviance residuals, $r_i^{D'}$.

$$r_i^L = \text{sign}(y_i - \widehat{\mu}_i)\left\{h_i(r_i^{P'})^2 + (1-h_i)(r_i^{D'})^2\right\}^{1/2} \tag{4.24}$$

4.5.9 Score residuals

These are the scores used in calculating the sandwich estimate of variance. The scores are related to the score function (estimating equation), which is optimized.

$$r_i^S = \frac{y_i - \widehat{\mu}_i}{\text{V}(\widehat{\mu}_i)}\left(\frac{\partial\eta}{\partial\mu}\right)_i^{-1} \tag{4.25}$$

4.6 Model statistics

Here we catalog several commonly used model fit statistics and criterion measures for comparing models. An excellent source for more information on these topics is Long

(1997). Most of the following measures are available to Stata users via the `fitstat` command implemented in Long and Freese (2000). Users may also use the official `estat` command, though the Bayesian information criterion (BIC) and Akaike information criterion (AIC) statistics available through these commands are not scaled by the number of observations.

4.6.1 Criterion measures

In the following definitions,

$$\begin{aligned} p &= \text{number of predictors} && (4.26)\\ n &= \text{number of observations} && (4.27)\\ L(M_k) &= \text{likelihood for model } k && (4.28)\\ \mathcal{L}(M_k) &= \text{log likelihood for model } k && (4.29)\\ D(M_k) &= \text{deviance of model } k && (4.30)\\ G^2(M_k) &= \text{likelihood-ratio test of model } k && (4.31) \end{aligned}$$

Next we provide the formulas for two criterion measures useful for model comparison. They include terms based on the log likelihood along with a penalty term based on the number of parameters in the model. In this way, the criterion measures seek to balance our competing desires for finding the best model (in terms of maximizing the likelihood) with model parsimony (including only those terms that significantly contribute to the model).

We introduce the two main model selection criterion measures below; one may also see Hansen and Yu (2001) for more considerations.

4.6.1.1 AIC

The Akaike (1973) information criterion may be used to compare competing models. A comparison may be made to nonnested models or models calculated across different samples. The criterion is such that the lower the value, the better fitting the model. Furthermore, a difference of greater than two indicates a marked preference for the model with the smaller criterion measure. The formula is given by

$$\text{AIC} = \frac{-2\mathcal{L}(M_k) + 2p}{n} \tag{4.32}$$

Including the term involving the number of parameters p is a penalty such that the criterion measure is especially amenable to comparing GLMs of the same link and variance function, but different covariate lists. That the penalty term specifies $2p$ as the penalty term is somewhat analogous to the choice of level in a formal hypothesis test. Some authors have suggested $3p$ or other values in comparing models, e.g., Atkinson (1988). Here we view the penalty term as the precision required to eliminate a candidate predictor from the model.

One should take care on how this statistic is calculated. The above definition includes the model log likelihood. In generalized linear models, one usually replaces this with a deviance-based calculation. However, the deviance does not involve the part of the density specification that is free of the parameter of interest, the term $c(y_i, \phi)$ in (3.3). Thus such calculations would involve different "penalties" if a comparison were made of two models from different families.

Another formulation of the AIC statistic is commonly found in statistical output. This form drops the n in the denominator. Although this formula does not adjust for the number of observations in the model, it is nevertheless a useful fit statistic. Both versions of the statistic are interpreted in the same manner, with the lower value indicating comparatively better fit. Care must be taken, though, to recognize which version is being used in a particular application and not to mix versions when comparing models.

4.6.1.2 BIC

Raftery (1996) presented an alternative information criterion based on a Bayesian comparison of models. Others have devised formulations of this fit statistic, but Raftery's is the earliest and most commonly applied. As with the AIC, this measure may be used to compare nonnested models as well as models calculated across different samples, i.e., comparing nested models. Again, like the AIC statistic, the better fitted model is the one having the smaller value. The BIC statistic is often negative; thus, the model having the most negative value is preferred.

The BIC statistic is particularly useful when comparing nonnested models. The statistic is defined as

$$\text{BIC} = D(M_k) - (\text{df}) \ln(n) \tag{4.33}$$

Contrary to the AIC, the BIC includes a penalty term that automatically becomes more severe as the sample size increases. This characteristic reflects the power available to detect significance. Investigations of the utility of this measure for comparisons of models appear in Bai et al. (1992) and for comparisons of distributions in Lindsey and Jones (1998).

When comparing nonnested models, one may assess the degree of model preference on the basis of the absolute value of the difference between the BIC statistics of the two models. The scale given by Raftery for determining the relative preference is

|difference|	Degree of preference
0–2	Weak
2–8	Positive
6–10	Strong
>10	Very strong

When we compare two models, A and B, if $\text{BIC}_A - \text{BIC}_B < 0$, then model A is preferred. If $\text{BIC}_A - \text{BIC}_B > 0$, then model B is preferred. The degree of preference is outlined above.

4.6.2 The interpretation of R² in linear regression

One of the most prevalent model measures is R^2. This statistic is usually discussed in introductory linear regression along with various ad hoc rules on its interpretation. A fortunate student is taught that there are many ways to interpret this statistic and that these interpretations have been generalized to areas outside linear regression.

For the linear regression model, we can define the R^2 measure in the following ways.

$$n = \text{number of observations} \tag{4.34}$$

$$p = \text{number of predictors} \tag{4.35}$$

$$M_\alpha = \text{model with only an intercept} \tag{4.36}$$

$$M_\beta = \text{model with intercept and predictors} \tag{4.37}$$

4.6.2.1 Percent variance explained

The most popular interpretation is that of the percent variance explained, where it can be shown that the R^2 statistic is equal to the ratio of variance of the fitted values and to the total variance.

$$\text{RSS} = \text{residual sum of squares} = \sum_{i=1}^{n}(y_i - \widehat{y}_i)^2 \tag{4.38}$$

$$\text{TSS} = \text{total sum of squares} = \sum_{i=1}^{n}(y_i - \bar{y})^2 \tag{4.39}$$

$$R^2 = \frac{\text{TSS} - \text{RSS}}{\text{TSS}} = 1 - \frac{\text{RSS}}{\text{TSS}} = 1 - \frac{\sum_{i=1}^{n}(y_i - \widehat{y}_i)^2}{\sum_{i=1}^{n}(y_i - \bar{y})^2} \tag{4.40}$$

4.6.2.2 The ratio of variances

Another interpretation originates from the equivalence of the statistic to the ratio of the variances.

$$R^2 = \frac{\widehat{\text{V}}(\widehat{y})}{\widehat{\text{V}}(y)} = \frac{\widehat{\text{V}}(\widehat{y})}{\widehat{\text{V}}(\widehat{y}) + \widehat{\text{V}}(\widehat{\epsilon})} \tag{4.41}$$

4.6.2.3 A transformation of the likelihood ratio

R^2 is equal to a transformation of the likelihood ratio between the full and null models. If the errors are assumed to be normal,

$$R^2 = 1 - \left\{ \frac{L(M_\alpha)}{L(M_\beta)} \right\}^{2/n} \tag{4.42}$$

4.6.2.4 A transformation of the F test

In assessing the fitted model, an F test may be performed to assess the adequacy of the additional parameters. The R^2 statistic can be shown to equal a transformation of this test.

The hypothesis that all the independent variables are zero may be tested with an F test and then R^2 may be written in terms of that statistic as

$$R^2 = \frac{Fp}{Fp + (n - k - 1)} \tag{4.43}$$

4.6.2.5 Squared correlation

The statistic is also equal to the squared correlation between the fitted and observed outcomes. This is the reason for the statistic's name.

$$R^2 = \{\text{Corr}(y, \widehat{y})\}^2 \tag{4.44}$$

4.6.3 Generalizations of linear regression R^2 interpretations

Starting from each of the interpretations in the preceding section, we may generalize the formula for use with models other than linear regression.

The most important fact to remember is that the extended or generalized versions of the R^2 statistic extend the particular interpretation that served as its genesis. As generalizations, these statistics are sometimes called pseudo-R^2 statistics. However, this name makes it to easy to forget which original formula was used in the derivation. Many people make the mistake that the use of any pseudo-R^2 statistic can be interpreted in the familiar and popular "percent variance explained" manner. Although the various interpretations in linear regression result in the same calculated value, the pseudo-R^2 scalar criteria generalized from different definitions do not result in the same value.

Just as there are adjusted R^2 measures for linear regression, there is active research in adjusting pseudo-R^2 criterion measures. We list only a few of the proposed adjusted measures.

4.6.3.1 Efron's pseudo-R^2

Efron (1978) defines the following measure as an extension to the regression model's "percent variance explained" interpretation:

$$R^2_{\text{Efron}} = 1 - \frac{\sum_{i=1}^{n}(y_i - \widehat{y}_i)^2}{\sum_{i=1}^{n}(y_i - \overline{y})^2} \tag{4.45}$$

Efron's presentation was directed at binary-outcome models and listed $\widehat{\pi}_i$ for $\widehat{y}_i$. The measure could be used for continuous models; the equation is the same as given in (4.40), and the measure is sometimes called the sum of squares R^2 or R^2_{SS}.

4.6.3.2 McFadden's likelihood-ratio index

Sometimes called the *likelihood-ratio index*, McFadden (1974) defines the following measure as another extension to the "percent variance explained" interpretation:

$$R^2_{\text{McFadden}} = 1 - \frac{\mathcal{L}(M_\beta)}{\mathcal{L}(M_\alpha)} \tag{4.46}$$

Note: Can be used for any model fitted by maximum likelihood.

4.6.3.3 Ben-Akiva and Lerman adjusted likelihood-ratio index

Ben-Akiva and Lerman (1985) extended McFadden's pseudo-R^2 measure to include an adjustment. Their adjustment is in the spirit of adjusted R^2 measure in linear regression and the formula is given by

$$R^2_{\text{Ben-Akiva\&Lerman}} = 1 - \frac{\mathcal{L}(M_\beta) - p}{\mathcal{L}(M_\alpha)} \tag{4.47}$$

where k is the number of parameters in the M_β model. Adjusted R^2 measures have been proposed in several forms. The aim of adjusting the calculation of the criterion is to address the fact that R^2 monotonically increases as terms are added to the model. Adjusted R^2 measures include penalty or shrinkage terms so that noncontributory terms will not significantly increase the criterion measure.

Note: Can be used for any model fitted by maximum likelihood.

4.6.3.4 McKelvey and Zavoina ratio of variances

McKelvey and Zavoina (1975) define the following measure as an extension of the "ratio of variances" interpretation:

$$R^2_{\text{McKelvey\&Zavoina}} = \frac{\widehat{\mathrm{V}}(\widehat{y}*)}{\widehat{\mathrm{V}}(y^*)} \tag{4.48}$$

$$= \frac{\widehat{\mathrm{V}}(\widehat{y}^*)}{\widehat{\mathrm{V}}(\widehat{y}^*) + \mathrm{V}(\epsilon)} \tag{4.49}$$

$$\mathrm{V}(\epsilon) = \begin{cases} 1 & \text{probit} \\ \pi^2/3 & \text{logit} \end{cases} \tag{4.50}$$

$$\widehat{\mathrm{V}}(\widehat{y}^*) = \widehat{\boldsymbol{\beta}}' \widehat{\mathrm{V}} \widehat{\boldsymbol{\beta}} \tag{4.51}$$

$$\widehat{\mathrm{V}} = \text{variance–covariance matrix of } \boldsymbol{\beta} \tag{4.52}$$

Note: Used for ordinal, binary, or censored outcomes.

4.6.3.5 Transformation of likelihood ratio

A maximum likelihood pseudo-R^2 measure extends the "transformation of the likelihood ratio" interpretation by using the formula

$$R^2_{\mathrm{ML}} = 1 - \left\{ \frac{L(M_\alpha)}{L(M_\beta)} \right\}^{2/n} \tag{4.53}$$

$$= 1 - \exp(-G^2/n) \tag{4.54}$$

$$G^2 = -2\ln\{L(M_\alpha)/L(M_\beta)\} \tag{4.55}$$

where the relationship to G^2 (a likelihood-ratio chi-squared statistic) is shown in Maddala (1983).

Note: Can be used for any model fitted by maximum likelihood.

4.6.3.6 Cragg and Uhler normed measure

Cragg and Uhler (1970) introduced a normed version of the transformation of the likelihood ratio pseudo-R^2 measure from the observation that as the fit of the two comparison models coincides, R^2_{ML} does not approach 1. The normed version is given by

$$R^2_{\text{Cragg\&Uhler}} = \frac{R^2_{\mathrm{ML}}}{\max R^2_{\mathrm{ML}}} \tag{4.56}$$

$$= \frac{1 - \{L(M_\alpha)/L(M_\beta)\}^{2/n}}{1 - L(M_\alpha)^{2/n}} \tag{4.57}$$

Note: Can be used for any model fitted by maximum likelihood.

4.6.4 More R^2 measures

4.6.4.1 The count R^2

A measure based on the correct classification of ordinal outcomes may be defined as in Maddala (1992) using

$$R^2_{\text{count}} = \frac{1}{n}\sum_j n_{jj} \tag{4.58}$$

$$n_{jj} = \text{number of observed } j \text{ correctly classified as } j \tag{4.59}$$

Note: For models with binary or ordinal outcome.

4.6.4.2 The adjusted count R^2

Because a poor model might be expected to correctly classify half of the observations, an adjusted measure accounts for the largest row margin.

$$R^2_{\text{count}} = \frac{\sum_j n_{jj} - \max(n_{r+})}{n - \max(n_{r+})} \tag{4.60}$$

$$n_{jj} = \text{number of observed } j \text{ correctly classified as } j \tag{4.61}$$

$$n_{r+} = \text{marginal sum for observed outcome } j \tag{4.62}$$

Note: For models with binary or ordinal outcome.

4.6.4.3 Veall and Zimmermann R^2

Veall and Zimmermann (1992) define a goodness-of-fit criterion measure for ordered-outcome models given by

$$R^2_{\text{Veall\&Zimmermann}} = \left(\frac{\delta - 1}{\delta - R^2_{\text{McFadden}}}\right) R^2_{\text{McFadden}} \tag{4.63}$$

$$\delta = \frac{n}{2\mathcal{L}(M_\alpha)} \tag{4.64}$$

Note: For models with ordinal outcome.

4.6.4.4 Cameron–Windmeijer R^2

Cameron and Windmeijer (1997) define a proportionate reduction in uncertainty measure valid for exponential families. The formula is given by

$$R^2_{\text{Cameron\&Windmeijer}} = 1 - \frac{K(y,\mu)}{K(y,\mu_0)} \tag{4.65}$$

where $K()$ is the Kullback–Leibler divergence, μ is the predicted values for the fitted model, and μ_0 is the estimated mean for the constant-only model. This measure may be solved for various families, and the associated formulas are given in tables A.12 and A.13. This measure assumes that the model includes a constant term.

Part II

Continuous-Response Models

5 The Gaussian family

Contents

In the 1800s, Karl Friedrich Gauss, the eponymous prince of mathematics, described the least-squares fitting method and the distribution that bears his name. Having a symmetric bell shape, the Gaussian density function is often referred to as the normal density. The normal cumulative distribution function is a member of the exponential family of distributions and so may be used as a basis for a GLM. Moreover, since the theory of GLMs was first conceived to be an extension to the normal OLS model, we will begin with an exposition of how OLS fits into the GLM framework. Understanding how other GLM models generalize this basic form should then be easier.

Regression models based on the Gaussian or normal distribution are commonly referred to as OLS models. This standard regression model is typically the first model taught in beginning statistics courses.

The Gaussian distribution is a continuous distribution of real numbers with support over the real line $\Re = (-\infty, +\infty)$. A model based on this distribution assumes that the response variable, called the *dependent variable* in the social sciences, takes the shape of the Gaussian distribution. Equivalently, we consider that the error term in the equation

$$y = \mathbf{x}\boldsymbol{\beta} + \epsilon \tag{5.1}$$

is normally or Gaussian distributed. In fact, GLIM, the original GLM software program, uses the term `ERROR` to designate the family or distribution of a model. Regardless, we may think of the underlying distribution as either that of the response variable or of the error term.

5.1 Derivation of the GLM Gaussian family

The various components of a particular GLM can be obtained from the base probability function. The manner in which the probability function is parameterized relates directly to the algorithm used for estimation, i.e., IRLS or Newton–Raphson (N–R). If we intend to estimate parameters by using the standard or traditional IRLS algorithm, then the probability function is parameterized in terms of the mean μ (estimated fitted value). On the other hand, if the estimation is to be performed in terms of N–R, parameterization is in terms of $\mathbf{x}\boldsymbol{\beta}$, the linear predictor. To clarify: the IRLS algorithm considers maximizing a function in terms of the mean where the introduction of covariates is delayed until specification of the link function. On the other hand, maximum likelihood typically writes the function to be maximized in terms of the specific link function so that derivatives may be clearly seen.

For the Gaussian distribution, μ is the same as $\mathbf{x}\boldsymbol{\beta}$ in its canonical form. We say that the canonical link is the identity link; i.e. there is a straightforward identity between the fitted value and the linear predictor. The canonical-link Gaussian regression model is the paradigm instance of a model with an identity link.

The Gaussian probability function, parameterized in terms of μ, is expressed as

$$f(y;\mu,\sigma^2) = \frac{1}{\sqrt{2\pi\sigma^2}}\exp\left\{-\frac{(y-\mu)^2}{2\sigma^2}\right\} \tag{5.2}$$

where $f()$ is the generic form of the density function of y, given parameters μ and σ^2; y is the response variable; μ is the mean parameter; and σ^2 is the scale parameter.

In GLM theory, the density function is reexpressed in exponential family form. That is, all density functions are derived from members of the exponential family of distributions and can be expressed in the form

$$f(y;\theta,\phi) = \exp\left\{\frac{y\theta - b(\theta)}{a(\phi)} + c(y;\phi)\right\} \tag{5.3}$$

where θ is the canonical link function, $b(\theta)$ is the cumulant, $a(\phi)$ is the scale parameter, and $c(y,\phi)$ is the normalization function. Normalization assures that the probability function sums or integrates to 1. Although the normalization term is a necessary part of maximum likelihood estimation, it plays no role in traditional GLMs using IRLS. The reason, which will be demonstrated later, is that IRLS defines the deviance. This value is proportional to the difference of log likelihoods of two nested ML models. Any normalization terms thus cancel out in the calculation of the deviance function; only those terms containing an instance of μ are essential to the calculation.

5.2 Derivation in terms of the mean

GLM theory traditionally derives the elements of the GLM algorithm in terms of μ. This derivation leads to the standard IRLS algorithm. We first detail this method.

Expressing the Gaussian density function in exponential-family form results in the specification

$$f(y;\mu,\sigma^2) = \exp\left\{-\frac{(y-\mu)^2}{2\sigma^2} - \frac{1}{2}\ln\left(2\pi\sigma^2\right)\right\} \tag{5.4}$$

$$= \exp\left\{\frac{y\mu-\mu^2/2}{\sigma^2} - \frac{y^2}{2\sigma^2} - \frac{1}{2}\ln\left(2\pi\sigma^2\right)\right\} \tag{5.5}$$

where we can easily determine that $\theta = \mu$ and $b(\theta) = \mu^2/2$.

The mean, expected value of the outcome y, is determined by taking the derivative of the cumulant, $b(\theta)$, with respect to θ. The variance is the second derivative. Hence,

$$b'(\theta) = \frac{\partial b}{\partial \theta} \tag{5.6}$$

$$= \frac{\partial b}{\partial \mu}\frac{\partial \mu}{\partial \theta} \tag{5.7}$$

$$= (\mu)(1) = \mu \tag{5.8}$$

$$b''(\theta) = \frac{\partial^2 b}{\partial \theta^2} \tag{5.9}$$

$$= \frac{\partial}{\partial \theta}\left[\frac{\partial b}{\partial \mu}\frac{\partial \mu}{\partial \theta}\right] \tag{5.10}$$

$$= \frac{\partial}{\partial \theta}(\mu) \tag{5.11}$$

$$= \frac{\partial}{\partial \mu}\mu\frac{\partial \mu}{\partial \theta} \tag{5.12}$$

$$= (1)(1) = 1 \tag{5.13}$$

since $\partial\mu/\partial\theta = 1$ for this case and b is understood to be the function $b(\theta)$.

The mean is simply μ and the variance is 1. The scale parameter, ϕ, is σ^2. The other important statistic required for the GLM–Gaussian algorithm is the derivative of the link, μ.

$$\frac{\partial \theta}{\partial \mu} = \frac{\partial \mu}{\partial \mu} = 1 \tag{5.14}$$

Finally, the deviance function is calculated as

$$D = 2\phi\left\{\mathcal{L}(y,\sigma^2;y) - \mathcal{L}(\mu,\sigma^2;y)\right\} \tag{5.15}$$

$\mathcal{L}(y,\sigma^2;y)$ represents the saturated log-likelihood function and is calculated by substituting the value of y_i for each instance of μ_i; the saturated model is one that perfectly reproduces the data, which is why $y_i = \mu_i$. When subtracting the observed $\mathcal{L}$ from the saturated function, we find that the normalization terms cancel from the final deviance equation. The observed log-likelihood function is given by

$$\mathcal{L}(\mu,\sigma^2;y) = \sum_{i=1}^{n}\left\{\frac{y_i\mu_i-\mu_i^2/2}{\sigma^2} - \frac{y_i^2}{2\sigma^2} - \frac{1}{2}\ln\left(2\pi\sigma^2\right)\right\} \tag{5.16}$$

The saturated log likelihood is

$$\mathcal{L}(y,\sigma^2;y) = \sum_{i=1}^{n}\left\{\frac{y_i^2 - y_i^2/2}{\sigma^2} - \frac{y_i^2}{2\sigma^2} - \frac{1}{2}\ln\left(2\pi\sigma^2\right)\right\} \tag{5.17}$$

$$= \sum_{i=1}^{n}\left\{-\frac{1}{2}\ln\left(2\pi\sigma^2\right)\right\} \tag{5.18}$$

The Gaussian deviance is thereby calculated as

$$D = 2\sigma^2\sum_{i=1}^{n}\left\{-\frac{1}{2}\ln\left(2\pi\sigma^2\right) - \frac{y_i\mu_i - \mu_i^2/2}{\sigma^2} + \frac{y_i^2}{2\sigma^2} + \frac{1}{2}\ln\left(2\pi\sigma^2\right)\right\} \tag{5.19}$$

$$= 2\sigma^2\sum_{i=1}^{n}\left\{\frac{y_i^2 - 2y_i\mu_i + \mu_i^2}{2\sigma^2}\right\} \tag{5.20}$$

$$= \sum_{i=1}^{n}(y_i - \mu_i)^2 \tag{5.21}$$

The Gaussian deviance is identical to the sum of squared residuals. This fact is unique to the Gaussian family. However, since the Gaussian variance is 1, the Pearson χ^2 statistic, defined as

$$\chi^2 = \sum_{i=1}^{n}\frac{(y_i - \mu_i)^2}{\mathrm{V}(\mu_i)} \tag{5.22}$$

is equal to the deviance. This too is a unique feature of the Gaussian family.

The IRLS algorithm for the canonical-link Gaussian family is shown in the generic form below. Of course, there are other algorithms, some of which use a `DO-WHILE` loop (e.g., S-Plus and XploRe) rather than a `WHILE` conditional loop (e.g., Stata). Nevertheless, all IRLS algorithms follow the basic logic illustrated here.

5.3 IRLS GLM algorithm (nonbinomial)

With summing understood, the generic IRLS form for nonbinomial models is

Listing 5.1: IRLS algorithm for nonbinomial models

```
1   μ = (y + mean(y))/2
2   η = g
3   WHILE (abs(ΔDev) > tolerance)  {
4       w = 1/(Vg'^2)
5       z = η + (y − μ)g'
6       β = (X^T WX)^{-1} X^T Wz
7       η = Xβ
8       μ = g^{-1}
9       OldDev = Dev
10      Dev = deviance function
11      ΔDev = Dev - OldDev
12  }
13  χ^2 = Σ(y − μ)^2/V
```

where V is the variance function, g is the link function, g' is the derivative of the link, and g^{-1} is the inverse link function. The tolerance specifies the sensitivity of when iterations are to cease. Typically, the tolerance is set to $1.0e-6$. Other methods of defining tolerance exist, including the measure of difference in estimating coefficients between iterations.

This measure provides a basis for the canonical Gaussian algorithm:

Listing 5.2: IRLS algorithm for Gaussian models

```
1   Dev = 1
2   μ = (y − mean(y))/2
3   η = μ                                        /* identity link */
4   WHILE  (abs(ΔDev) > tolerance) {
5       β = (X^T X)^{-1} X^T y
6       η = xβ
7       μ = η
8       OldDev = Dev
9       Dev = Σ(y − μ)^2
10      ΔDev = Dev - OldDev
11  }
12  χ^2 = Σ(y − μ)^2
```

Again, in the canonical form, we do not typically use the GLM routine for fitting model parameters. The reason is clear from looking at the above algorithm. The GLM estimation of $\boldsymbol{\beta}$ is simplified in the canonical Gaussian instance because w, the weight, equals 1 and z, the working response, is the same as y, the actual response. Thus

$$(\mathbf{X}^{\mathrm{T}}\mathbf{W}\mathbf{X})^{-1}\mathbf{X}^{\mathrm{T}}\mathbf{W}\mathbf{z} = (\mathbf{X}^{\mathrm{T}}\mathbf{X})^{-1}\mathbf{X}^{\mathrm{T}}\mathbf{y} \tag{5.23}$$

In fact, the above algorithm can be further reduced because of the identity between μ and η.

Listing 5.3: IRLS algorithm (reduced) for Gaussian models

```
1   Dev = 1
2   μ = (y − mean(y))/2
3   WHILE (abs(ΔDev) > tolerance)  {
4       β = (X^T X)^{-1} X^T y
5       μ = Xβ
6       OldDev = Dev
7       Dev = Σ(y − μ)^2
8       ΔDev = Dev - OldDev
9   }
10  χ^2 = Σ(y − μ)^2
```

Moreover, iterations to find estimates are not necessary because estimates can be determined using only one step. Hence, in actuality, only the line estimating $\boldsymbol{\beta}$ is necessary. This too is unique among GLMs.

When modeling the canonical Gaussian, it is perhaps best to use the standard linear regression procedure common to most statistical packages. However, using a GLM routine typically allows access to a variety of diagnostic residuals, as well as to goodness-of-fit (GOF) statistics. Using GLM when analyzing the Gaussian model also allows easy comparison to other Gaussian links, as well as to other GLM families.

The χ^2 statistic is traditionally called the Pearson χ^2 GOF statistic. It is the ratio of the squared residuals to the variance. Because the Gaussian family variance is 1, the deviance and Pearson χ^2 statistics are identical. This is not the case for other families. However, on this account, the deviance-based dispersion is the same as the χ^2-based dispersion.

Dispersion statistics are obtained by dividing the residual degrees of freedom from the deviance or Pearson χ^2 statistic, respectively. The residual degrees of freedom is calculated as $n - p$, where n is the number of cases in the model and p is the number of model predictors (possibly including a constant).

The statistics described above are important to the assessment of model fit. For the canonical-link Gaussian case, the values are identical to the residual statistics displayed in an analysis of variance (ANOVA) table, which usually accompanies a linear regression printout.

$$\text{ERROR SS} = \text{deviance} = \text{Pearson } \chi^2 \tag{5.24}$$

$$\text{MEAN SE} = \text{dispersion (both deviance and } \chi^2\text{)} \tag{5.25}$$

The scale is defined as 1 when using the IRLS algorithm. The reason for this is simple. A traditional GLM, using the IRLS algorithm, provides only mean parameter estimates; it does not directly estimate the scale or ancillary parameter. If we desired a separate estimate of the scale, we could use an ML algorithm. However, the deviance dispersion can be used to estimate the scale value, though, without estimating an associated standard error (which would be obtained in an ML approach).

5.4 Maximum likelihood estimation

Maximum likelihood (ML) estimation of the canonical-link Gaussian model requires specifying the Gaussian log likelihood in terms of $\mathbf{x}\boldsymbol{\beta}$. Parameterized in terms of $\mathbf{x}\boldsymbol{\beta}$, the Gaussian density function is given by

$$f(y;\boldsymbol{\beta},\sigma^2) = \frac{1}{\sqrt{2\pi\sigma^2}}\exp\left\{-\frac{1}{2\sigma^2}(y-\mathbf{x}\boldsymbol{\beta})^2\right\} \tag{5.26}$$

After we express the joint density in exponential family form, the log likelihood becomes

$$\mathcal{L}(\mu,\sigma^2;y) = \sum_{i=1}^{n}\left\{\frac{y_i\mathbf{x}_i\boldsymbol{\beta}-(\mathbf{x}_i\boldsymbol{\beta})^2/2}{\sigma^2} - \frac{y_i^2}{2\sigma^2} - \frac{1}{2}\ln\left(2\pi\sigma^2\right)\right\} \tag{5.27}$$

This result implies that

$$\begin{aligned}
\theta &= \mathbf{x}\boldsymbol{\beta} && (5.28)\\
b(\theta) &= (\mathbf{x}\boldsymbol{\beta})^2/2 = \theta^2/2 && (5.29)\\
b'(\theta) &= \mathbf{x}\boldsymbol{\beta} && (5.30)\\
b''(\theta) &= 1 && (5.31)\\
a(\phi) &= \sigma^2 && (5.32)
\end{aligned}$$

The ML algorithm estimates σ^2 in addition to standard parameters and linear predictor estimates. The estimated value of σ^2 should be identical to the dispersion values, but ML provides a standard error and confidence interval. Both estimates differ from the usual OLS estimates, which specifies estimators for the regression model without assuming normality. The OLS estimated standard error uses $(n-p)$ in the denominator, whereas the ML and GLM estimates use (n); the numerator is the sum of the squared residuals. This slight difference (which makes no difference asymptotically and makes no difference in the estimated regression coefficients) is most noticeable in the estimated standard errors for models based on small samples.

The (partial) derivatives of the canonical-link Gaussian log likelihood are given by

$$\begin{aligned}
\frac{\partial\mathcal{L}}{\partial\beta_j} &= \sum_{i=1}^{n}\frac{1}{\sigma^2}(y_i-\mathbf{x}_i\boldsymbol{\beta})x_{ji} && (5.33)\\
\frac{\partial\mathcal{L}}{\partial\sigma} &= \sum_{i=1}^{n}\frac{1}{\sigma}\left[\left(\frac{y_i-\mathbf{x}_i\boldsymbol{\beta}}{\sigma}\right)^2-1\right] && (5.34)
\end{aligned}$$

and the second (partial) derivatives are given by

$$\frac{\partial^2 \mathcal{L}}{\partial \beta_j \partial \beta_k} = -\sum_{i=1}^{n} \frac{1}{\sigma^2} x_{ji} x_{ki} \tag{5.35}$$

$$\frac{\partial^2 \mathcal{L}}{\partial \beta_j \partial \sigma} = -\sum_{i=1}^{n} \frac{2}{\sigma^3} (y_i - \mathbf{x}_i \boldsymbol{\beta}) x_{ji} \tag{5.36}$$

$$\frac{\partial^2 \mathcal{L}}{\partial \sigma \partial \sigma} = -\sum_{i=1}^{n} \frac{1}{\sigma^2} \left\{ 3 \left(\frac{y_i - \mathbf{x}_i \boldsymbol{\beta}}{\sigma} \right)^2 - 1 \right\} \tag{5.37}$$

5.5 GLM log-normal models

An important and perhaps the foremost reason for using GLM as a framework for model construction is the ability to easily adjust models to fit particular response data situations. The canonical-link Gaussian model assumes a normally distributed response. Although the normal model is robust to moderate deviations from this assumption, it is nevertheless the case that many data situations are not amenable to or appropriate for normal models.

Unfortunately, many researchers have used the canonical-link Gaussian for data situations that do not meet the assumptions on which the Gaussian model is based. Until recently, few software packages allowed users to model data by means other than the normal model. Granted, many researchers had little training in nonnormal modeling, but most popular software packages now have GLM capabilities or at least implement many of the most widely used GLM procedures, e.g., logistic, probit, and Poisson regression.

The log-normal model is based on the Gaussian distribution. It uses the log rather than the (canonical) identity link. The log link is generally used for response data that can take only positive values on the continuous scale, or values greater than 0. The data must be such that nonpositive values are not only absent but also theoretically precluded.

Before GLM, researchers usually modeled positive-only data with the normal model. However, they first took the natural log of the response prior to modeling. In so doing, they explicitly acknowledged the need to normalize the response relative to the predictors, thus accommodating one of the assumptions of the Gaussian model. The problem with this method is one of interpretation. Fitted values, as well as parameter estimates, are in terms of the log response. This obstacle often proves to be inconvenient.

A better approach is to internalize within the model itself the log transformation of the response. The log link exponentiates the linear predictor, or $\mathbf{x}\boldsymbol{\beta}$, rather than log-transforming the response to linearize the relationship between the response and predictors. This procedure, implicit within the GLM algorithm, allows easy interpretation of estimates, as well as of fitted values.

The implementation of a log link within the ML algorithm is straightforward—simply substitute $\exp(\mathbf{x}\boldsymbol{\beta})$ for each instance of $\mathbf{x}\boldsymbol{\beta}$ in the log-likelihood function. The canonical-link Gaussian log-likelihood function is thus transformed to a log-normal model. The log-likelihood function is then written in exponential-family form:

$$\mathcal{L}(\mu; y, \sigma^2) = \sum_{i=1}^{n} \left\{ \frac{y_i \exp(\mathbf{x}_i\boldsymbol{\beta}) - \{\exp(\mathbf{x}_i\boldsymbol{\beta})\}^2/2}{\sigma^2} - \frac{y_i^2}{2\sigma^2} - \frac{1}{2}\ln\left(2\pi\sigma^2\right) \right\} \tag{5.38}$$

Creating a log-normal, or log-linked Gaussian, model using the standard IRLS algorithm is only a bit more complicated. Recalling that parameterization is in terms of μ, we must change the link function from $\eta = \mu$ to $\eta = \ln(\mu)$ and the inverse link function from $\mu = \eta$ to $\mu = \exp(\eta)$.

Finally, we have a value for the derivative of the link. In the canonical form, the derivative is 1, hence excluding it from the algorithm. The derivative of the log link is $1/\mu$ and is placed in the algorithm at line 4 in listing 5.1. All other aspects of the Gaussian-family algorithm remain the same, except that we must extend the algorithm to its standard form.

5.6 Expected versus observed information matrix

To reiterate our discussion from chapter 3, OIM standard errors are more conservative than those based on EIM, and the difference becomes increasingly noticeable in datasets with fewer observations. Again referring to the discussions of the third chapter, it was demonstrated that the observed information matrix reduces to the naive expected information matrix for the canonical link. The log link is not the canonical link for the Gaussian. Hence, it may be better to model noncanonical linked data by using a method that provides standard errors based on the observed information matrix (especially in smaller datasets).

One can adjust the IRLS estimating algorithm such that the observed information matrix is used rather than the expected matrix. One need only create a function, $\mathbf{W}_o$, which calculates the observed information matrix given in (3.59).

$$\mathbf{W}_o = \mathbf{W} + (y - \mu)\left\{ \frac{\mathrm{V}g'' + \mathrm{V}'g'}{\mathrm{V}^2(g')^3} \right\} \tag{5.39}$$

where $g' = g'(\mu)$, the derivative of the link; $g'' = g''(\mu)$, the second derivative of the link; $\mathrm{V} = \mathrm{V}(\mu)$, the variance function; V^2 = variance squared; and $\mathbf{W}$ is the standard IRLS weight function given in (3.47).

The generic form of the GLM IRLS algorithm, using the observed information matrix, is then given by

Listing 5.4: IRLS algorithm for GLM using OIM

```
1   μ = (y + mean(y))/2
2   η = g
3   WHILE (abs(ΔDev) > tolerance) {
4       W = 1/(Vg'^2)
5       z = η + (y − μ)g' − offset
6       W_o = W + (y − μ)(Vg'' + V'g')/(V^2(g')^3)
7       β = (X^T W_o X)^{-1} X^T W_o z
8       η = Xβ + offset
9       μ = g^{-1}
10      OldDev = Dev
11      Dev = ∑ (deviance function)
12      ΔDev = Dev - OldDev
13  }
14  χ^2 = ∑(y − μ)^2/V
```

where V' is the derivative with respect to μ of the variance function.

Using this algorithm, which we can call the IRLS–OIM algorithm, estimates and standard errors will be identical to those produced using maximum likelihood. The scale, estimated by the ML routine, is estimated as the deviance-based dispersion. However, unlike ML output, the IRLS–OIM algorithm fails to provide standard errors for the scale. Of course, the benefit to using IRLS–OIM is that the typical ML problems with starting values are nearly always avoided. Also ML software usually takes a bit longer to converge than does a straightforward IRLS–OIM algorithm.

Finally, we need a value for the derivative of the link. In canonical form the derivative is 1, hence being in effect excluded from the algorithm. The derivative of the log link, $\ln(\mu)$, is $1/\mu$ and is appropriately placed into the algorithm below at line 4.

Listing 5.5: IRLS algorithm for log-normal models using OIM

```
1   μ = (y + mean(y))/2
2   η = ln(μ)
3   WHILE (abs(ΔDev) > tolerance {
4       W = 1/(1(1/μ)^2) = μ^2
5       z = η + (y − μ)μ − offset
6       W_o = W + (y − μ)μ^2
7       β = (X^T W_o X)^{-1} X^T W_o z
8       η = Xβ + offset
9       μ = exp(η)
10      OldDev = Dev
11      Dev = ∑(y − μ)^2
12      ΔDev = Dev - OldDev
13  }
14  χ^2 = ∑(y − μ)^2
```

5.7 Other Gaussian links

The standard or normal linear model, OLS regression, uses the identity link, which is the canonical form of the distribution. It is robust to most minor violations of Gaussian distributional assumptions but is used in more data modeling situations than is appropriate. In fact, it was used to model binary, proportional, and discrete count data. Appropriate software was unavailable and there was not widespread understanding of the statistical problems inherent in such an approach.

There is no theoretical reason why researchers using the Gaussian family for a model should be limited to using only the identity or log links. The reciprocal link, $1/\mu$, has been used to model data with a rate response. It also may be desirable to use the power links with the Gaussian model. In doing so, we approximate the Box–Cox ML algorithm, which is used as a normality transform. Some software packages do not have Box–Cox capabilities; therefore, using power links with the Gaussian family may assist in achieving the same purposes.

5.8 Example: Relation to OLS

We provide a simple example showing the relationship of OLS regression output to that of GLM. The example we use comes from a widely used study on prediction of low-birth-weight babies (Hosmer and Lemeshow 2000). The response `bwt` is the birth weight of the baby. The predictors are `lwt`, the weight of the mother; `race`, where a value of 1 indicates white, 2 indicates black, and 3 indicates other; a binary variable `smoke`, which indicates whether the mother has a history of smoking; another binary variable `ht`, which indicates whether the mother has a history of hypertension; and a final binary variable `ui`, which indicates whether the mother has a history of uterine irritability. We generated indicator variables for use in model specification `race2` indicating black and `race3` indicating other. The standard Stata regression output is given by

```
. use http://www.stata-press.com/data/hh2/lbw
(Hosmer & Lemeshow data)

. regress bwt lwt race2 race3 smoke ht ui

      Source |       SS       df       MS              Number of obs =     189
-------------+------------------------------           F(  6,   182) =    9.59
       Model |  23992342.5     6  3998723.74           Prob > F      =  0.0000
    Residual |  75922956.1   182  417159.099           R-squared     =  0.2401
-------------+------------------------------           Adj R-squared =  0.2151
       Total |  99915298.6   188  531464.354           Root MSE      =  645.88

------------------------------------------------------------------------------
         bwt |      Coef.   Std. Err.      t    P>|t|     [95% Conf. Interval]
-------------+----------------------------------------------------------------
         lwt |   4.238846   1.675414     2.53   0.012     .9331126    7.544579
       race2 |    -475.22    145.589    -3.26   0.001    -762.4792   -187.9607
       race3 |   -349.669   112.3456    -3.11   0.002    -571.3363   -128.0018
       smoke |  -355.7326   103.4345    -3.44   0.001    -559.8176   -151.6475
          ht |  -584.6279   199.6238    -2.93   0.004    -978.5024   -190.7533
          ui |  -523.8869    134.664    -3.89   0.000    -789.5903   -258.1835
       _cons |   2837.343   243.6806    11.64   0.000      2356.54    3318.145
------------------------------------------------------------------------------
```

GLM output:

```
. glm bwt lwt race2 race3 smoke ht ui
Iteration 0:   log likelihood = -1487.5585

Generalized linear models                      No. of obs      =        189
Optimization     : ML                          Residual df     =        182
                                               Scale parameter =   417159.1
Deviance         =  75922956.11                (1/df) Deviance =   417159.1
Pearson          =  75922956.11                (1/df) Pearson  =   417159.1

Variance function: V(u) = 1                    [Gaussian]
Link function    : g(u) = u                    [Identity]

                                               AIC             =   15.81543
Log likelihood   = -1487.558492                BIC             =   7.59e+07

------------------------------------------------------------------------------
             |                 OIM
         bwt |      Coef.   Std. Err.      z    P>|z|     [95% Conf. Interval]
-------------+----------------------------------------------------------------
         lwt |   4.238846    1.675414     2.53   0.011      .9550942    7.522598
       race2 |    -475.22     145.589    -3.26   0.001     -760.5691   -189.8708
       race3 |   -349.669    112.3456    -3.11   0.002     -569.8623   -129.4758
       smoke |  -355.7326    103.4345    -3.44   0.001     -558.4606   -153.0046
          ht |  -584.6279    199.6238    -2.93   0.003     -975.8834   -193.3723
          ui |  -523.8869     134.664    -3.89   0.000     -787.8235   -259.9503
       _cons |   2837.343    243.6806    11.64   0.000      2359.738    3314.948
------------------------------------------------------------------------------
```

Note the similarity of estimate and standard error values. Parameter p-values differ because the standard OLS method displays p-values based on the Student's t distribution, whereas GLM p-values are in terms of the z or normal distribution. Note also the comparative values of the deviance and Pearson statistics with the residual sum of squares, as well as between the scaled deviance and Pearson statistics and the ANOVA mean squared error, or residual mean square (MS).

Modeling with GLM provides an estimate of 417,159.1 for the scale value, σ^2. If we used the `irls` option with the `glm` program, having estimation depend solely on Fisher scoring, then the scale is not estimated directly and is displayed as 1 on the output heading. Of course, the deviance scale will provide the same information under both methods of estimation, but this will not always be the case with other GLM families.

```
. glm bwt lwt race2 race3 smoke ht ui, irls
Iteration 1:   deviance =  7.59e+07
Generalized linear models                          No. of obs      =       189
Optimization     : MQL Fisher scoring              Residual df     =       182
                   (IRLS EIM)                      Scale parameter =         1
Deviance         =  75922956.11                    (1/df) Deviance =  417159.1
Pearson          =  75922956.11                    (1/df) Pearson  =  417159.1
Variance function: V(u) = 1                        [Gaussian]
Link function    : g(u) = u                        [Identity]
                                                   BIC             =  7.59e+07

------------------------------------------------------------------------------
             |                 EIM
         bwt |      Coef.   Std. Err.      z    P>|z|     [95% Conf. Interval]
-------------+----------------------------------------------------------------
         lwt |   4.238846   1.675414     2.53   0.011      .9550942    7.522598
       race2 |    -475.22    145.589    -3.26   0.001     -760.5691   -189.8708
       race3 |   -349.669   112.3456    -3.11   0.002     -569.8623   -129.4758
       smoke |  -355.7326   103.4345    -3.44   0.001     -558.4606   -153.0046
          ht |  -584.6279   199.6238    -2.93   0.003     -975.8834   -193.3723
          ui |  -523.8869    134.664    -3.89   0.000     -787.8235   -259.9503
       _cons |   2837.343   243.6806    11.64   0.000      2359.738    3314.948
------------------------------------------------------------------------------
```

Finally, looking at the output log (whether we use the IRLS algorithm) demonstrates what was said before. The canonical-link Gaussian algorithm iterates just once, which is unique among GLM specifications.

5.9 Example: Beta-carotene

http://lib.stat.cmu.edu/datasets/Plasma_Retinol is a URL where (as of writing this text) data from an unpublished study are located.[1]

We repeat analyses on 315 observations of dietary intake. The goal of the study is to investigate whether low dietary intake or low plasma concentrations of retinol, beta-carotene, or other carotenoids might be associated with certain cancers. The investigators collected information over a 3-year period from patients who elected to have a surgical biopsy or removal of a lesion that was found to be noncancerous. Such procedures retrieved tissue from the lung, colon, breast, skin, ovary, or uterus.

The outcome variable `betap` is the plasma concentration reading of beta-carotene. The independent variables of interest include the age of the patient in years, `age`; an indicator of female sex, `female`; an indicator of whether the patient takes vitamins, `pills`; the average daily dietary intake in kilocalories, `kcal`; the average daily dietary intake of fiber, `fiber`; the average daily dietary intake of alcohol, `alcohol`; a reading of the patient's cholesterol, `chol`; and the average daily intake of dietary beta-carotene, `betad`. We fit a linear model, obtain the Pearson residuals from the fitted model, and draw a histogram, which clearly shows the skewness of the distribution.

1. Data and descriptions of the data are used with permission.

```
. glm betap age female pills kcal fat fiber alcohol chol betad
Iteration 0:   log likelihood =  -2062.328

Generalized linear models                          No. of obs      =        315
Optimization     : ML                              Residual df     =        305
                                                   Scale parameter =   29393.52
Deviance         =  8965024.147                    (1/df) Deviance =   29393.52
Pearson          =  8965024.147                    (1/df) Pearson  =   29393.52

Variance function: V(u) = 1                        [Gaussian]
Link function    : g(u) = u                        [Identity]

                                                   AIC             =   13.15764
Log likelihood   = -2062.327961                    BIC             =    8963270

------------------------------------------------------------------------------
             |                 OIM
       betap |      Coef.   Std. Err.      z    P>|z|     [95% Conf. Interval]
-------------+----------------------------------------------------------------
         age |   .9515906   .7453438     1.28   0.202    -.5092565    2.412438
      female |   31.27419   31.99056     0.98   0.328    -31.42615    93.97454
       pills |   71.89446   21.07482     3.41   0.001     30.58858    113.2003
        kcal |  -.0358134   .0517117    -0.69   0.489    -.1371666    .0655397
         fat |   .1546087   .8174872     0.19   0.850    -1.447637    1.756854
       fiber |   7.771203     2.8335     2.74   0.006     2.217644    13.32476
     alcohol |   1.228665   1.249511     0.98   0.325    -1.220331    3.677661
        chol |  -.1388087   .1079521    -1.29   0.199    -.3503909    .0727736
       betad |   .0162778   .0075557     2.15   0.031     .0014689    .0310867
       _cons |   15.61038   67.45794     0.23   0.817    -116.6048    147.8255
------------------------------------------------------------------------------

. predict double res, pearson

. histogram res
(bin=17, start=-286.79243, width=82.533625)
```

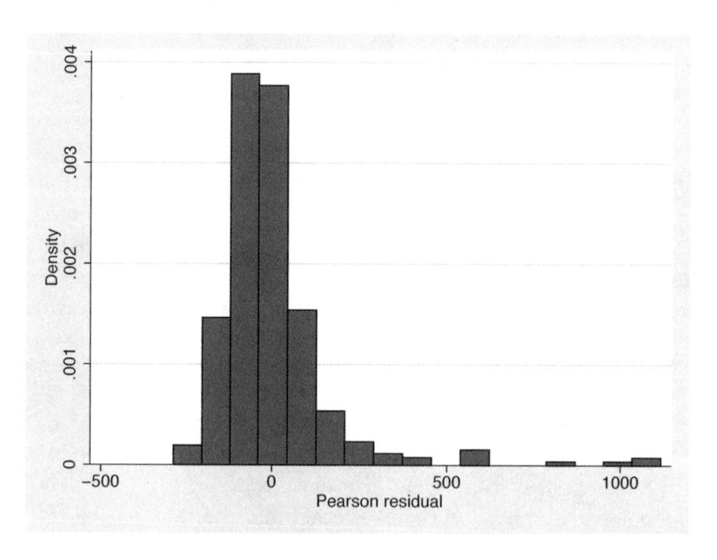

Figure 5.1: Pearson residuals obtained from linear model

In addition to the histogram, we can plot the ordered Pearson residuals versus normal probability points to see how well the two agree. If the Pearson residuals follow a normal distribution, the plot should look like a line along the diagonal from the bottom left of the plot to the top right. This plot, figure 5.2, emphasizes the conclusions already reached looking at figure 5.1. We also illustrate the Pearson residuals versus one of the covariates, the daily caloric intake, as an example of how one investigates whether the variance is related to the value of the independent variables.

```
. sort res
. gen nscore = invnormal(_n/316)
. twoway scatter nscore res
```

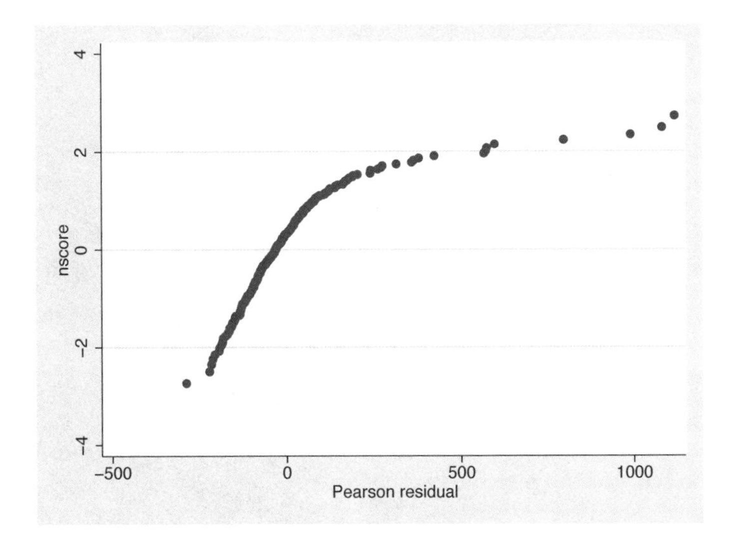

Figure 5.2: Normal scores versus sorted Pearson residuals obtained from linear model

Figure 5.3 demonstrates that the variance does not depend on the dietary caloric intake. However, the positive residuals seem to be much larger (in absolute value) than the negative values. There also appear to be 2 observations with unusually large daily caloric intake. In a full analysis, one would calculate Cook's distance measures to investigate leverage and would more fully investigate any suspected outliers.

(Continued on next page)

```
. twoway scatter res kcal, yline(0)
```

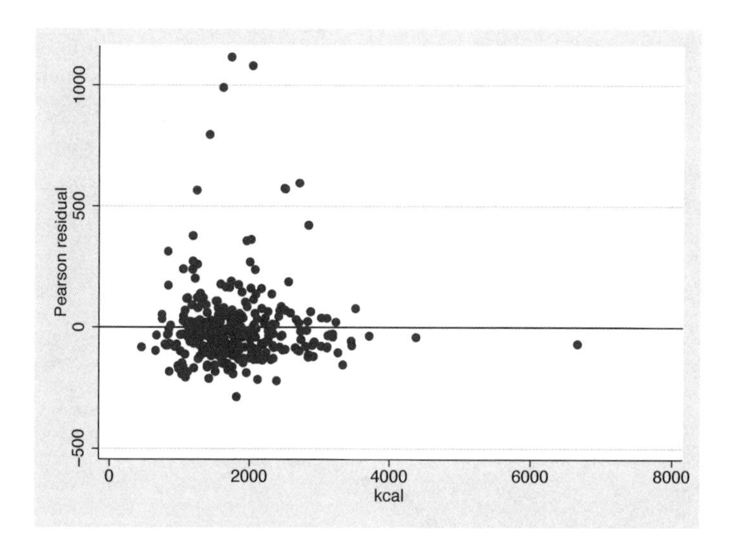

Figure 5.3: Pearson residuals versus kilocalories; Pearson residuals obtained from linear model

For our quick illustration, we will drop from further models the 2 observations with unusually large average daily intake of kilocalories. Here we switch from a linear regression model to a log-normal model, one that uses the log link to transform the linear predictor to the scale of the observed values.

```
. drop if kcal > 4000
(2 observations deleted)

. glm betap age female pills kcal fat fiber alcohol chol betad, link(log) nolog

Generalized linear models                         No. of obs      =        313
Optimization     : ML                             Residual df     =        303
                                                  Scale parameter =    28332.2
Deviance         =  8584655.935                   (1/df) Deviance =    28332.2
Pearson          =  8584655.935                   (1/df) Pearson  =    28332.2

Variance function: V(u) = 1                       [Gaussian]
Link function    : g(u) = ln(u)                   [Log]

                                                  AIC             =   13.12106
Log likelihood   = -2043.445673                   BIC             =    8582915

------------------------------------------------------------------------------
             |                 OIM
       betap |      Coef.   Std. Err.      z    P>|z|     [95% Conf. Interval]
-------------+----------------------------------------------------------------
         age |   .0056749   .0035317     1.61   0.108    -.0012471    .0125969
      female |   .3368722   .1966366     1.71   0.087    -.0485284    .7222728
       pills |   .4761023   .1340172     3.55   0.000     .2134334    .7387712
        kcal |  -.0000842   .0002087    -0.40   0.687    -.0004933    .0003249
         fat |   -.000546   .0037247    -0.15   0.883    -.0078462    .0067543
       fiber |    .042346   .0098439     4.30   0.000     .0230524    .0616397
     alcohol |   .0222378   .0102478     2.17   0.030     .0021525    .0423232
        chol |  -.0007851   .0006329    -1.24   0.215    -.0020256    .0004553
       betad |    .000062   .0000269     2.30   0.021     9.23e-06    .0001147
       _cons |   3.911164   .3715337    10.53   0.000     3.182971    4.639357
------------------------------------------------------------------------------

. predict double betaphat, mu

. predict double resloglink, pearson

. histogram resloglink
(bin=17, start=-274.24709, width=80.920857)

. twoway scatter resloglink betaphat, yline(0)
```

Investigating residual plots, including figure 5.4 and figure 5.5, demonstrate that the residuals are still not close to the normal distribution.

(Continued on next page)

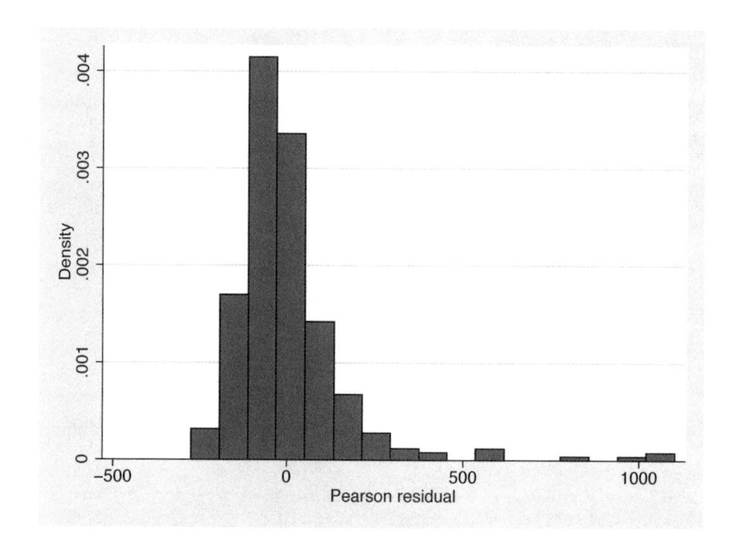

Figure 5.4: Pearson residuals obtained from log-normal model (two outliers removed)

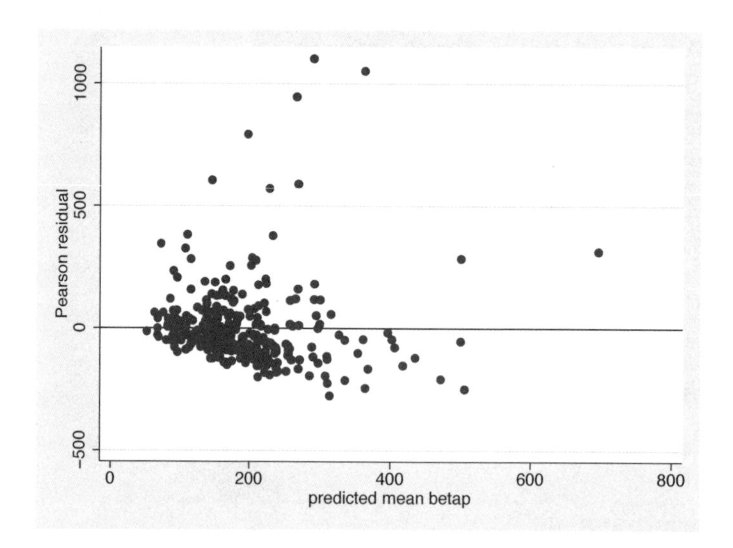

Figure 5.5: Pearson residuals versus fitted values from log-normal model (two outliers removed)

We now switch our focus from transforming the linear predictor to transforming the outcome variable to see if a better model may be found. In this model, we transform the observed variable and leave the linear predictor on the identity link. We call this the lognormal model. However, we admit that this can easily be confused with the log-normal (log-*hyphen*-normal) model. The lognormal model is derived from the lognormal distribution with an unspecified link function whereas the log-normal model is derived from the normal distribution incorporating the log-link.

```
. gen double logbetap = log(betap)
(1 missing value generated)
. glm logbetap age female pills kcal fat fiber alcohol chol betad
Iteration 0:   log likelihood = -321.38508
Generalized linear models                          No. of obs      =        312
Optimization     : ML                              Residual df     =        302
                                                   Scale parameter =   .4746678
Deviance         =  143.3496683                    (1/df) Deviance =   .4746678
Pearson          =  143.3496683                    (1/df) Pearson  =   .4746678
Variance function: V(u) = 1                        [Gaussian]
Link function    : g(u) = u                        [Identity]
                                                   AIC             =   2.124263
Log likelihood   = -321.3850781                    BIC             =  -1591.037

------------------------------------------------------------------------------
             |                 OIM
    logbetap |      Coef.   Std. Err.      z    P>|z|     [95% Conf. Interval]
-------------+----------------------------------------------------------------
         age |   .0062599   .0030183     2.07   0.038     .0003442    .0121756
      female |   .2596354   .1316948     1.97   0.049     .0015184    .5177524
       pills |   .3338829   .0849924     3.93   0.000     .1673008    .5004649
        kcal |  -.0002552   .0002092    -1.22   0.223    -.0006652    .0001548
         fat |   .0016595   .0033158     0.50   0.617    -.0048393    .0081583
       fiber |   .0397854   .0114408     3.48   0.001     .0173619    .0622089
     alcohol |   .0102508   .0088814     1.15   0.248    -.0071565     .027658
        chol |  -.0004819   .0004566    -1.06   0.291    -.0013769     .000413
       betad |    .000042   .0000305     1.38   0.169    -.0000179    .0001019
       _cons |   4.019137   .2835851    14.17   0.000      3.46332    4.574953
------------------------------------------------------------------------------
. predict double logbetaphat, mu
. predict double reslognormal, pearson
(1 missing value generated)
. histogram reslognormal
(bin=17, start=-2.4730175, width=.26211838)
```

This model shows much better results for the residuals in terms of achieving normality with constant variance. Figure 5.6 shows a histogram of the Pearson residuals from the model. The histogram is much more symmetric than that obtained from earlier models.

(Continued on next page)

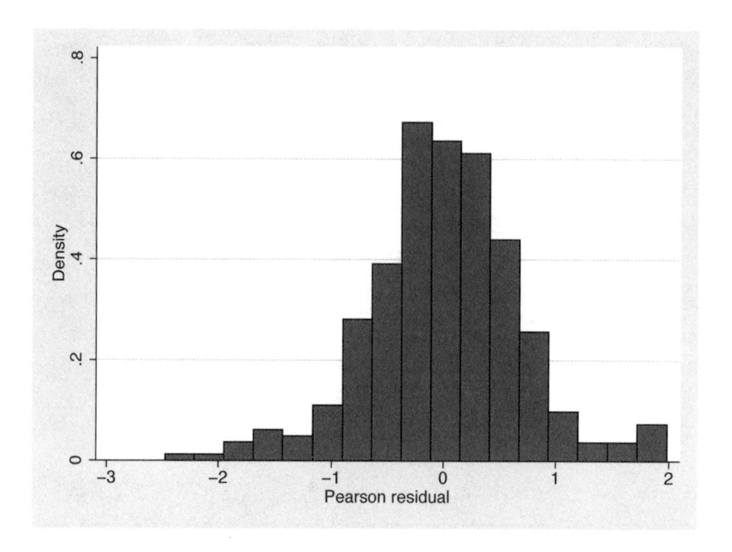

Figure 5.6: Pearson residuals from lognormal model (log-transformed outcome, two outliers removed, and zero outcome removed)

The standard residual plot of the Pearson residuals versus the fitted values illustrated in figure 5.7 shows the classic band of random values with no discernible features.

```
. twoway scatter reslognormal logbetaphat, yline(0)
```

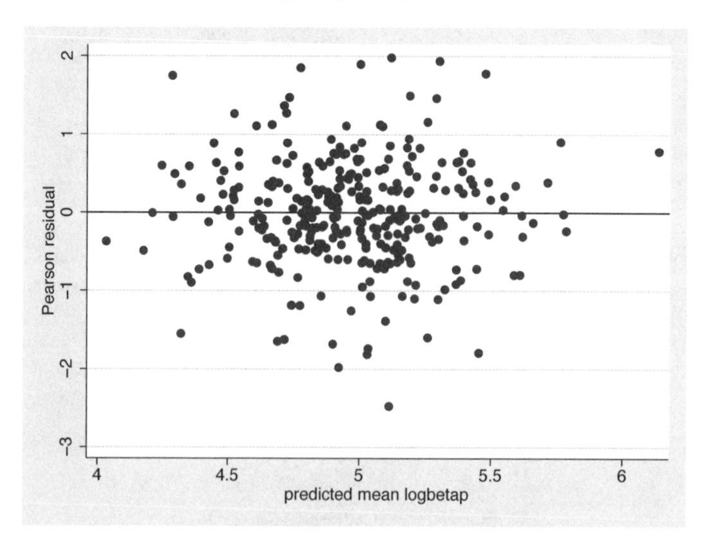

Figure 5.7: Pearson residuals versus fitted values from lognormal model (log-transformed outcome, two outliers removed, and zero outcome removed)

Figure 5.8 of the ordered Pearson residuals versus percentiles from the normal distribution shows better agreement (though the results are not randomly around the reference line), whereas figure 5.9 shows no discernible relationship of the Pearson residuals with the average daily kilocaloric intake.

```
. sort reslognormal
. gen nscorelb = invnormal(_n/315)
. twoway scatter nscorelb reslognormal || line reslognormal reslognormal
. twoway scatter reslognormal kcal, yline(0)
```

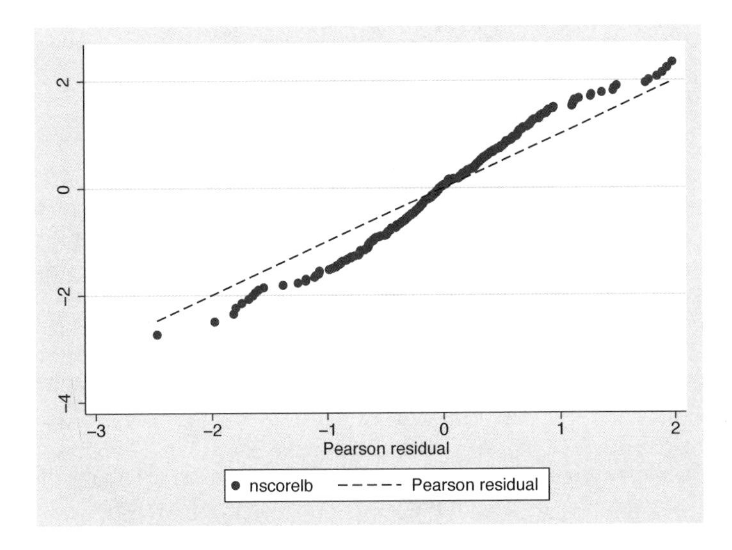

Figure 5.8: Normal scores versus sorted Pearson residuals obtained from lognormal model (log-transformed outcome, two outliers removed, and zero outcome removed)

(*Continued on next page*)

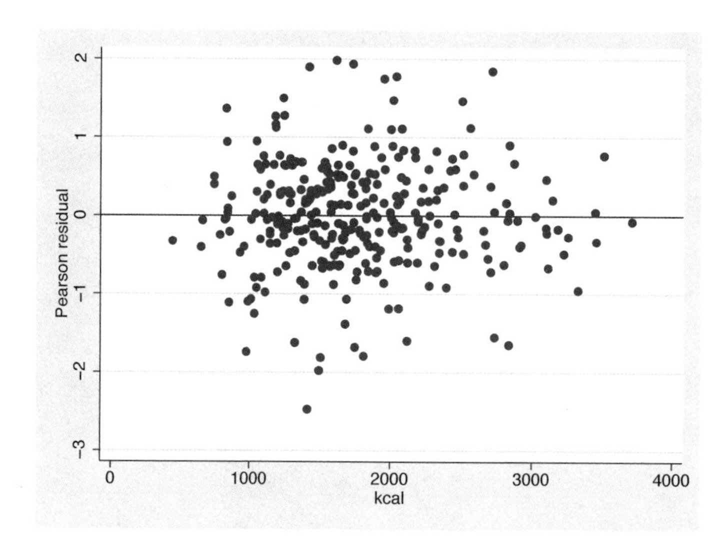

Figure 5.9: Pearson residuals versus kilocalories; Pearson residuals obtained from lognormal model (log-transformed outcome, two outliers removed, and zero outcome removed)

6 The gamma family

Contents

The gamma model is used for data situations in which the response can take only values greater than or equal to 0. Used primarily with continuous-response data, the GLM gamma family can be used with count data where there are many different count results that in total take the shape of a gamma distribution.

Ideally, the gamma model is best used with positive responses having a constant coefficient of variation. However, the model is robust to wide deviations from the latter criterion. In fact, because the shape of the two-parameter gamma distribution is pliable and can be parameterized to fit many response shapes, it may be preferable over the Gaussian model for many strictly positive response data situations.

The traditional GLM framework limits models to one parameter, that of the mean or μ. Scale parameters are constrained to the value of 1 when part of either the variance or link function, or they may be assigned a value by the user. However, in the latter case the model is no longer a straightforward GLM but rather a quasilikelihood model. A prime example of the latter is the traditional GLM version of the negative binomial model. Examples of the former include the Gaussian, gamma, and inverse Gaussian models.

Although the traditional IRLS GLM algorithm does not formally estimate the scale or ancillary parameter as would an FIML algorithm, the scale may be estimated from the IRLS Pearson-based dispersion or scale statistic. That is, the value for the scale parameter in the gamma regression model using FIML is nearly the same as that for the IRLS Pearson-based estimate.

We demonstrate various GLM-type parameterizations for the gamma model in this chapter. We willl not, however, discuss the three-parameter model, or give substantial discussion to gamma models that incorporate censoring. However, the GLM gamma model can be used to directly model exponential regression, but we relegate those types of discussions to the final section in this chapter.

6.1 Derivation of the gamma model

The base density function for the gamma distribution is

$$f(y;\mu,\phi) = \frac{1}{y\Gamma(1/\phi)}\left(\frac{y}{\mu\phi}\right)^{1/\phi}\exp\left(-\frac{y}{\mu\phi}\right) \tag{6.1}$$

In exponential-family form, the above probability density appears as

$$f(y;\mu,\phi) = \exp\left\{\frac{y/\mu-(-\ln\mu)}{-\phi}+\frac{1-\phi}{\phi}\ln y-\frac{\ln\phi}{\phi}-\ln\Gamma\left(\frac{1}{\phi}\right)\right\} \tag{6.2}$$

This equation provides us with the link and the cumulant given by

$$\theta = 1/\mu \tag{6.3}$$
$$b(\theta) = -\ln(\mu) \tag{6.4}$$

From this we may derive the mean and the variance

$$b'(\theta) = \frac{\partial b}{\partial\mu}\frac{\partial\mu}{\partial\theta} \tag{6.5}$$
$$= \left(-\frac{1}{\mu}\right)(-\mu^2) \tag{6.6}$$
$$= \mu \tag{6.7}$$
$$b''(\theta) = \frac{\partial^2 b}{\partial\mu^2}\left(\frac{\partial\mu}{\partial\theta}\right)+\frac{\partial b}{\partial\mu}\frac{\partial^2\mu}{\partial\theta^2} \tag{6.8}$$
$$= (1)(-\mu^2) \tag{6.9}$$
$$= -\mu^2 \tag{6.10}$$

The variance here is an ingredient of the variance of y, which is found using $b''(\theta)a(\phi) = -\mu^2(-\phi) = \phi\mu^2$.

From the density function, in exponential-family form, we derive the log-likelihood ($\mathcal{L}$) function by dropping the exp and braces from (6.2). The deviance function may be calculated from $\mathcal{L}$:

$$D = 2\phi\left\{\mathcal{L}(y;y)-\mathcal{L}(\mu;y)\right\} \tag{6.11}$$
$$= 2\phi\sum_{i=1}^{n}\frac{1+\ln(y_i)-y_i/\mu_i-\ln(\mu_i)}{-\phi} \tag{6.12}$$
$$= 2\sum_{i=1}^{n}\left\{\frac{y_i-\mu_i}{\mu_i}-\ln\left(\frac{y_i}{\mu_i}\right)\right\} \tag{6.13}$$

The other important statistic for the estimating algorithm is the derivative of the gamma link function, or g':

$$\frac{\partial g(\mu)}{\partial \mu} = \frac{\partial \theta}{\partial \mu} = -\frac{1}{\mu^2} \tag{6.14}$$

θ is often parameterized as $-1/\mu$ or as $-y/\mu$. We have placed the negation outside the link to be associated with the scale to provide a direct relationship between θ and the canonical reciprocal link.

The important statistics that distinguish one GLM from another, in the canonical form, are listed below.

$$\text{link}: \quad \frac{1}{\mu} \tag{6.15}$$

$$\text{inverse link}: \quad \frac{1}{\eta} \tag{6.16}$$

$$\text{variance}: \quad \mu^2 \tag{6.17}$$

$$g': \quad -\frac{1}{\mu^2} \tag{6.18}$$

$$\text{deviance}: \quad 2\sum_{i=1}^{n}\left\{\frac{y_i - \mu_i}{\mu_i} - \ln\left(\frac{y_i}{\mu_i}\right)\right\} \tag{6.19}$$

Inserting these functions into the canonical GLM algorithm yields

Listing 6.1: IRLS algorithm for gamma models

```
1   μ = (y + mean(y))/2
2   η = 1/μ
3   WHILE (abs(ΔDev) > tolerance) {
4       W = μ^2
5       z = η + (y − μ)/μ^2
6       β = (X^T W X)^(−1) X^T W z
7       η = Xβ
8       μ = 1/η
9       OldDev = Dev
10      Dev = 2∑((y − μ)/μ − ln(y/μ))
11      ΔDev = Dev - OldDev
12  }
13  χ^2 = ∑(y − μ)^2/μ^2
```

An OIM-based maximum likelihood algorithm will give identical answers to those produced by the standard IRLS algorithm because the OIM of the former reduces to the expected information matrix of the latter. We have previously mentioned that only for the noncanonical link does this make a difference. We have also shown how one can amend the IRLS algorithm to allow use of the OIM. Stata's `glm` command uses a default Newton–Raphson algorithm where standard errors are based on the OIM. This is a modification of the standard EIM-based IRLS algorithm and is similar to the default hybrid algorithm used by SAS software. Refer to chapter 3 for details.

6.2 Example: Reciprocal link

We now present a rather famous example of a reciprocal-linked gamma dataset. The example first gained notoriety in McCullagh and Nelder (1989) and later was given a full examination in Hilbe and Turlach (1995). The example deals with car insurance claims (`claims.dta`) and models average claims for damage to an owner's car on the basis of the policy holder's age group (PA 1–8), the vehicle age group (VA 1–4), and the car group (CG 1–4). A frequency weight is given, called `number`, which represents the number of identical covariate patterns related to a particular outcome.

The criterion of coefficient of variation constancy across cell groups is assumed and has been validated in previous studies. Again the gamma model is robust to deviation from this criterion, but not so much that it should not be assessed. This is particularly the case with the canonical-link model.

A schema of the model given the reciprocal link is given by

$$\mu = \sigma w(\beta_0 + \beta_1 \text{PA} + \beta_2 \text{CG} + \beta_3 \text{VA})^{-1} \tag{6.20}$$

The main-effects model is displayed below. The levels for the age group, vehicle age group, and car group are included in the model through automated production of indicator variables by using Stata's `xi` command.

```
. use http://www.stata-press.com/data/hh2/claims

. xi: glm y i.pa i.cg i.va [fw=number], family(gamma) irls
i.pa              _Ipa_1-8            (naturally coded; _Ipa_1 omitted)
i.cg              _Icg_1-4            (naturally coded; _Icg_1 omitted)
i.va              _Iva_1-4            (naturally coded; _Iva_1 omitted)

Iteration 1:   deviance =  135.5555
Iteration 2:   deviance =  124.8845
Iteration 3:   deviance =  124.7828
Iteration 4:   deviance =  124.7828
Iteration 5:   deviance =  124.7828

Generalized linear models                          No. of obs      =      8942
Optimization     : MQL Fisher scoring              Residual df     =      8928
                   (IRLS EIM)                      Scale parameter =         1
Deviance         =  124.7827519                    (1/df) Deviance =  .0139766
Pearson          =  131.7861885                    (1/df) Pearson  =   .014761

Variance function: V(u) = u^2                      [Gamma]
Link function    : g(u) = 1/u                      [Reciprocal]

                                                   BIC             = -81106.76

------------------------------------------------------------------------------
             |                 EIM
           y |      Coef.   Std. Err.      z    P>|z|     [95% Conf. Interval]
-------------+----------------------------------------------------------------
      _Ipa_2 |   .0001014   .0000482     2.10   0.035     6.91e-06    .0001959
      _Ipa_3 |     .00035   .0000456     7.68   0.000     .0002607    .0004393
      _Ipa_4 |   .0004623   .0000454    10.19   0.000     .0003734    .0005512
      _Ipa_5 |     .00137   .0000463    29.58   0.000     .0012792    .0014608
      _Ipa_6 |   .0009695   .0000447    21.68   0.000     .0008819    .0010571
      _Ipa_7 |   .0009164   .0000451    20.33   0.000     .0008281    .0010048
      _Ipa_8 |   .0009201   .0000459    20.03   0.000       .00083    .0010101
      _Icg_2 |   .0000377   .0000186     2.02   0.043     1.12e-06    .0000742
      _Icg_3 |  -.0006139   .0000188   -32.68   0.000    -.0006507   -.0005771
      _Icg_4 |  -.0014206     .00002   -71.20   0.000    -.0014597   -.0013815
      _Iva_2 |   .0003663   .0000111    32.87   0.000     .0003445    .0003881
      _Iva_3 |   .0016512   .0000251    65.90   0.000     .0016021    .0017003
      _Iva_4 |   .0041537   .0000489    84.98   0.000     .0040579    .0042495
       _cons |   .0034105   .0000462    73.85   0.000       .00332     .003501
------------------------------------------------------------------------------
```

Using `predict`, we can generate both the Anscombe residuals and the expected value of the model. Using the predicted value, we can generate the log of the variance and illustrate a useful residual plot in figure 6.1.

```
. predict double anscombe, anscombe
(5 missing values generated)

. predict double logvar, mu

. replace logvar = log(logvar*logvar)
(128 real changes made)

. label var logvar "Log(variance)"

. twoway scatter anscombe logvar, yline(0)
```

(Continued on next page)

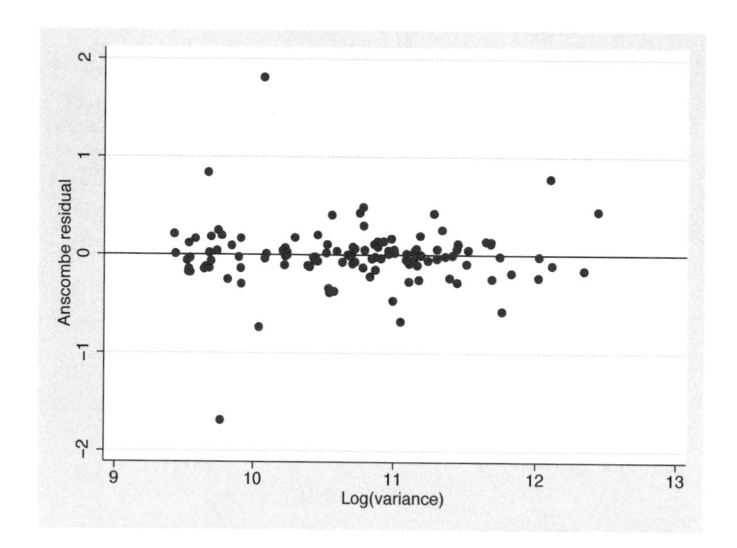

Figure 6.1: Anscombe residuals versus log (variance)

A check, in figure 6.1, of the Anscombe residuals versus the log of the variance for the gamma family indicates no problems.

The results may be interpreted as the rate at which each pound sterling (the outcome) is paid for an average claim over a unit of time, which is unspecified in the data. Because the algorithm uses the IRLS method of estimation, the scale is shown as 1. However, as we have seen before, the value of the scale is point-estimated as the Pearson dispersion, or, in this instance, 0.015.

The interactions discussed in the above referenced sources may also be evaluated by using the Stata command

```
. xi: glm y i.pa i.cg i.va i.pa*i.cg i.pa*i.va i.cg*i.va [fw=number],
> family(gamma) irls
```

The above needs to be on one line when issued in Stata.

Both of our reference sources found that the extra interactions in the model were nonproductive, adding no more predictive value to the model. The manner in which the authors showed this to be the case underlies appropriate modeling techniques and warrants discussion here.

Typically, we assess the worth of a predictor, or group of predictors, on the basis of the p-value from a Wald or likelihood-ratio test statistic. However, particularly for large-data situations, many or most of the predictors appear statistically significant. The goal, of course, is to discover the most parsimonious model possible that still possesses maximal explanatory value. The standard methods above are of no help. However, creating a deviance table, described by the references cited above, aids in achieving parsimony while differentiating explanatory power for candidate models.

The method is as follows. The deviance from each successive addition of main effect term or interaction term is displayed in a vertical row, with values of the difference between deviances, degrees of freedom, and mean differences placed in corresponding rows to the right. Of course, the order the predictors enter the model does matter and manipulation may be required.

The following table for the `claims` dataset is reproduced here in part from McCullagh and Nelder (1989) and Hilbe and Turlach (1995):

Model	Deviance	Difference	DF	Mean Difference
Constant	649.9			
Main effects				
PA	567.7	82.2	7	11.7
PA CG	339.4	228.3	3	76.1
PA CG VA	124.8	214.7	3	71.6
Interaction terms				
+PA*CG	90.7	34.0	21	1.62
+PA*VA	71.0	19.7	21	0.94
+CG*VA	65.6	5.4	9	1.13

We have shown the output for the main effects model above. The header portion for the results from the full interaction model appears as

```
Generalized linear models                         No. of obs       =       8942
Optimization      : MQL Fisher scoring            Residual df      =       8877
                    (IRLS EIM)                    Scale param      =          1
Deviance          =  65.58462                     (1/df) Deviance  =   .0073882
Pearson           =  64.19831                     (1/df) Pearson   =    .007232
```

The header information is consistent with the deviance table.

Each additional main effect term results in a substantial difference in deviance values. This is especially the case when adjusted for degrees of freedom. The addition of interaction terms results in a markedly reduced value of the mean differences between hierarchical levels. For this reason, we can accept the main effects model as effectively representing a parsimonious and sufficiently explanatory model. Interactions are not needed.

6.3 Maximum likelihood estimation

For ML estimation of a model, the log-likelihood function (rather than the deviance) plays a paramount role in estimation. It is the log-likelihood function that is maximized, hence the term "maximum likelihood". However, when we use the traditional GLM IRLS algorithm, whether we modify it such that the OIM is created and used to calculate standard errors, the algorithm effectively specifies a limited information maximum-likelihood (LIML) algorithm when equivalent link and variance functions are specified; it is LIML because the scale parameter is ancillary (estimated using method of moments with no estimated standard error). However, the ML algorithm may estimate all parameters (FIML) or may treat the scale parameter as ancillary (LIML).

We may illustrate the log-likelihood function for a particular link function (parameterization) by substituting the inverse link function $g^{-1}(\eta)$ for μ. For example, the gamma canonical link (reciprocal function) is

$$\mu = \frac{1}{\mathbf{x}\boldsymbol{\beta}} \tag{6.21}$$

For each instance of a μ in the gamma $\mathcal{L}$ function, we replace μ with the inverse link of the linear predictor $1/\mathbf{x}\boldsymbol{\beta}$.

$$\begin{aligned}
\mathcal{L} &= \sum_{i=1}^{n} \left\{ \frac{y_i/\mu_i - (-\ln \mu_i)}{-\phi} + \frac{1-\phi}{\phi} \ln y_i - \frac{\ln \phi}{\phi} - \ln \Gamma\left(\frac{1}{\phi}\right) \right\} && (6.22) \\
&= \sum_{i=1}^{n} \left\{ \frac{y_i \mathbf{x}_i\boldsymbol{\beta} - \ln(\mathbf{x}_i\boldsymbol{\beta})}{-\phi} + \frac{1-\phi}{\phi} \ln y_i - \frac{\ln \phi}{\phi} - \ln \Gamma\left(\frac{1}{\phi}\right) \right\} && (6.23)
\end{aligned}$$

In the previous section we presented an example related to car insurance claims. We fitted the model using the traditional IRLS algorithm. The maximum likelihood approach results in the same estimates and standard errors when the canonical link is involved. This was the case in the discussion of the Gaussian family of models. We also saw how the IRLS approach accurately estimated the scale parameter, even though the IRLS algorithm assumed a value of 1. In the main-effects model for the `claims` dataset, the scale was estimated as 0.014761. To show the scale when modeled using the default modified N–R algorithm for LIML, we have the following (header) results:

```
. xi: glm y i.pa i.cg i.va [fw=number], family(gamma) notable nolog
i.pa              _Ipa_1-8            (naturally coded; _Ipa_1 omitted)
i.cg              _Icg_1-4            (naturally coded; _Icg_1 omitted)
i.va              _Iva_1-4            (naturally coded; _Iva_1 omitted)

Generalized linear models                    No. of obs      =        8942
Optimization     : ML                        Residual df     =        8928
                                             Scale parameter =     .014761
Deviance         =  124.7827519              (1/df) Deviance =    .0139766
Pearson          =  131.7861915              (1/df) Pearson  =     .014761

Variance function: V(u) = u^2                [Gamma]
Link function    : g(u) = 1/u                [Reciprocal]

                                             AIC             =    12.91782
Log likelihood   = -57741.59316              BIC             =   -81106.76
```

6.4 Log-gamma models

We mentioned before that the reciprocal link models the rate per unit of the model response, given a specific set of explanatory variables or predictors. The log-linked gamma represents the log-rate of the response. This model specification is identical to exponential regression. Such a specification, of course, models data with a negative exponential decline. However, unlike the exponential models found in survival analysis, we cannot use the log-gamma model with censoring effects. We see, though, that non-censored exponential models can be fitted with GLM specifications. We leave that to the end of this chapter.

The log-gamma model, like its reciprocal counterpart, is used with data in which the response is greater than 0. Examples can be found in nearly every discipline. For instance, in health analysis, length of stay (LOS) can generally be modeled using log-gamma regression because stays are always constrained to be positive. LOS data are generally modeled using Poisson or negative binomial regression since the elements of LOS are discrete. However, when there are many LOS elements, i.e., many different LOS values, many researchers find the gamma or inverse Gaussian models to be acceptable and preferable.

Before GLM, data that are now modeled using log-gamma techniques were generally modeled using Gaussian regression with a log-transformed response. Although the results are usually similar between the two methods, the log-gamma, which requires no external transformation, is easier to interpret and comes with a set of residuals with which to evaluate the worth of the model. Hence, the log-gamma is finding increased use among researchers who once used Gaussian techniques.

The IRLS algorithm for the log-gamma model is the same as that for the canonical-link model except that the link and inverse link become $\ln(\mu)$ and $\exp(\eta)$, respectively, and the derivative of g is now $1/\mu$. The ease with which we can change between models is one of the marked beauties of GLM. However, because the log link is not canonical, the IRLS and modified ML algorithms will give different standard errors. But, except in extreme cases, differences in standard errors are usually minimal. Except perhaps when working with small datasets, which method of estimation is used generally makes little inferential difference.

An example will point this out. We use the cancer dataset that comes packaged with Stata, `cancer.dta`. The variable `age` is a continuous predictor and `drug` is factored or leveled. We generated indicator variables specifying `tab drug, gen(drug)`, where `drug1`, the placebo, serves as the reference level. The IRLS model is fitted as

```
. sysuse cancer, clear
(Patient Survival in Drug Trial)
. gen drug2 = drug==2
. gen drug3 = drug==3
```

(Continued on next page)

```
. glm studytime age drug2 drug3, family(gamma) link(log) irls nolog
Generalized linear models                          No. of obs      =        48
Optimization     : MQL Fisher scoring              Residual df     =        44
                   (IRLS EIM)                      Scale parameter =         1
Deviance         =  16.17463555                    (1/df) Deviance =  .3676054
Pearson          =  13.99281815                    (1/df) Pearson  =  .3180186
Variance function: V(u) = u^2                      [Gamma]
Link function    : g(u) = ln(u)                    [Log]
                                                   BIC             = -154.1582
```

studytime	Coef.	EIM Std. Err.	z	P>|z|	[95% Conf.	Interval]
age	-.0447763	.014732	-3.04	0.002	-.0736505	-.0159021
drug2	.5743482	.1969367	2.92	0.004	.1883594	.9603371
drug3	1.052072	.1977145	5.32	0.000	.6645583	1.439585
_cons	4.645976	.8353019	5.56	0.000	3.008814	6.283138

Estimation using the default modified ML algorithm displays the following table:

```
. glm studytime age drug2 drug3, family(gamma) link(log) nolog
Generalized linear models                          No. of obs      =        48
Optimization     : ML                              Residual df     =        44
                                                   Scale parameter =  .3180529
Deviance         =  16.17463553                    (1/df) Deviance =  .3676054
Pearson          =  13.99432897                    (1/df) Pearson  =  .3180529
Variance function: V(u) = u^2                      [Gamma]
Link function    : g(u) = ln(u)                    [Log]
                                                   AIC             =  7.403608
Log likelihood   = -173.6866032                    BIC             = -154.1582
```

studytime	Coef.	OIM Std. Err.	z	P>|z|	[95% Conf.	Interval]
age	-.0447789	.015112	-2.96	0.003	-.0743979	-.01516
drug2	.5743689	.1986342	2.89	0.004	.185053	.9636847
drug3	1.0521	.1965822	5.35	0.000	.6668056	1.437394
_cons	4.646108	.8440093	5.50	0.000	2.99188	6.300336

The standard errors are slightly different, as was expected. And a scale parameter is estimated and is provided in the output header. That the parameter estimates differ as well may seem strange. They do only because of different optimization criteria. If the tolerance is altered in either model, the parameters take slightly different values. If the starting values are different, as they are here, then estimates may vary slightly. Since the differences in parameter estimates are typically less than .01, the statistical information remains unchanged.

For a matter of comparison, we can model the same data by using a Stata two-parameter log-gamma program called `lgamma` from Hilbe (2000). It uses a full ML routine for estimation, with no ties to GLM and no IRLS method. `lgamma` treats the additional parameter in an equivalent manner as the regression coefficient parameters. Thus this command is an FIML command for fitting the log-gamma model.

```
. lgamma studytime age drug2 drug3, nolog
Log-gamma model                                   Number of obs   =         48
                                                  LR chi2(3)      =      28.00
Log likelihood = -160.91053                       Prob > chi2     =     0.0000

------------------------------------------------------------------------------
   studytime |      Coef.   Std. Err.      z    P>|z|     [95% Conf. Interval]
-------------+----------------------------------------------------------------
         age |  -.0447789   .0151597    -2.95   0.003    -.0744914   -.0150665
       drug2 |   .5743689   .1992613     2.88   0.004      .183824    .9649138
       drug3 |     1.0521   .1972027     5.34   0.000     .6655893     1.43861
       _cons |   4.646108   .8466737     5.49   0.000     2.986658    6.305558
-------------+----------------------------------------------------------------
     /ln_phi |  -1.139231   .1942152    -5.87   0.000    -1.519886   -.7585759
-------------+----------------------------------------------------------------
         phi |   .3200651   .0621615                      .2187369    .4683329
------------------------------------------------------------------------------
```

Here we see that the parameter estimates are identical to those produced using GLM with the modified ML algorithm. Standard errors are also similar to those of the GLM model. The scale estimate is the same to within .002 of the previous estimate. Again different starting values or different tolerances can account for slight differences in output. They typically have no bearing on statistical accuracy or on statistical interpretation.

The log-likelihood function for the log-gamma model, parameterized in terms of $\mathbf{x}\boldsymbol{\beta}$, is

$$\mathcal{L} = \sum_{i=1}^{n} \left\{ \frac{y_i/\exp(\mathbf{x}_i\boldsymbol{\beta}) + \mathbf{x}_i\boldsymbol{\beta}}{-\phi} + \frac{\phi+1}{\phi}\ln y_i - \frac{\ln\phi}{\phi} - \ln\Gamma\left(\frac{1}{\phi}\right)\right\} \tag{6.24}$$

Above, we compared the estimated parameters and standard errors for the various estimation methods. A rather striking difference: the GLM model fitted with the modified ML algorithm produced a log-likelihood value of -173.69, whereas the FIML `lgamma` command produced a value of -160.91. Why are the values so different? The answer lies in how the GLM model alters the deviance-based optimization into one for the log likelihood. The deviance functions used in the GLM algorithms do not include normalization factors or scaling factors, and so these terms are lost to calculations for the (limited information) log-likelihood values.

Here the scaled log likelihood is $-173.6866032/.3180529 = -546.09344$. The normalization term $\{(\phi+1)/\phi\}\ln y_i - \ln\phi/\phi - \ln\Gamma(1/\phi)$ would be 385.18238 such that the comparable log likelihood is $-546.09344 + 385.18238 = -160.91106$.

6.5 Identity-gamma models

The identity link, with respect to the gamma family, models duration data. It assumes that there is a one-to-one relationship between μ and η. In this relationship, it is like the canonical-link Gaussian model. However, here it is a noncanonical link.

Choosing between different links of the same model family may sometimes be hard. McCullagh and Nelder (1989) support the method of minimal deviances. They also check the residuals to observe closeness of fit and the normality and independence of the standardized residuals themselves. In this example, we have two noncanonical links that we are interested in comparing: log and identity links. All other factors the same, choosing the model having the least value for the deviance is preferable. Here it is the identity-linked model. We give both the IRLS and ML outputs below. The log-gamma model has a deviance of 16.17, whereas the identity-gamma model has a deviance of 15.09—a reduction of 1.08 or 6.68%.

We can use other tests to assess differential model worth among links. These tests include the BIC and AIC. Models having lower values of these criteria are preferable. With respect to BIC, if the absolute difference between the BIC for two models is less than 2, then there is only weak evidence that the model with the smaller BIC is preferable. Absolute differences between 2 and 6 give positive support for the model with the smaller BIC, whereas absolute differences between 6 and 10 offer strong support. Absolute differences more than 10 are strong support for the model with the smaller BIC.

```
. glm studytime age drug2 drug3, family(gamma) link(id) irls nolog  /* IRLS */
Generalized linear models                          No. of obs      =         48
Optimization     : MQL Fisher scoring              Residual df     =         44
                   (IRLS EIM)                      Scale parameter =          1
Deviance         =  15.09376233                    (1/df) Deviance =   .3430401
Pearson          =  12.98530838                    (1/df) Pearson  =   .2951206

Variance function: V(u) = u^2                      [Gamma]
Link function    : g(u) = u                        [Identity]

                                                   BIC             = -155.2391

------------------------------------------------------------------------------
             |                 EIM
   studytime |      Coef.   Std. Err.      z    P>|z|     [95% Conf. Interval]
-------------+----------------------------------------------------------------
         age |  -.5369706   .1334085    -4.03   0.000    -.7984464   -.2754948
       drug2 |   6.472958   2.134955     3.03   0.002     2.288522    10.65739
       drug3 |   16.20353   3.874013     4.18   0.000     8.610607    23.79646
       _cons |   38.96542    8.19225     4.76   0.000      22.9089    55.02193
------------------------------------------------------------------------------
```

```
. glm studytime age drug2 drug3, family(gamma) link(id) nolog        /* N-R  */

Generalized linear models                         No. of obs      =         48
Optimization     : ML                             Residual df     =         44
                                                  Scale parameter =   .2951209
Deviance         =  15.09376233                   (1/df) Deviance =   .3430401
Pearson          =  12.98531951                   (1/df) Pearson  =   .2951209

Variance function: V(u) = u^2                     [Gamma]
Link function    : g(u) = u                       [Identity]

                                                  AIC             =    7.38109
Log likelihood   = -173.1461666                   BIC             =  -155.2391

------------------------------------------------------------------------------
             |                 OIM
   studytime |      Coef.   Std. Err.      z    P>|z|     [95% Conf. Interval]
-------------+----------------------------------------------------------------
         age |   -.536976    .1256999    -4.27   0.000    -.7833433   -.2906087
       drug2 |    6.47296    2.131972     3.04   0.002     2.294372    10.65155
       drug3 |   16.20346    3.804618     4.26   0.000     8.746548    23.66038
       _cons |   38.96574    7.747343     5.03   0.000     23.78123    54.15025
------------------------------------------------------------------------------
```

6.6 Using the gamma model for survival analysis

We mentioned earlier that exponential regression may be modeled using a log-linked gamma regression. Using the default ML version of `glm` provides the necessary means by which the observed information matrix is used to calculate standard errors. This is the same method used by typical ML implementations of exponential regression. See Stata's `streg`, which replaced the older `ereg` command, for a summary of the exponential regression model.

Why are the exponential regression results the same as the log-gamma model results? The answer rests in the likelihood functions. The exponential probability distribution has the form

$$f\left(\frac{y}{\mu}\right) = \frac{1}{\mu}\exp\left(-\frac{y}{\mu}\right) \tag{6.25}$$

where μ is parameterized as

$$\mu = \exp(\mathbf{x}\boldsymbol{\beta}) \tag{6.26}$$

such that the function appears as

$$f\left(\frac{y}{\mathbf{x}\boldsymbol{\beta}}\right) = \frac{1}{\exp(\mathbf{x}\boldsymbol{\beta})}\exp\left\{-\frac{y}{\exp(\mathbf{x}\boldsymbol{\beta})}\right\} \tag{6.27}$$

The exponential log likelihood is thus

$$\mathcal{L}(\boldsymbol{\beta}; y) = \sum_{i=1}^{n}\left\{-\mathbf{x}_i\boldsymbol{\beta} - \frac{y_i}{\exp(\mathbf{x}_i\boldsymbol{\beta})}\right\} \tag{6.28}$$

Now recall the log-gamma log likelihood given in (6.24). Drop references to the scale parameter since the exponential distribution assumes a scale of 1. We find that the log-gamma log likelihood in that form is identical to that of the exponential log likelihood, causing the similarity in output. However, the log link is noncanonical. To obtain the same standard errors between the two models, we must fit the log-gamma by using the modified Newton–Raphson method and explicitly specify a scale of one. This is the default parameterization when using Stata.

Below we display output for an exponential model using first GLM and then a standard exponential regression model.

```
. glm studytime age drug2 drug3, family(gamma) link(log) scale(1) nolog

Generalized linear models                     No. of obs      =          48
Optimization     : ML                         Residual df     =          44
                                              Scale parameter =           1
Deviance         =  16.17463553               (1/df) Deviance =    .3676054
Pearson          =  13.99432897               (1/df) Pearson  =    .3180529

Variance function: V(u) = u^2                 [Gamma]
Link function    : g(u) = ln(u)               [Log]

                                              AIC             =    7.403608
Log likelihood   = -173.6866032               BIC             =   -154.1582

------------------------------------------------------------------------------
             |                 OIM
   studytime |      Coef.   Std. Err.      z    P>|z|     [95% Conf. Interval]
-------------+----------------------------------------------------------------
         age |  -.0447789   .0267961    -1.67   0.095    -.0972983    .0077404
       drug2 |   .5743689   .3522116     1.63   0.103    -.1159532    1.264691
       drug3 |     1.0521    .348573     3.02   0.003      .368909     1.73529
       _cons |   4.646108    1.49657     3.10   0.002     1.712886     7.57933
------------------------------------------------------------------------------
(Standard errors scaled using dispersion equal to square root of 1)

. stset studytime

     failure event:  (assumed to fail at time=studytime)
obs. time interval:  (0, studytime]
 exit on or before:  failure

------------------------------------------------------------------------------
       48  total obs.
        0  exclusions
------------------------------------------------------------------------------
       48  obs. remaining, representing
       48  failures in single record/single failure data
      744  total analysis time at risk, at risk from t =         0
                             earliest observed entry t =         0
                                  last observed exit t =        39
```

```
. streg age drug2 drug3, nohr dist(exp) nolog
         failure _d:  1 (meaning all fail)
   analysis time _t:  studytime
Exponential regression -- log relative-hazard form
No. of subjects =           48                    Number of obs    =        48
No. of failures =           48
Time at risk    =          744
                                                  LR chi2(3)       =     11.75
Log likelihood  =   -56.087318                    Prob > chi2      =    0.0083

------------------------------------------------------------------------------
          _t |      Coef.   Std. Err.      z    P>|z|     [95% Conf. Interval]
-------------+----------------------------------------------------------------
         age |   .0447789    .0267961     1.67   0.095    -.0077404    .0972983
       drug2 |  -.5743689    .3522116    -1.63   0.103    -1.264691    .1159532
       drug3 |    -1.0521     .348573    -3.02   0.003     -1.73529    -.368909
       _cons |  -4.646108     1.49657    -3.10   0.002     -7.57933   -1.712886
------------------------------------------------------------------------------
```

Likewise, the regression parameters are the same because the GLM algorithms are limited information and have the same partial derivative as exponential regression—the two models have the same estimating equation for the regression parameters; they differ only in sign. The log-likelihood values differ because the models contain different normalization and scaling terms.

The gamma model (when estimated via GLM specification) cannot incorporate censoring into its routine. However, a right-censored exponential regression model may be duplicated with GLM methodology, as we show in section 12.5.

Gamma regression, in all its parameterizations and forms, has enjoyed little use when compared with the standard OLS or Gaussian model. Researchers have often used OLS regression when they should have used one of the gamma models. Certainly, one of the historic reasons was the lack of software. There has also been a corresponding lack of understanding of the gamma model and all its variations. We suspect that this will change.

7 The inverse Gaussian family

Contents

The inverse Gaussian is the most rarely used model of all traditional GLMs. Primary sources nearly always list the inverse Gaussian in tables of GLM families, but it is rarely discussed. Even McCullagh and Nelder (1989) give only passing comment of its existence.

The inverse Gaussian distribution does have a history of use in reliability studies. Typically, however, these are limited to models having only one predictor, much like the traditional nonlinear regression model. We endeavor to add more information so that researchers may realize that the inverse Gaussian does in fact have practical application.

7.1 Derivation of the inverse Gaussian model

The inverse Gaussian probability distribution is a continuous distribution having two parameters given by

$$f(y;\mu,\sigma^2) = \frac{1}{\sqrt{2\pi y^3\sigma^2}} \exp\left\{-\frac{(y-\mu)^2}{2(\mu\sigma)^2 y}\right\} \tag{7.1}$$

In exponential form, the inverse Gaussian distribution is given by

$$\begin{aligned} f(y;\mu,\sigma^2) &= \exp\left\{-\frac{(y-\mu)^2}{2y(\mu\sigma)^2} - \frac{1}{2}\ln\left(2\pi y^3\sigma^2\right)\right\} & (7.2)\\ &= \exp\left\{\frac{y/\mu^2 - 2/\mu}{-2\sigma^2} + \frac{1/y}{-2\sigma^2} + \frac{\sigma^2}{-2\sigma^2}\ln\left(2\pi y^3\sigma^2\right)\right\} & (7.3) \end{aligned}$$

The log-likelihood function may be derived from the exponential-family form by dropping the exp and its associated braces.

$$\mathcal{L} = \sum_{i=1}^{n} \left\{ \frac{y_i/(2\mu_i^2) - 1/\mu_i}{-\sigma^2} + \frac{1}{-2y_i\sigma^2} - \frac{1}{2}\ln\left(2\pi y_i^3\sigma^2\right) \right\} \tag{7.4}$$

GLM theory provides that, in canonical form, the link and cumulant functions are

$$\theta = \frac{1}{2\mu^2} = \frac{1}{2}\mu^{-2} \tag{7.5}$$

$$b(\theta) = \frac{1}{\mu} \tag{7.6}$$

$$a(\phi) = -\sigma^2 \tag{7.7}$$

The sign and coefficient value are typically dropped from the inverse Gaussian link function when inserted into the GLM algorithm. It is normally given the value of $1/\mu^2$ and the inverse link function the value of $1/\sqrt{\eta}$.

The inverse Gaussian mean and variance functions are derived from the cumulant

$$b'(\theta) = \frac{\partial b}{\partial \mu}\frac{\partial \mu}{\partial \theta} = \left(\frac{-1}{\mu^2}\right)\left(-\mu^3\right) = \mu \tag{7.8}$$

$$b''(\theta) = \frac{\partial^2 b}{\partial \mu^2}\left(\frac{\partial \mu}{\partial \theta}\right)^2 + \frac{\partial b}{\partial \mu}\frac{\partial^2 \mu}{\partial \theta^2} \tag{7.9}$$

$$= \left(\frac{2}{\mu^3}\right)\left(\mu^6\right) + \left(\frac{-1}{\mu^2}\right)\left(3\mu^5\right) = 2\mu^3 - 3\mu^3 = -\mu^3 \tag{7.10}$$

The derivative of the link, $g(\mu)$, is written $g'(\mu)$, or simply g'. Its derivation is given by

$$g' = \frac{\partial \theta}{\partial \mu} = \frac{\partial}{\partial \mu}\frac{1}{2\mu^2} = -\mu^{-3} \tag{7.11}$$

Finally, the deviance function is calculated from the saturated and model log-likelihood formulas

$$D = 2\sigma^2\left\{\mathcal{L}(y,\sigma^2;y) - \mathcal{L}(\mu,\sigma^2;y)\right\} \tag{7.12}$$

$$= 2\sigma^2 \sum_{i=1}^{n}\left\{\frac{y_i/(2y_i^2) - 1/y_i}{-\sigma^2} - \frac{y_i/(2\mu_i^2) - 1/\mu_i}{-\sigma^2}\right\} \tag{7.13}$$

$$= \sum_{i=1}^{n}\left\{\frac{y_i}{\mu_i^2} - \frac{2}{\mu_i} + \frac{1}{y_i}\right\} \tag{7.14}$$

$$= \sum_{i=1}^{n}\frac{(y_i - \mu_i)^2}{y_i\mu_i^2} \tag{7.15}$$

The IRLS GLM algorithm calculates the scale value, σ^2, as the Pearson-based dispersion statistic.

7.2 The inverse Gaussian algorithm

Inserting the appropriate inverse Gaussian formulas into the standard GLM algorithm, and excluding prior weights and subscripts, we have

Listing 7.1: IRLS algorithm for inverse Gaussian models

```
1      μ = (y − mean(y))/2
2      η = 1/μ^2
3      WHILE (abs(ΔDev) > tolerance) {
4            W = μ^3
5            z = η + (y − μ)/μ^3
6            β = (X^T W X)^{-1} X^T W z
7            η = Xβ
8            μ = 1/√η
9            OldDev = Dev
10           Dev = Σ(y − μ)^2/(yμ^2)
11           ΔDev = Dev - OldDev
12     }
13     χ^2 = σ Σ(y − μ)^2/μ^3
```

We have mentioned before that the IRLS algorithm may be amended such that the log-likelihood function is used as the basis of iteration rather than the deviance. However, when this is done, the log-likelihood function itself is usually amended. The normalization term is dropped such that terms with no instance of the mean parameter, μ, are excluded. Of course, prior weights are kept. For the inverse Gaussian, the amended log-likelihood function is usually given as

$$\mathcal{L}(\mu, \sigma^2; y) = \sum_{i=1}^{n} \frac{1}{\sigma^2} \left\{ -\frac{y_i}{2\mu_i^2} + \frac{1}{\mu_i} \right\} \tag{7.16}$$

7.3 Maximum likelihood algorithm

Maximum likelihood usually uses some modification of the Newton–Raphson (N–R) algorithm. We have seen before that the traditional GLM IRLS algorithm may be modified in such a manner that results in identical values to those produced using standard ML methods. Essentially, the IRLS algorithm is modified to produce standard errors from the observed, rather than the expected, information matrix (see the discussion on this topic in chap. 3). Parameter estimates remain unchanged. Moreover, for the canonical link, the two methods mathematically coincide.

Maximum likelihood (ML) relies on the log-likelihood function. Using full ML methods, not those that are amendments to the IRLS algorithm, parameter estimates and standard errors that are based on the observed information matrix are obtained as a direct result of maximization. The method requires use of the full log-likelihood function, which includes the scale parameter. And, in the process, the scale is estimated by ML techniques in the same manner as are other parameter estimates.

We show how the inverse Gaussian algorithm can be amended to simulate ML estimation when we address noncanonical links. We also provide more details about the full ML algorithm.

7.4 Example: The canonical inverse Gaussian

As mentioned earlier, the inverse Gaussian has found limited use in certain reliability studies. In particular, it seems most appropriate when modeling a nonnegative response having a high initial peak, rapid drop, and long right tail. If a discrete response has many different values together with the same properties, then the inverse Gaussian may be appropriate for this type of data as well. A variety of other shapes may also be modeled using inverse Gaussian regression.

We can create a synthetic inverse Gaussian dataset using tools available in Stata. This will allow us to observe some of the properties and relationships we have thus far discussed.

Shown below, we create a dataset of 25,000 cases with two random normal predictors having coefficients of 0.5 and −0.25, respectively, and a constant of 1. We also provide the model with a scale parameter of 0.5. A program for generating random numbers, `rndivgx` from Hilbe and Linde-Zwirble (1995), is used to create the dataset. We set the random number seed to allow recreation of these results. The version prefix is necessary because the `rndivgx` program relies on older facilities in the Stata program.

```
. set seed 12345
. version 3: set seed 12345
. set obs 25000
obs was 0, now 25000
. generate x1 = abs(invnormal(uniform()))
. generate x2 = abs(invnormal(uniform()))
. generate eta = 1 + .5*x1 -.25*x2
. generate mu = 1/sqrt(eta)
. rndivgx mu, s(0.5)
( Generating .................... )
Variable xig created.
. glm xig x1 x2, family(ig) nolog
Generalized linear models                          No. of obs      =      25000
Optimization     : ML                              Residual df     =      24997
                                                   Scale parameter =   .2845213
Deviance         =    7114.2851                    (1/df) Deviance =   .2846056
Pearson          =  7112.178056                    (1/df) Pearson  =   .2845213
Variance function: V(u) = u^3                      [Inverse Gaussian]
Link function    : g(u) = 1/(u^2)                  [Power(-2)]
                                                   AIC             =    1.56014
Log likelihood   = -19498.7551                     BIC             = -246021.1
```

xig	Coef.	OIM Std. Err.	z	P>|z|	[95% Conf.	Interval]
x1	.4797303	.0138798	34.56	0.000	.4525263	.5069343
x2	-.2298438	.0101248	-22.70	0.000	-.2496881	-.2099995
_cons	.9931728	.0151022	65.76	0.000	.963573	1.022773

σ^2 is calculated using the Pearson statistic listed in the output by typing

```
. display sqrt(.2845213)
0.53340538
```

This value is close to that specified to the random number generator program of 0.5.

7.5 Noncanonical links

The log and identity links are the primary noncanonical links associated with the inverse Gaussian distribution. This is similar to the gamma model.

We usually do not refer to a link function with full ML. The log-likelihood function is simply adjusted such that the inverse link, parameterized in terms of $x\beta$, is substituted for each instance of μ in the log-likelihood function. Hence, for the log-inverse Gaussian, we have

$$\mathcal{L} = \sum_{i=1}^{n} \left[\frac{y_i/\{2\exp(\mathbf{x}_i\boldsymbol{\beta})^2\} - 1/\exp(\mathbf{x}_i\boldsymbol{\beta})}{-\sigma^2} + \frac{1}{-2y_i\sigma^2} - \frac{1}{2}\ln\left(2\pi y_i^3\sigma^2\right) \right] \tag{7.17}$$

Using the IRLS approach, we can shape the canonical form of the inverse Gaussian to the log link by making the following substitutions:

	Canonical	Substitution
link	$-1/(2\mu^2)$	$\ln(\mu)$
inverse link	$(-2\eta)^{-1/2}$	$\exp(\eta)$
$g'(\mu)$	$1/\mu^3$	$1/\mu$

All other aspects of the algorithm remain the same.

If we wish to amend the basic IRLS algorithm so that the observed information matrix is used, hence making it similar to ML output, then another adjustment must be made. This has to do with a modification of the weight function, w. The log-inverse Gaussian algorithm, implementing a modified N–R approach, appears as

Listing 7.2: IRLS algorithm for log-inverse Gaussian models using OIM

```
1    μ = (y − mean(y))/2
2    η = ln(μ)
3    WHILE (abs(ΔDev) > tolerance) {
4         W = 1/(μ^3(1/μ)^2) = 1/μ
5         z = η + (y − μ)/μ
6         W_o = W + 2(y − μ)/μ^2
7         β = (X^T W_o X)^{-1} X^T W_o z
8         η = Xβ
9         μ = 1/exp(η)
10        OldDev = Dev
11        Dev = Σ(y − μ)^2/(yσ^2μ^2)
12        ΔDev = Dev - OldDev
13   }
14   χ^2 = Σ(y − μ)^2/μ^3
```

For an example, we use a dataset from the U.S. Medicare database called MedPar. We model length of stay (LOS) on the basis of type of admission, admitype, which is zero if the admission is emergency or urgent and one if it is elective, and codes, which records the number of ICD-9 codes recorded (up to 9). The data are from one hospital and are for one diagnostic code (DRG). The goal is to see whether the type of admission has any bearing on the length of stay, adjusted for the number of codes.

Since inverse Gaussian models fit positively valued outcome measures, the log link is a useful alternative to the canonical link.

```
. use http://www.stata-press.com/data/hh2/medpar1
. glm los admitype codes, family(ig) link(log) nolog
Generalized linear models                          No. of obs      =       3676
Optimization     : ML                              Residual df     =       3673
                                                   Scale parameter =   .1024074
Deviance         =  304.271883                     (1/df) Deviance =   .0828402
Pearson          = 376.1423892                     (1/df) Pearson  =   .1024074
Variance function: V(u) = u^3                      [Inverse Gaussian]
Link function    : g(u) = ln(u)                    [Log]
                                                   AIC             =   6.619001
Log likelihood   = -12162.72308                    BIC             = -29849.52

------------------------------------------------------------------------------
             |                 OIM
         los |      Coef.   Std. Err.      z    P>|z|     [95% Conf. Interval]
-------------+----------------------------------------------------------------
    admitype |  -.1916237   .0259321    -7.39   0.000    -.2424496   -.1407977
       codes |   .0992712   .0055994    17.73   0.000     .0882964    .1102459
       _cons |   1.057156   .0452939    23.34   0.000     .9683816     1.14593
------------------------------------------------------------------------------
```

Here we treat codes as a classification variable and investigate the log link, identity link, and the canonical link.

```
. xi: glm los admitype i.codes, family(ig) link(log) nolog
i.codes           _Icodes_1-9         (naturally coded; _Icodes_1 omitted)

Generalized linear models                          No. of obs      =      3676
Optimization     : ML                              Residual df     =      3666
                                                   Scale parameter =  .0984565
Deviance         =  298.4689189                    (1/df) Deviance =  .0814154
Pearson          =  360.9415865                    (1/df) Pearson  =  .0984565

Variance function: V(u) = u^3                      [Inverse Gaussian]
Link function    : g(u) = ln(u)                    [Log]

                                                   AIC             =   6.62123
Log likelihood   = -12159.8216                     BIC             = -29797.85

------------------------------------------------------------------------------
             |                 OIM
         los |      Coef.   Std. Err.      z    P>|z|     [95% Conf. Interval]
-------------+----------------------------------------------------------------
    admitype |  -.2027366   .0253892    -7.99   0.000    -.2524985   -.1529747
   _Icodes_2 |   .3201292   .2065013     1.55   0.121    -.0846058    .7248643
   _Icodes_3 |   .3337039   .2013326     1.66   0.097    -.0609008    .7283086
   _Icodes_4 |   .3521312   .1977577     1.78   0.075    -.0354667    .7397291
   _Icodes_5 |   .4764281   .1968441     2.42   0.016     .0906208    .8622353
   _Icodes_6 |   .4754921   .1964791     2.42   0.016     .0904001    .8605841
   _Icodes_7 |   .4551619   .1956977     2.33   0.020     .0716014    .8387224
   _Icodes_8 |   .5848797   .1948331     3.00   0.003      .203014    .9667455
   _Icodes_9 |   .9008458   .1924717     4.68   0.000     .5236082    1.278083
       _cons |   1.103884    .192179     5.74   0.000     .7272204    1.480548
------------------------------------------------------------------------------

. xi: glm los admitype i.codes, family(ig) link(id) nolog
i.codes           _Icodes_1-9         (naturally coded; _Icodes_1 omitted)

Generalized linear models                          No. of obs      =      3676
Optimization     : ML                              Residual df     =      3666
                                                   Scale parameter =  .0971354
Deviance         =  296.9162798                    (1/df) Deviance =  .0809919
Pearson          =  356.0982589                    (1/df) Pearson  =  .0971354

Variance function: V(u) = u^3                      [Inverse Gaussian]
Link function    : g(u) = u                        [Identity]

                                                   AIC             =  6.620808
Log likelihood   = -12159.04528                    BIC             = -29799.41

------------------------------------------------------------------------------
             |                 OIM
         los |      Coef.   Std. Err.      z    P>|z|     [95% Conf. Interval]
-------------+----------------------------------------------------------------
    admitype |  -1.145623   .1267899    -9.04   0.000    -1.394127   -.8971193
   _Icodes_2 |   1.013613   .5289403     1.92   0.055    -.0230911    2.050317
   _Icodes_3 |   .9906395   .5086064     1.95   0.051    -.0062107     1.98749
   _Icodes_4 |   1.058073   .4912463     2.15   0.031     .0952479    2.020898
   _Icodes_5 |   1.479535   .4960473     2.98   0.003     .5073004     2.45177
   _Icodes_6 |   1.600658   .4943341     3.24   0.001     .6317811    2.569535
   _Icodes_7 |   1.442416   .4878528     2.96   0.003     .4862419     2.39859
   _Icodes_8 |   2.065912   .4903178     4.21   0.000     1.104907    3.026917
   _Icodes_9 |   3.943232   .4748458     8.30   0.000     3.012551    4.873913
       _cons |   3.414347   .4690347     7.28   0.000     2.495056    4.333638
------------------------------------------------------------------------------
```

```
. xi: glm los admitype i.codes, family(ig) nolog
i.codes          _Icodes_1-9        (naturally coded; _Icodes_1 omitted)

Generalized linear models                          No. of obs      =      3676
Optimization     : ML                              Residual df     =      3666
                                                   Scale parameter =  .1011533
Deviance         =  301.6265109                    (1/df) Deviance =  .0822767
Pearson          =  370.8279408                    (1/df) Pearson  =  .1011533

Variance function: V(u) = u^3                      [Inverse Gaussian]
Link function    : g(u) = 1/(u^2)                  [Power(-2)]

                                                   AIC             =  6.622089
Log likelihood   = -12161.4004                     BIC             =  -29794.7

------------------------------------------------------------------------------
             |                 OIM
         los |      Coef.   Std. Err.      z    P>|z|     [95% Conf. Interval]
-------------+----------------------------------------------------------------
    admitype |   .0084326   .0016685     5.05   0.000     .0051624    .0117028
   _Icodes_2 |  -.0616845   .0549764    -1.12   0.262    -.1694363    .0460673
   _Icodes_3 |   -.067647   .0543874    -1.24   0.214    -.1742444    .0389504
   _Icodes_4 |  -.0699813   .0541194    -1.29   0.196    -.1760534    .0360909
   _Icodes_5 |  -.0858411   .0539192    -1.59   0.111    -.1915208    .0198385
   _Icodes_6 |  -.0831355   .0539198    -1.54   0.123    -.1888164    .0225454
   _Icodes_7 |  -.0829273   .0538908    -1.54   0.124    -.1885513    .0226967
   _Icodes_8 |  -.0939159   .0538107    -1.75   0.081     -.199383    .0115512
   _Icodes_9 |  -.1111008   .0537499    -2.07   0.039    -.2164487   -.0057529
       _cons |   .1297928   .0537508     2.41   0.016     .0244432    .2351424
------------------------------------------------------------------------------
```

The identity link may also be used with the inverse Gaussian. It is appropriate when modeling duration-type data. It may also be used when there is evidence of a relationship of identity between the fit and linear predictor. Usually, we can determine which model is preferable, between the log and identity links, from the value of the deviance function. Of course, we assume that all other aspects of the data and model are the same. If there is a significant difference between the deviance values between two models, then the model with the lower deviance is preferred. If there is little difference between the two, then either may be used. The same logic may be applied with respect to BIC or AIC statistics, as mentioned in chapter 6. Residual analysis may help in determining model preference as well. For the hospital data modeled above, the identity link is nearly the same, in terms of BIC and AIC statistics, as the results of the log model. Neither is preferred over the other.

8 The power family and link

Contents

8.1 Power links

The links that are associated with members of the continuous family of GLM distributions may be thought of in terms of powers. In fact, except for the standard binomial links and the canonical form of the negative binomial, all links are powers of μ. The relationships are listed in the following table:

Link	Function	Power function
Identity	μ	μ^1
Log	$\ln(\mu)$	μ^0
Reciprocal	$1/\mu$	μ^{-1}
Inverse quadratic	$1/\mu^2$	μ^{-2}

A generic power link function can thus be established as

$$\text{Power}(a) = \begin{cases} \mu^a & \text{if } a \neq 0 \\ \ln(\mu) & \text{if } a = 0 \end{cases} \tag{8.1}$$

The corresponding generic inverse link function is

$$\text{Power}(a) = \begin{cases} \eta^{-a} & \text{if } a \neq 0 \\ \exp(\eta) & \text{if } a = 0 \end{cases} \tag{8.2}$$

Variance functions for the continuous distributions as well as for the Poisson distribution can also be thought of in terms of powers. The following table displays the relationships:

Family	Link	Power function
Gaussian	Identity	$\mu^0 = 1$
Poisson	Log	$\mu^1 = \mu$
Gamma	Square	μ^2
Inverse Gaussian	Cube	μ^3

The power link can play an important role in assessing the fit of models based on continuous distributions. For instance, checking the differences in log likelihood or deviance function for various links within a particular GLM family can help the researcher construct the optimal model. With the gamma models, we can use the canonical or inverse link (power $= -1$), the log link (power $= 0$), or the identity link (power $= 1$). The model having the lower deviance value, or greater log-likelihood value, is the preferred model (all other aspects of the model being equal). However, the optimal model may fall between that of the log and identity links, e.g., a link of $\mu^{0.5}$, or the square root link. If the fit of residuals is better using this link, and the deviance, BIC, and AIC statistics are less than those of other links, then the choice of the gamma model with power link of $\mu^{0.5}$ is preferred over other models. Of course, the square root link is not one of the standard links, but it is nevertheless valid.

When one specifies a power link within a given family, the algorithm makes the appropriate adjustments to the link, inverse link, and derivative of the link, g'. No adjustments need be made for the deviance or log-likelihood functions or for the variance function. The derivative of the link is given in table A.2. Some of the important power relationships are outlined in table A.14.

8.2 Example: Power link

We use a power analysis on the `claims` data discussed in chapter 6 on the gamma family. The canonical reciprocal link was used to model the data. In power link terms, we used power $= -1$. Using power links, we can determine whether the canonical inverse reciprocal link is optimal. We can also ascertain whether another distribution may be preferable for the modeling of the data at hand.

We modeled the main effects for the claims dataset as

```
. xi: glm y i.pa i.cg i.va [fw=number], family(gamma)
```

The algorithm used the default canonical inverse reciprocal link. Using the power link option, we may obtain the same result by specifying

```
. xi: glm y i.pa i.cg i.va [fw=number], family(gamma) link(power -1)
```

Link	Gamma deviance	Inverse Gaussian deviance
−2.0	130.578	0.656
−1.5	126.826	0.638
−1.0	*124.783	0.628
−0.5	124.801	*0.626
0	127.198	0.634
0.5	132.228	0.665
1.0	139.761	0.687

This table shows us that the most preferable link for the gamma family is the canonical reciprocal link. If the data are modeled with the inverse Gaussian family, then the most preferable link is the inverse square root. We do not compare deviance values across families. To make comparisons of models across families, we use the BIC or AIC statistics. One must also evaluate the significance of the predictors. Here the significance of predictors is nearly identical.

The AIC and BIC statistics for the canonical gamma model are 12.9 and −2.6, respectively. For the inverse square root–linked inverse Gaussian model, the values are 18.2 and −126.7. The AIC and BIC statistics may be used to decide whether one family of models is preferable to another for a given set of data. Here the gamma model appears more appropriate than the inverse Gaussian for the main-effects claims data.

8.3 The power family

The power function may be extended so that it is regarded as a family of distributions. That is, we may model data as a power model, like a gamma or logit model. The difference is that the power distribution is not directly based on a probability distribution that is a member of the exponential family of distributions.

Although the log likelihood and deviance functions were derived directly from the exponential family form of a probability distribution, one can derive a working quasideviance from the variance function. See the chapter on quasilikelihood for more details. The quasideviance for the power family is

$$\frac{2y}{(1-a)\,(y^{1-a}-\mu^{1-a})} - \frac{2}{(2-a)\,(y^{2-a}-\mu^{2-a})} \tag{8.3}$$

One can then create a power algorithm that uses the above deviance function, together with the power link, the power inverse link, the derivative of the power link, and the power variance functions. This is an example of a quasideviance model, one that is an extended member of the GLM family.

One use of power links is to assess optimal models within a GLM distribution family. One can use the link function to check various statistics resulting from changes to a in the power link. This gives a wide variety of possible links for each continuous family.

However, we may need to use a general power link model to assess models between families. Of course, we may simply create a series of models using a continuum of links within each of the Gaussian ($a = 1$), Poisson ($a = 0$), gamma ($a = -1$), and inverse Gaussian ($a = -2$) families. If all other aspects of the models are the same, then the model having the lower deviance value and best-fitting residuals will be the model of choice. However, we are thereby giving wide leniency to the notion of "aspects".

The power family can model data with links that are different from those used in the standard GLM repertoire. Models with positive power values can be created, e.g., powers of 0.5 (square root), 2 (square), 2.5, or 3 (cube). Power links greater than 1 may be appropriate for data having a response with a sharp increase of values.

We know no notable use of the power family algorithm. Typically, power analysis rests with comparative links, as we demonstrated in section 8.2. However, one can use a generalized power family algorithm to model data, but the literature has no evidence that one can derive substantial benefit therefrom.

Part III

Binomial Response Models

9 The binomial–logit family

Contents

Except for the canonical Gaussian model, binomial models are used more often than any other member of the GLM family. Because the normal model is rarely used in GLM, we are left with the binomial model.

Binomial regression models are used in analyses having discrete (corresponding to the number of successes for k trials) or proportional responses. Models of binary responses are actually based on the Bernoulli distribution. The Bernoulli distribution is a degenerate case of the binomial where the number of trials is equal to 1.

Table 9.1: Binomial regression models

Response	Model
Binary: $\{0,1\}$	Bernoulli = Binomial(1)
Proportional: $\{0,1,\ldots,k\}$	Binomial(k)

Proportional-response data involve two variables identified by the number of successes, y_i, of a population of k_i trials, where both variables are indexed by i since proportional data do not require that the number of trials be the same for each observation. Hence, a binomial model can be considered as a weighted binary response model, or simply as a two-part response relating the proportion of successes to trials. We discuss the relationship between these two methods of dealing with discrete data. However, binary data can be modeled as proportional, but the reverse is not always possible.

Binary responses take the form of 0 or 1, with 1 typically representing a success or some other positive result, whereas 0 is taken as a failure or some negative result. Responses may also take the form of 1/2, a/b, or a similar pair of values. However, values other than 0 and 1 are translated to a 0/1 format by the binary regression algorithm before estimation. The reason for this will be apparent when we describe the probability function. Also, in Stata, a 0 indicates failure and *any other nonzero nonmissing value* represents a success in Bernoulli models.

The binomial family is associated with several links where the logit link is the canonical link. The links are associated with either the Bernoulli or the full binomial distribution. The common binomial links (there are others) are given in table 9.2.

Table 9.2: Common binomial link functions

logit	probit	log-log	complementary log-log
identity	log	inverse	log-complement

See table A.2 for the associated formulas.

We begin our exposition of the binomial model by describing it in its canonical form. Since the binary is included within the proportional model, we begin by examining the latter. Conveniently, the canonical link is also the most well known of GLM models (the logistic or logit regression model).

9.1 Derivation of the binomial model

The binomial probability function is expressed as

$$f(y;k,p) = \binom{k}{y} p^y (1-p)^{k-y} \tag{9.1}$$

where p is the probability of success, y is the proportional numerator, and k is the denominator. p is sometimes referred to π in the literature when the focus is solely on the logit model; however, we use p throughout this text. It is the sole parameter for all binomial models (including the Bernoulli).

In exponential family form, the binomial distribution is written

$$\begin{aligned} f(y;k,p) &= \exp\left\{y\ln(p) + k\ln(1-p) - y\ln(1-p) + \ln\binom{k}{y}\right\} & (9.2)\\ &= \exp\left\{y\ln\left(\frac{p}{1-p}\right) + k\ln(1-p) + \ln\binom{k}{y}\right\} & (9.3)\end{aligned}$$

Identifying the link and cumulant is easy when a member of the exponential family of probability distributions is expressed in exponential form as above. The link is θ and the cumulant is $b(\theta)$. The link, θ, and the cumulant, $b(\theta)$, functions are then given by

$$\theta = \ln\left(\frac{p}{1-p}\right) \tag{9.4}$$
$$b(\theta) = -k\ln(1-p) \tag{9.5}$$

and $\phi = 1$ (the variance has no scale component).

The mean and variance functions can be calculated as the first and second derivatives, respectively, of the cumulant function.

$$\begin{aligned}
b'(\theta) &= \frac{\partial b}{\partial p}\frac{\partial p}{\partial \theta} && (9.6)\\
&= \frac{k}{1-p}p(1-p) && (9.7)\\
&= kp && (9.8)\\
b''(\theta) &= \frac{\partial^2 b}{\partial p^2}\left(\frac{\partial p}{\partial \theta}\right)^2 + \frac{\partial b}{\partial p}\frac{\partial^2 p}{\partial \theta^2} && (9.9)\\
&= \frac{k}{(1-p)^2}(1-p)^2p^2 + \frac{k}{1-p}(1-p)p(1-2p) && (9.10)\\
&= kp^2 + kp(1-2p) && (9.11)\\
&= kp(1-p) && (9.12)
\end{aligned}$$

The relationship of p and μ is thus given by

$$\mu = kp \tag{9.13}$$

and the variance is given by

$$\mathrm{V}(\mu) = \mu\left(1 - \frac{\mu}{k}\right) \tag{9.14}$$

The linear predictor η can also be expressed in terms of μ. θ is the same as η for the canonical link. Here the linear predictor is also the canonical link function. Hence,

$$\theta = \eta = g(\mu) = g = \ln\left(\frac{\mu}{k-\mu}\right) \tag{9.15}$$

The inverse link can easily be derived from the above as

$$\mu = g^{-1}(\eta) = \frac{k}{1+\exp(-\eta)} = \frac{k\exp(\eta)}{1+\exp(\eta)} \tag{9.16}$$

Both forms of inverse link are acceptable. We use primarily the first form, but there is no inherent reason to prefer one over the other, and you will see both forms in the literature.

The derivative of the link is also necessary to the binomial algorithm

$$g'(\mu) = g' = \frac{\partial}{\partial\mu}\ln\left(\frac{\mu}{k-\mu}\right) = \frac{k}{\mu(k-\mu)} \tag{9.17}$$

Finally, to have all the functions required for the binomial regression algorithm, we need to determine the deviance, which is calculated as

$$D = 2\left\{\mathcal{L}(y;y) - \mathcal{L}(\mu;y)\right\} \tag{9.18}$$

when $\phi = 1$.

The log-likelihood function is thus required for calculating the deviance. The log likelihood is also important for calculating various fit statistics, and it takes the place of the deviance function when using maximum likelihood methods to estimate model parameters.

We can obtain the log-likelihood function directly from the exponential-family form of the distribution. Removing the exp and braces, we have

$$\mathcal{L} = \sum_{i=1}^{n}\left\{y_i \ln\left(\frac{\mu_i}{1-\mu_i}\right) + k_i\ln(1-\mu_i) + \ln\binom{k_i}{y_i}\right\} \tag{9.19}$$

Given the log likelihood, the deviance is calculated as

$$\begin{aligned} D &= 2\sum_{i=1}^{n}\left\{y_i\ln\left(\frac{y_i}{1-y_i}\right) + k_i\ln(1-y_i) - y_i\ln\left(\frac{\mu_i}{1-\mu_i}\right) + k_i\ln(1-\mu_i)\right\} && (9.20)\\ &= 2\sum_{i=1}^{n}\left\{y\ln\left(\frac{y_i}{\mu_i}\right) + (k_i-y_i)\ln\left(\frac{k_i-y_i}{k_i-\mu_i}\right)\right\} && (9.21)\end{aligned}$$

The normalization term drops out of the deviance formula altogether. This dropout occurs for all members of the GLM family and is a prime reason why researchers code the IRLS algorithm by using the (simpler) deviance rather than the ML algorithm with the log likelihood.

The binomial deviance function is typically given a multiple form depending on the values of y and k. We list the appropriate calculations of the deviance for specific observations. See also table A.11.

Bernoulli

$$D_i(k_i = 1; y_i = 0) = -2\ln(1-\mu_i) \tag{9.22}$$

$$D_i(k_i = 1; y_i = 1) = -2\ln(\mu_i) \tag{9.23}$$

Binomial(k)

$$D_i(k_i > 1; y_i = 0) = 2k_i \ln\left(\frac{k_i}{k_i - \mu_i}\right) \tag{9.24}$$

$$D_i(k_i > 1; y_i = k_i) = 2k_i \ln\left(\frac{k_i}{\mu_i}\right) \tag{9.25}$$

$$D_i(k_i > 1; 0 < y_i < k_i) = 2y_i \ln\left(\frac{y_i}{\mu_i}\right) + 2(k_i - y_i)\ln\left(\frac{k_i - y_i}{k_i - \mu_i}\right) \tag{9.26}$$

Most commercial software algorithms use the above reductions, but to our knowledge Stata is the only one that documents the full extent of the limiting values.

9.2 Derivation of the Bernoulli model

The binary or Bernoulli response probability distribution is simplified from the binomial distribution. The binomial denominator k is equal to 1 and the outcomes y are constrained to the values $\{0, 1\}$. Again y may initially be represented in your dataset as 1 or 2, or some other alternative set. If so, you must change your data to the above description.

However, some programs use the opposite default behavior: they use 0 to denote a success and 1 to denote a failure. Transferring data unchanged between Stata and a package that uses this other codification will result in fitted models with reversed signs on the estimated coefficients.

The Bernoulli response probability function is

$$f(y; p) = p^y(1-p)^{1-y} \tag{9.27}$$

The binomial normalization (combination) term has disappeared, which makes the function comparatively simple.

In canonical (exponential) form, the Bernoulli distribution is written

$$f(y; p) = \exp\left\{y \ln\left(\frac{p}{1-p}\right) + \ln(1-p)\right\} \tag{9.28}$$

Below you will find the various functions and relationships that are required to complete the Bernoulli algorithm, in canonical form. Because the canonical form is commonly referred to as the logit model, these functions can be thought of as those of the binary logit or logistic regression algorithm.

$$\theta = \ln\left(\frac{p}{1-p}\right) = \eta = g(\mu) = \ln\left(\frac{\mu}{1-\mu}\right) \tag{9.29}$$

$$g^{-1}(\theta) = \frac{1}{1+\exp(-\eta)} = \frac{\exp(\eta)}{1+\exp(\eta)} \tag{9.30}$$

$$b(\theta) = -\ln(1-p) = -\ln(1-\mu) \tag{9.31}$$

$$b'(\theta) = p = \mu \tag{9.32}$$

$$b''(\theta) = p(1-p) = \mu(1-\mu) \tag{9.33}$$

$$g'(\mu) = \frac{1}{\mu(1-\mu)} \tag{9.34}$$

The Bernoulli log likelihood and deviance functions are

$$\mathcal{L}(\mu; y) = \sum_{i=1}^{n}\left\{y_i \ln\left(\frac{\mu_i}{1-\mu_i}\right) + \ln(1-\mu_i)\right\} \tag{9.35}$$

$$D = 2\sum_{i=1}^{n}\left\{y_i \ln\left(\frac{y_i}{\mu_i}\right) + (1-y_i)\ln\left(\frac{1-y_i}{1-\mu_i}\right)\right\} \tag{9.36}$$

As we observed for the binomial probability function, separate values are often given for the log likelihood and deviance when $y = 0$. When this is done, a separate function is generally provided in case $y = 1$. We list these below for individual observations.

$$\mathcal{L}_i(\mu_i; y_i = 0) = \ln(1-\mu_i) \tag{9.37}$$

$$\mathcal{L}_i(\mu_i; y_i = 1) = \ln(\mu_i) \tag{9.38}$$

$$D_i(y_i = 0) = -2\ln(1-\mu_i) \tag{9.39}$$

$$D_i(y_i = 1) = -2\ln(\mu_i) \tag{9.40}$$

9.3 The binomial regression algorithm

The canonical binomial algorithm is commonly referred to as logistic or logit regression. Traditionally, binomial models have three commonly used links: logit, probit, and complementary log-log (or clog-log for short). There are other links that we will discuss. However, statisticians typically refer to a GLM-based regression by its link function, hence the still-used reference to probit or clog-log regression. For the same reason, statisticians generally referred to the canonical form as logit regression. This terminology is still used.

Over time, some researchers began referring to *logit regression* as *logistic regression*. They made a distinction based on the type of predictors in the model. A logit model comprised factor variables. The logistic model, on the other hand, had at least one continuous variable as a predictor. Although this distinction has now been largely discarded, we still find reference to it in older sources. Logit and logistic refer to the same basic model.

In the previous section we provided all the functions required to construct the binomial algorithm. Since this is the canonical form, it is also the algorithm for logistic regression. We first give the proportional-response form, since it encompasses the simpler model.

Listing 9.1: IRLS algorithm for proportional logistic regression

```
1   Dev = 0
2   μ = (y + 0.5)/(k + 1)              /* initialization of μ */
3   η = ln(μ/(k − μ))                  /* initialization of η */
4   WHILE (abs(ΔDev) > tolerance ) {
5       W = μ(k − μ)
6       z = η + k(y − μ)/(μ(k − μ))
7       β = (X^T W X)^{-1} X^T W z
8       η = Xβ
9       μ = k/(1 + exp(−η))
10      OldDev = Dev
11      Dev = 2∑ y ln(y/μ) − (k − y) ln((k − y)/(k − μ))
12      ΔDev = Dev - OldDev
13  }
14  χ^2 = ∑(y − μ)^2/(μ(1 − μ/k))
```

The Bernoulli or binary algorithm can be constructed by setting k to the value of 1. Since the binary model often reduces the deviance on the basis of the value of y, we show the entire algorithm.

Listing 9.2: IRLS algorithm for binary logistic regression

```
1   Dev = 0
2   μ = (y + 0.5)/2                    /* initialization of μ */
3   η = ln(μ/(1 − μ))                  /* initialization of η */
4   WHILE (abs(ΔDev) > tolerance ) {
5       W = μ(1 − μ)
6       z = η + (y − μ)/(μ(1 − μ))
7       β = (X^T W X)^{-1} X^T W z
8       η = Xβ
9       μ = 1/(1 + exp(−η))
10      OldDev = Dev
11      Dev = 2∑ ln(1/μ) if (y = 1)
12      Dev = 2∑ ln(1/(1 − μ)) if (y = 0)
13      ΔDev = Dev - OldDev
14  }
15  χ^2 = ∑(y − μ)^2/(μ(1 − μ))
```

There are several ways to construct the various GLM algorithms. Other algorithms may implement a `DO-WHILE` loop rather than the type we have shown here. Others may use the log-likelihood function as the basis for iteration; i.e., iteration stops when there is a defined minimal difference in log-likelihood values. Other algorithms iterate until there is minimal change in parameter estimates. One can find each of these methods in commercial GLM software, but they all result in statistically similar results. The variety we show here is the basis for the Stata IRLS `glm` procedure.

9.4 Example: Logistic regression

Our logistic regression example uses a medical dataset. The data come from a subset of the National Canadian Registry of Cardiovascular Disease (FASTRAK), sponsored by Hoffman-La Roche Canada. The goal is to model death within 48 hours by various explanatory variables. For this study we are interested in the comparative risk for patients who enter a hospital with an anterior infarct (heart attack) rather than an inferior infarct. The terms `anterior` and `inferior` refer to the site on the heart where damage has occurred. The model is adjusted by `hcabg`, history of having had a cardiac bypass surgery (CABG); `age1-age4`, levels of age; and Killip class, levels of risk where the higher the level, the greater the risk. Levels of each variable in the model are shown in the table below.

Table 9.3: Variables for `heart` data

`death = 1`	death within 48 hours of myocardial infarction onset
`death = 0`	no death
`anterior = 1`	anterior infarction
`anterior = 0`	inferior infarction
`hcabg = 1`	history of CABG
`hcabg = 0`	no history of CABG
`kk1 = 1`	killip class 1
`kk2 = 1`	killip class 2
`kk3 = 1`	killip class 3
`kk4 = 1`	killip class 4
`age1 = 1`	age less than 60 years
`age2 = 1`	age between 60 and 69 years
`age3 = 1`	age between 70 and 79 years
`age4 = 1`	age greater than or equal to 80 years

We model the data twice, first displaying the coefficients and then showing coefficients (and standard errors) converted to odds ratios. To demonstrate the similarity of results, we also use the slower N–R algorithm for the first model and IRLS for the second. The z statistics are identical for both models, which would not necessarily be the case if we used a noncanonical link.

The z-values are produced by dividing each coefficient by its standard error. When coefficients are exponentiated and standard errors are converted via the delta method to their odds ratio equivalent, it is a Stata convention that z-values remain the same.

GLM theory assumes that the predictors enter into the model with normal standard errors. Some commercial software routines will display t-values instead, in a manner similar to that produced in normal linear regression. Slight variances for the p-values will result. However, the differences are usually not great enough to matter.

9.4.1 Model producing logistic coefficients: The heart data

```
. use http://www.stata-press.com/data/hh2/heart01
. glm death anterior hcabg kk2 kk3 kk4 age2-age4, family(bin) nolog
Generalized linear models                          No. of obs      =       4483
Optimization     : ML                              Residual df     =       4474
                                                   Scale parameter =          1
Deviance         =  1273.251066                    (1/df) Deviance =    .284589
Pearson          =  4220.194937                    (1/df) Pearson  =   .9432711
Variance function: V(u) = u*(1-u)                  [Bernoulli]
Link function    : g(u) = ln(u/(1-u))              [Logit]
                                                   AIC             =   .2880328
Log likelihood   = -636.6255331                    BIC             = -36344.35

------------------------------------------------------------------------------
             |                 OIM
       death |      Coef.   Std. Err.      z    P>|z|     [95% Conf. Interval]
-------------+----------------------------------------------------------------
    anterior |   .6425552   .1675538     3.83   0.000     .3141557    .9709547
       hcabg |   .7444459    .352956     2.11   0.035     .0526648    1.436227
         kk2 |   .8116998    .180502     4.50   0.000     .4579223    1.165477
         kk3 |   .7756968   .2690602     2.88   0.004     .2483486    1.303045
         kk4 |   2.659656   .3559947     7.47   0.000     1.961919    3.357393
        age2 |   .4930224   .3101877     1.59   0.112    -.1149343    1.100979
        age3 |   1.511168   .2662236     5.68   0.000     .9893793    2.032957
        age4 |   2.185288   .2718383     8.04   0.000     1.652494    2.718081
       _cons |  -5.052069   .2586081   -19.54   0.000    -5.558931   -4.545206
------------------------------------------------------------------------------
```

9.4.2 Model producing logistic odds ratios

```
. glm death anterior hcabg kk2 kk3 kk4 age2-age4, family(bin) irls ef nolog
Generalized linear models                          No. of obs      =       4483
Optimization     : MQL Fisher scoring              Residual df     =       4474
                   (IRLS EIM)                      Scale parameter =          1
Deviance         =  1273.251066                    (1/df) Deviance =    .284589
Pearson          =  4220.186292                    (1/df) Pearson  =   .9432692

Variance function: V(u) = u*(1-u)                  [Bernoulli]
Link function    : g(u) = ln(u/(1-u))              [Logit]
                                                   BIC             = -36344.35

------------------------------------------------------------------------------
             |                 EIM
       death | Odds Ratio   Std. Err.      z    P>|z|     [95% Conf. Interval]
-------------+----------------------------------------------------------------
    anterior |   1.901333   .3185755     3.83   0.000     1.369103    2.640464
       hcabg |   2.105275   .7430692     2.11   0.035     1.054076    4.204801
         kk2 |   2.251732   .4064421     4.50   0.000     1.580786    3.207453
         kk3 |   2.172105   .5844269     2.88   0.004     1.281907    3.680486
         kk4 |   14.29137   5.087652     7.47   0.000     7.112966    28.71423
        age2 |    1.63726   .5078572     1.59   0.112     .8914272    3.007112
        age3 |   4.532029    1.20653     5.68   0.000     2.689572    7.636636
        age4 |   8.893222   2.417514     8.04   0.000     5.219999    15.15123
------------------------------------------------------------------------------
```

We find that patients who enter Canadian hospitals with an anterior infarct have nearly twice the odds of death within 48 hours of the onset of their symptoms as patients who have inferior infarcts. This result is in keeping with what experience has taught cardiologists. Moreover, those who have had a previous CABG appear to have twice the risk as those who have not had the procedure. Those who enter the hospital with a higher Killip-level risk score, or those who are older, are also at greater risk. In particular, those who are 80 and over are at greatest risk. The outlook appears bleak for someone who enters the hospital with an anterior infarct, experienced a previous CABG, is at Killip-level 4, and is over age 80. In fact, we can calculate the probability of `death` for such a patient directly from parameter results.

The probability of `death` can be calculated for each observation in the dataset by summing the various estimate terms. First, multiply the value of each appropriate predictor by its coefficient; then sum across terms. The sum is called the linear predictor, or η. Thereafter, apply the inverse link transform to η. Doing so produces a μ value for each observation, which is the probability in the case of binary logistic regression. In Stata, we need only type

```
. predict mu, mu
```

or

```
. predict mu
```

to obtain probabilities after modeling.

The same principle applies to covariate patterns that are not actually in the model dataset, like our high-risk patient above. We can obtain the probability of `death` for such a patient by implementing the same methods. First, calculate the linear predictor, and then use the inverse link transform. For our high-risk patient we have

$$\begin{aligned}\eta &= .6425552(1) + .7444459(1) + 2.659656(1) + 2.185288(1) - 5.052069(1) \quad (9.41)\\ &= 1.1798761 \quad (9.42)\end{aligned}$$

Applying the inverse link gives

$$\mu = \frac{1}{1 + \exp(-1.1798761)} = .7649 \tag{9.43}$$

Hence, a patient who has all the highest risk factors in the model has a 76.5% probability of death within 48 hours of symptom onset, regardless of treatment. Fortunately, no one in the actual data had all these risk factors.

We can also calculate the odds of a person dying within the given time framework by applying the following formula:

$$\text{Odds} = \frac{p}{1-p} \tag{9.44}$$

We can thereby calculate the odds of death within 48 hours of symptom onset for a patient entering a Canadian hospital with all of the highest risk factors as

$$\text{Odds} = \frac{.7649}{1 - .7649} = 3.25 \tag{9.45}$$

Of course, there are perhaps other important confounding explanatory variables that were not a part of the model. If this were the case, there would probably be a change in the probability and odds ratio. But if the model is well fitted and correlates well with the population as a whole then the added predictors will have little effect on the probability.

9.5 GOF statistics

Models can easily be constructed without thought to how well the model actually fits the data. All too often this is seen in publications—logistic regression results, with parameter estimates, and standard errors together with corresponding p-values without associated GOF statistics.

Models are simply that: *models*. They are not aimed at providing a one-to-one fit, called overfitting. The problem with overfitting is that the model is difficult to adapt to related data situations outside the actual data modeled. If the `heart` model were overfitted, it would have little value in generalizing to other heart patients.

One measure of fit is called the *confusion matrix*. The origins of the term may be more confusing than the information provided in the table. The confusion matrix simply classifies the number of instances where

- $y = 1$, predicted $y = 1$
- $y = 1$, predicted $y = 0$
- $y = 0$, predicted $y = 1$
- $y = 0$, predicted $y = 0$

Predicted values can be calculated by obtaining the linear predictor, $\mathbf{x}\boldsymbol{\beta}$, or η, and converting it by means of the logistic transform, which is the inverse link function. Both of these statistics are available directly after modeling. In Stata, we can obtain the values after modeling by typing

```
. predict xb, xb    /* linear predictor */
```

or

```
. predict mu, mu    /* fitted value   */
```

We may also obtain the linear predictor and then compute the fit for them:

```
. gen mu = 1/(1+exp(-xb))
```

To create a confusion matrix, we simply tabulate the actual 0/1 response with values of $\widehat{\mu}$. The matrix may be cut at different values depending on the distribution of the data. One method that has been used is to cut at the mean value of $\widehat{\mu}$.

Stata's `logistic` command for fitting maximum likelihood logistic regression models includes many useful postestimation commands. We refit the previous model by using the command

```
. logistic death anterior hcabg kk2 kk3 age2-age4, eform nolog
```

Various statistics for this model are then presented in summary table form by using the `estat classification` command. This command is useful for investigating the areas in which models are strong or weak in fitting data.

We find a 96% rate of correct classification for this model.

```
. estat classification
Logistic model for death

              ---------- True ----------
Classified |         D            ~D  |      Total
-----------+--------------------------+-----------
     +     |         4             7  |         11
     -     |       172          4300  |       4472
-----------+--------------------------+-----------
   Total   |       176          4307  |       4483
Classified + if predicted Pr(D) >= .5
True D defined as death != 0
--------------------------------------------------
Sensitivity                     Pr( +| D)    2.27%
Specificity                     Pr( -|~D)   99.84%
Positive predictive value       Pr( D| +)   36.36%
Negative predictive value       Pr(~D| -)   96.15%
--------------------------------------------------
False + rate for true ~D        Pr( +|~D)    0.16%
False - rate for true D         Pr( -| D)   97.73%
False + rate for classified +   Pr(~D| +)   63.64%
False - rate for classified -   Pr( D| -)    3.85%
--------------------------------------------------
Correctly classified                        96.01%
--------------------------------------------------
```

Area under the receiver operating characteristic (ROC) curve is another measure of fit. SAS calls this value a (Harrell's) C statistic. Regardless, higher values indicate a better fit. The Stata documentation for the `logistic` and `lroc` commands provide excellent discussion of this topic; see the respective entries in the *Stata Base Reference Manual*.

```
. lroc, nograph
Logistic model for death
number of observations =     4483
area under ROC curve   =   0.7965
```

A third measure of fit exploits the notion of the actual response versus predicted value that was given in the confusion matrix. The Hosmer–Lemeshow GOF statistic, together with a related table, divides the range of probability values, μ, into groups. The authors recommend 10 groups for relatively large datasets, but the number can be amended to fit the data situation. The point is to provide a count, for each respective group, of actually observed ones, fitted or expected ones, observed zeros, and expected zeros. Theoretically, observed and expected counts should be close. Diagnostically, it is easy to identify groups where deviations from the expected occur. When this is the case, we can go back to the data to see the covariate pattern(s) that gave rise to the discrepancy.

A fit statistic is provided that measures the overall correspondence of counts. Based on the χ^2 distribution, a Hosmer–Lemeshow (H–L) statistic with a p-value greater than 0.05 is considered a good fit. The lower the H–L statistic, the less variance in fit, and the greater the p-value.

```
. lfit, group(10) table
Logistic model for death, goodness-of-fit test
  (Table collapsed on quantiles of estimated probabilities)
```

Group	Prob	Obs_1	Exp_1	Obs_0	Exp_0	Total
1	0.0064	5	4.7	731	731.3	736
2	0.0104	5	5.0	481	481.0	486
3	0.0120	4	6.5	537	534.5	541
4	0.0137	1	0.5	34	34.5	35
5	0.0195	5	8.3	451	447.7	456
6	0.0282	19	21.3	782	779.7	801
7	0.0429	4	6.4	147	144.6	151
8	0.0538	30	26.8	479	482.2	509
9	0.0976	37	26.3	295	305.7	332
10	0.6072	66	70.2	370	365.8	436

```
      number of observations =      4483
            number of groups =        10
     Hosmer-Lemeshow chi2(8) =         9.49
                 Prob > chi2 =         0.3025
```

There has been discussion concerning the utility of this test in the general case. The H–L statistic creates 10 groups in the dataset, forming them on the deciles of the fitted values. When there are ties (multiple equal outcomes), choosing the decile induces a randomness to the statistic when the independent variables are different over observations producing the ties. We should ensure that this randomness does not affect a particular application of the test. The original authors have also advised that the proposed test should not be treated as a sole terminal test but that the test be part of a more complete investigation. One concern with the test is that the groups may contain subjects with widely different values of the covariates. As such, one can find situations where one set of fixed groups shows that the model fits, whereas another set of fixed groups shows that the model does not fit.

We provide more discussion of grouped or proportional binomial models in chapter 11, where we discuss problems of overdispersion. We demonstrate how certain binary response models can be converted to grouped datasets, as well as how to use a grouped binomial model to assess the fit of its binary case counterpart.

Researchers also evaluate the fit of a model by observing a graph of its residuals. Several residuals are available. Most are listed in the early chapters of this text. One of the most well-used methods of graphing GLM residuals is the graph of standardized deviance versus the fitted value. In it, points should be randomly displayed.

9.6 Interpretation of parameter estimates

We have provided an overview of various methods used to assess the logistic model fit. Now we turn to the actual interpretation of parameter estimates.

We know from basic statistical knowledge that linear regression coefficients are all interpreted in the same way. The coefficient represents the amount of change to the response for a one-unit change in the variable associated with the coefficient, with other predictors held to their same values. The same is the case for odds ratios.

The table listing of odds ratios for the heart dataset is

```
. ereturn disp, eform(Odds Ratio)
```

death	Odds Ratio	EIM Std. Err.	z	P>|z|	[95% Conf.	Interval]
anterior	1.901333	.3185755	3.83	0.000	1.369103	2.640464
hcabg	2.105275	.7430692	2.11	0.035	1.054076	4.204801
kk2	2.251732	.4064421	4.50	0.000	1.580786	3.207453
kk3	2.172105	.5844269	2.88	0.004	1.281907	3.680486
kk4	14.29137	5.087652	7.47	0.000	7.112966	28.71423
age2	1.63726	.5078572	1.59	0.112	.8914272	3.007112
age3	4.532029	1.20653	5.68	0.000	2.689572	7.636636
age4	8.893222	2.417514	8.04	0.000	5.219999	15.15123

Now consider the odds ratio for anterior of 1.90. It can be interpreted as

$$\frac{\Pr(\texttt{anterior}+1)/\{1-\Pr(\texttt{anterior}+1)\}}{\Pr(\texttt{anterior})/\{1-\Pr(\texttt{anterior})\}} \tag{9.46}$$

or

$$\frac{\Pr(p+1)/\{1-\Pr(p+1)\}}{\Pr(p)/\{1-\Pr(p)\}} \tag{9.47}$$

This is the meaning of the odds ratio. It means the same as the odds of the variable p being incremented by a value of 1, p+1, divided by p itself. The odds ratio of the coefficient is the resulting value. Calculating this in Stata, we find

```
. predict double p                    /* p = predicted probability    */
(option mu assumed; predicted mean death)
(905 missing values generated)

. replace anterior=anterior+1         /* increment anterior            */
(4696 real changes made)

. predict double p1                   /* p1= pred prob with anterior+1 */
(option mu assumed; predicted mean death)
(905 missing values generated)

. gen double or=(p1/(1-p1))/(p/(1-p)) /* Calculate ratio of odds       */
(905 missing values generated)

. summ or                             /* display odds ratio (anterior) */
```

Variable	Obs	Mean	Std. Dev.	Min	Max
or	4483	1.901333	5.41e-16	1.901333	1.901333

Incrementing the predictor anterior by 1 and calculating the new odds ratio using the formula shown displays the odds ratio observed in the parameter table. This relationship holds for all predictors and shows that the common interpretation of coefficients is the same when dealing with odds ratios.

Sometimes researchers wish to know the comparative worth of a predictor to the model. Just looking at the odds ratio of coefficient value does little to that end. A common method used to assess the impact of a predictor on the model is to look at the standardized value of its coefficient. The Stata `lstand` command (available for download by using `findit`) can be used to obtain standardized coefficients, together with partial correlations.

Standardized coefficients are calculated as

$$\frac{\beta s}{\sqrt{\pi^2/3}} \tag{9.48}$$

β is the parameter estimate or coefficient, and s is the standard deviation of the predictor. The denominator is the standard deviation of the underlying logistic distribution. Large values of the standardized coefficient indicate greater impact or worth to the model. Here the higher age groups are the most important predictors to the model.

Partial correlations, otherwise known as Atkinson's R, are calculated from

$$R = \sqrt{\frac{W-2}{2|\mathcal{L}_o|}} \tag{9.49}$$

where W is the Wald statistic, or t^2, and $|\mathcal{L}_o|$ is the absolute value of the intercept-only log-likelihood function. A partial correlation is the correlation of the response with a predictor, adjusted by the other predictors in the model.

```
. lstand
Table of Predictor Estimates:
Standardized Coefficients and Partial Correlations
No.  Var          Coef         OR      St.Coef      PartCorr    Prob(z)
========================================================================
0   Constant   -5.0521
1   anterior    0.6426     1.9013      0.1763        0.0925      0.000
2   hcabg       0.7444     2.1053      0.0740        0.0406      0.035
3   kk2         0.8117     2.2517      0.1789        0.1108      0.000
4   kk3         0.7757     2.1721      0.0989        0.0652      0.004
5   kk4         2.6597    14.2914      0.1624        0.1904      0.000
6   age2        0.4930     1.6373      0.1191        0.0188      0.112
7   age3        1.5112     4.5320      0.3680        0.1427      0.000
8   age4        2.1853     8.8932      0.3973        0.2054      0.000
========================================================================
```

The data below are from a study conducted by Milicer and Szczotka (1966) on young girls in Warsaw, Poland. Information was collected on the age of each girl on reaching menarche (for 3,918 girls). The dataset is compressed such that the age range of the girls was divided into 25 different categories. For each category (midpoint of the age interval) in the `age` variable, the number of girls having already reached menarche is recorded in the variable `menarche` and the total number of girls in the respective age range is recorded in the `total` variable. Thus the dataset is a good example of proportional data.

```
. use http://www.stata-press.com/data/hh2/warsaw, clear
. list
```

	age	total	menarche
1.	9.21	376	0
2.	10.21	200	0
3.	10.58	93	0
4.	10.83	120	2
5.	11.08	90	2
6.	11.33	88	5
7.	11.58	105	10
8.	11.83	111	17
9.	12.08	100	16
10.	12.33	93	29
11.	12.58	100	39
12.	12.83	108	51
13.	13.08	99	47
14.	13.33	106	67
15.	13.58	105	81
16.	13.83	117	88
17.	14.08	98	79
18.	14.33	97	90
19.	14.58	120	113
20.	14.83	102	95
21.	15.08	122	117
22.	15.33	111	107
23.	15.58	94	92
24.	15.83	114	112
25.	17.58	1049	1049

```
. generate proportion = menarche/total
. twoway scatter proportion age
```

The above two commands produces the illustration in figure 9.1 for the proportion of girls having reached menarche for each of the 25 age intervals. The figure shows a classic S-shaped curve common to many logistic regression problems.

(*Continued on next page*)

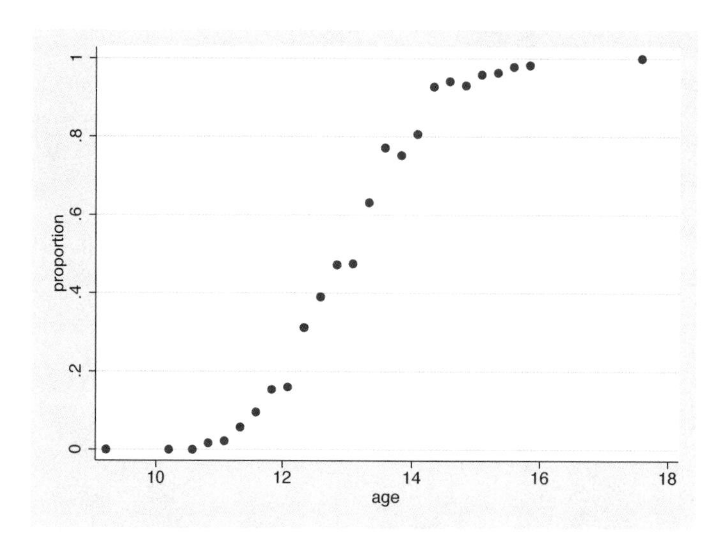

Figure 9.1: Sample proportions of girls reaching menarche for each age category

```
. glm menarche age, family(bin total)
Iteration 0:   log likelihood = -55.642964
Iteration 1:   log likelihood = -55.377952
Iteration 2:   log likelihood = -55.377628
Iteration 3:   log likelihood = -55.377628
Generalized linear models                          No. of obs      =        25
Optimization     : ML                              Residual df     =        23
                                                   Scale parameter =         1
Deviance         =  26.70345269                    (1/df) Deviance =   1.16102
Pearson          =  21.86985435                    (1/df) Pearson  =  .9508632
Variance function: V(u) = u*(1-u/total)            [Binomial]
Link function    : g(u) = ln(u/(total-u))          [Logit]
                                                   AIC             =   4.59021
Log likelihood   = -55.37762768                    BIC             = -47.33069

------------------------------------------------------------------------------
             |                 OIM
    menarche |      Coef.   Std. Err.      z    P>|z|     [95% Conf. Interval]
-------------+----------------------------------------------------------------
         age |   1.631968   .0589532    27.68   0.000     1.516422    1.747514
       _cons |  -21.22639   .7706859   -27.54   0.000    -22.73691   -19.71588
------------------------------------------------------------------------------
```

Figures 9.2 and 9.3 show two illustrations of the fit of the model to the data. To produce the initial plot, we specify

```
. predict double phat
(option mu assumed; predicted mean menarche)
. replace phat = phat/total
(25 real changes made)
. label var phat "Predicted proportion reaching menarche"
```

```
. twoway (scatter phat age, s(o)) (scatter proportion age, s(.))
```

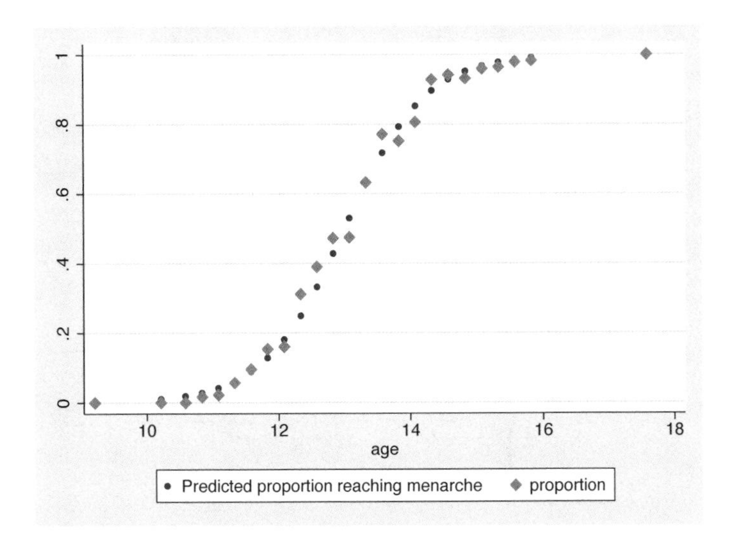

Figure 9.2: Predicted probabilities of girls reaching menarche for each age category

The second illustration first generates a variable that is used to create a diagonal line on the figure. Then the sample proportions are plotted versus the predicted probabilities. The diagonal line is a reference for equality of the two axis-defining values.

```
. generate equal = (_n-1)/24
. twoway (scatter phat proportion) (line equal equal),
> ytitle("Predicted proportion") xtitle("Observed proportion")
```

(Continued on next page)

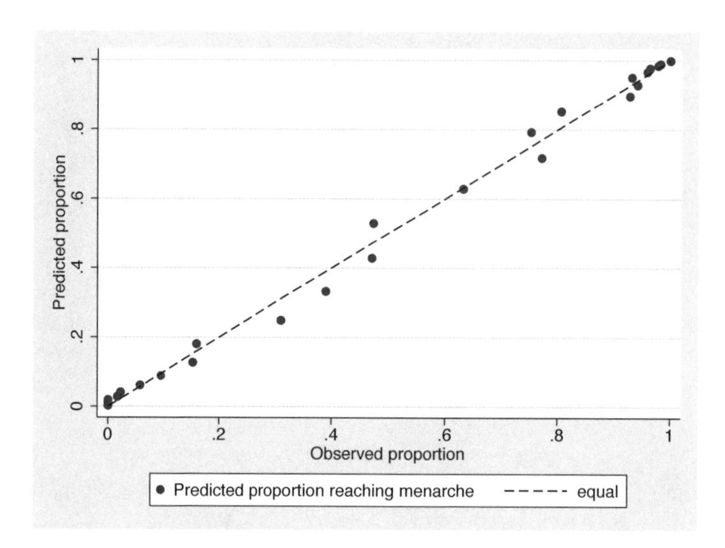

Figure 9.3: Predicted probabilities and sample proportions of girls reaching menarche for each age category

The next sample uses data from a survey on attitudes on entering the European Economic Community (EEC) are used in the next example. These data include whether the respondent views entry into the EEC favorably, `yes`; whether the respondent is not a union member, `unionnone`; whether the respondent is working class, `classwork`; whether the respondent has a minimum of schooling, `schoolmin`; and whether the respondent's attitude has not been constant, `consnot`.

```
. use http://www.stata-press.com/data/hh2/eec, clear
. glm yes consnot schoolmin unionnone classwork, family(bin total)
Iteration 0:   log likelihood = -44.599133
Iteration 1:   log likelihood = -44.577733
Iteration 2:   log likelihood = -44.577733
Generalized linear models                   No. of obs      =         16
Optimization     : ML                       Residual df     =         11
                                            Scale parameter =          1
Deviance         =  15.31299646             (1/df) Deviance =   1.392091
Pearson          =  15.02841402             (1/df) Pearson  =   1.366219
Variance function: V(u) = u*(1-u/total)     [Binomial]
Link function    : g(u) = ln(u/(total-u))   [Logit]
                                            AIC             =   6.197217
Log likelihood   =  -44.5777329             BIC             =  -15.18548
```

yes	Coef.	OIM Std. Err.	z	P>|z|	[95% Conf.	Interval]
consnot	-.9705364	.1244174	-7.80	0.000	-1.21439	-.7266827
schoolmin	-.2872516	.1125091	-2.55	0.011	-.5077654	-.0667377
unionnone	.1199609	.1107706	1.08	0.279	-.0971455	.3370673
classwork	-.4568704	.1239988	-3.68	0.000	-.6999036	-.2138372
_cons	1.42455	.1457412	9.77	0.000	1.138903	1.710198

Now we generate the interaction between working class and not constantly working to see whether such interaction might result in a better fitting model.

```
. gen consnotwork = consnot*classwork
. glm yes consnot schoolmin unionnone classwork consnotwork, family(bin total)
Iteration 0:   log likelihood = -44.013849
Iteration 1:   log likelihood = -43.997732
Iteration 2:   log likelihood = -43.997732
Generalized linear models                          No. of obs      =        16
Optimization     : ML                              Residual df     =        10
                                                   Scale parameter =         1
Deviance         =  14.15299404                    (1/df) Deviance =  1.415299
Pearson          =  13.89108844                    (1/df) Pearson  =  1.389109
Variance function: V(u) = u*(1-u/total)            [Binomial]
Link function    : g(u) = ln(u/(total-u))          [Logit]
                                                   AIC             =  6.249716
Log likelihood   = -43.99773169                    BIC             = -13.57289
```

yes	Coef.	OIM Std. Err.	z	P>|z|	[95% Conf.	Interval]
consnot	-.8076723	.1949602	-4.14	0.000	-1.189787	-.4255574
schoolmin	-.289237	.1126959	-2.57	0.010	-.5101169	-.068357
unionnone	.1210439	.1109298	1.09	0.275	-.0963745	.3384622
classwork	-.2832926	.2029896	-1.40	0.163	-.6811448	.1145596
consnotwork	-.2685823	.2493185	-1.08	0.281	-.7572376	.220073
_cons	1.334048	.1660547	8.03	0.000	1.008587	1.659509

The coefficient table lists the associated large-sample Wald-type test for consnotwork along with a p-value of 0.281. An analysis of deviance may also be performed. This analysis is similar to a likelihood-ratio test.

The model with the interaction has deviance 14.153 and the model without the interaction has deviance 15.313. The difference in deviance is 1.16 and the difference in the models is 1 degree of freedom (one covariate difference). This difference in deviance is a $\chi^2(1)$ variate with associated p-value given by

```
. display chi2tail(1,1.216)
.27014708
```

The resulting analysis of deviance p-value is close to the Wald test p-value listed in the table of parameter estimates. The interaction term does not significantly contribute to the model and therefore should not be in the final model.

10 The general binomial family

Contents

10.1 Noncanonical binomial models

In the previous chapter, we presented the canonical binomial model, including its derivation and separation into proportional and binary parameterizations. Here we discuss many of the noncanonical links associated with the binomial family.

Noncanonical links that have been used with the binomial family include the following:

Table 10.1: Common binomial noncanonical link functions

probit	log-log	clog-log	identity
log	inverse	log-complement	

See table A.2 for the associated formulas.

The first two, probit and clog-log, are perhaps the most often used. In fact, the clog-log regression model began in 1922, in the same year that likelihood theory was formulated by Fisher (1922). Twelve years later, Fisher (1934) discovered the exponential family of distributions and identified its more well-known characteristics.

That same year, Bliss (1934) used the new maximum likelihood approach to develop a probit model for bioassay analysis. It was not until 1944 that Berkson (1944) designed the proportional logit model to analyze bioassay data. Later, Dyke and Patterson (1952) used the same model to analyze cancer survey data.

Rasch (1960) then developed the Bernoulli or binary parameterization of the logit model. He used it in his seminal work on item analysis. Later in that decade and in the early 1970s, statisticians like Lindsey, Nelder, and Wedderburn began unifying various GLM. The last two theorists, Nelder and Wedderburn (1972), published a unified theory of GLM and later led a group that designed the first GLM software program, called GLIM. The software went through four major versions before development was discontinued. Version 4 came out in 1993. This was the same period that GENSTAT and S-Plus added support for GLM procedures and when both Stata and SAS were developing their respective GLM programs.

The GLM framework spurred more links that may not have been conceived outside its scope. For example, the exponential link was provided as an option to later versions of GLIM. It is defined as

$$\eta = (r_2 + \mu)^{r_1} \tag{10.1}$$

where r_1 and r_2 are user-specified constants. Because the link has found little use in the literature, we only mention it here. It is a modification to the power link that we discussed in an earlier chapter. Instead, some have used the odds power link. Wacholder (1986) brought log-binomial and identity-binomial links to the forefront where Hardin and Cleves (1999) introduced the techniques into Stata. They have been used for health risk and risk difference analysis. For example, several major infectious disease studies have effectively used the log-binomial model.

Again traditional GLM uses the IRLS algorithm to determine parameter estimates and associated standard errors. We have shown how one can amend the basic algorithm to produce standard errors based on the observed information matrix, and we have shown how one can model GLMs by using standard maximum likelihood techniques. We now investigate how noncanonical binomial models fit into this framework.

10.2 Noncanonical binomial links (binary form)

We begin by providing a table of noncanonical links, inverse links, and derivatives. We display them again to allow easy access to them when considering their position within the binomial algorithm.

Table 10.2: Noncanonical binomial link functions ($\eta = \mathbf{x}\boldsymbol{\beta} + \text{offset}$)

Link name	Link function $\eta = g(\mu)$	Inverse link $\mu = g^{-1}(\eta)$	First derivatives $\nabla = \partial\mu/\partial\eta$
Identity	μ	η	1
Log	$\ln(\mu)$	$\exp(\eta)$	μ
Log-complement	$\ln(1-\mu)$	$1-\exp(\eta)$	$\mu - 1$
Log-log	$-\ln\{-\ln(\mu)\}$	$\exp\{-\exp(-\eta)\}$	$-\mu\ln(\mu)$
Clog-log	$\ln\{-\ln(1-\mu)\}$	$1-\exp\{-\exp(\eta)\}$	$(\mu-1)\ln(1-\mu)$
Probit	$\Phi^{-1}(\mu)$	$\Phi(\eta)$	$\phi(\eta)$
Odds power ($\alpha = 0$)	$\ln\left(\frac{\mu}{1-\mu}\right)$	$\frac{e^\eta}{1+e^\eta}$	$\mu(1-\mu)$
Odds power ($\alpha \neq 0$)	$\frac{\mu/(1-\mu)^\alpha - 1}{\alpha}$	$\frac{(1+\alpha\eta)^{1/\alpha}}{1+(1+\alpha\eta)^{1/\alpha}}$	$\frac{\mu(1-\mu)}{1+\alpha\eta}$

You can incorporate table 10.2's formulas into the basic GLM algorithm to provide different linked models. However, noncanonical links offer a problem with standard errors. When modeling datasets having relatively few observations, we have found that it is better to either adjust the weights in the IRLS algorithm to construct an observed information matrix or to use ML methods to model the data (which also uses the observed information matrix). If neither adjustment is done, we are left with EIM standard errors based on the expected information matrix V_{EH} given in (3.66).

Using the expected information matrix to produce standard errors will have no significant impact on standard errors when the number of observations in the model is relatively large. Just how large depends on a variety of factors including the number of predictors in the model and how the data are balanced, e.g., the comparative number of ones to zeros for binary responses. However, even for smaller unbalanced datasets having many predictors, the resulting standard errors will usually not vary too much between the two estimating processes. Because the enhanced accuracy of standard errors is a paramount concern to researchers, using OIM as the basis of standard errors, when possible, is preferable.

We demonstrate in each of the following sections how one incorporates these formulas into the basic GLM algorithm. We also show how one can construct maximum likelihood algorithms to model several of the above links. Finally, we will give examples in each section to demonstrate how to use and interpret the models.

10.3 The probit model

The probit model was first used in bioassay or quantal analysis. Typically the probability of death was measured against the log of some toxin, e.g., an insecticide. Whether death occurred depended on the tolerance the subject had to the toxic agent. Subjects with a lower tolerance were more likely to die. Tolerance was assumed to be distributed normally, hence the use of the probit transform, $\mu = \Phi(\mathbf{x}\boldsymbol{\beta})$. Researchers still use probit for bioassay analysis, and they often use it to model other data situations.

The probit link function is the inverse of Φ, the cumulative normal distribution defined as

$$\mu = \int_{-\infty}^{\eta} \phi(u)du \tag{10.2}$$

where $\phi(u)$ is the standard normal density function

$$\frac{1}{\sqrt{2\pi}} \exp\left(-\frac{1}{2}u^2\right) \tag{10.3}$$

Using a probit regression model on binary or grouped binomial data typically results in output similar to that from logistic regression. Logistic regression is usually preferred because of the wide variety of fit statistics associated with the model. Moreover, exponentiated probit coefficients cannot be interpreted as odds ratios, as in logistic models (see sec. 10.6.2). However, if normality is involved in the linear relationship, as it often is in bioassay, then probit may be the appropriate model. It may also be used when the researcher is not interested in odds but rather in prediction or classification. Then, if the deviance of a probit model is significantly lower than that of the corresponding logit model, then the former is preferred. This dictum holds when comparing any of the links within the binomial family.

Both the logit and probit transforms produce a sigmoidal curve mapping the probability of μ to the linear predictor, η. Of course, the probability interval is $(0, 1)$, and the linear predictor is in $(-\infty, \infty)$. Both functions are symmetric, with a mean of 0.5.

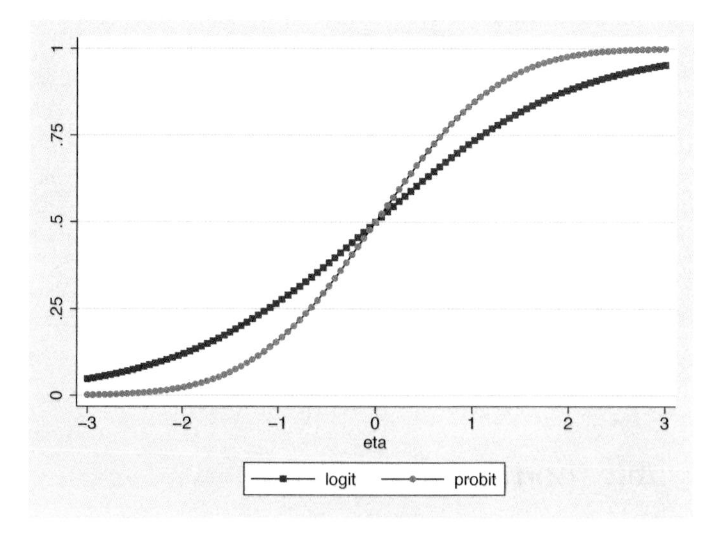

Figure 10.1: Probit and logit functions

One can easily change the traditional Fisher scoring IRLS logit algorithm to a probit algorithm. However, doing so implies that we maintain the same basis for producing standard errors. The substitution of the probit link, inverse link, and derivative of the link into the algorithm is all that is required. Since the link is a bit more complicated than that of the logit link, we add an initial line before calculating w. The probit algorithm, parameterized for binary response models, is

Listing 10.1: IRLS algorithm for binary probit regression

1 `Dev = 0`
2 $\mu = (y + 0.5)/2$ `/* initialization of` μ `*/`
3 $\eta = \Phi^{-1}(\mu)$ `/* initialization of` η `*/`
4 `WHILE (abs(`Δ`Dev) > tolerance ) {`
5     $\phi = \exp(-\eta^2/2)/\sqrt{2\pi}$ `/* standard normal density */`
6     $W = \phi^2/(\mu(1-\mu))$
7     $z = \eta + (y-\mu)/\phi$
8     $\beta = (X^{\mathrm{T}}WX)^{-1}X^{\mathrm{T}}Wz$
9     $\eta = X\beta$
10     $\mu = \Phi(\eta)$
11   `OldDev = Dev`
12   `Dev =` $2\sum \ln(1/\mu)$ `if` $(y=1)$
13   `Dev =` $2\sum \ln(1/(1-\mu))$ `if` $(y=0)$
14   Δ`Dev = Dev - OldDev`
15 `}`
16 $\chi^2 = \sum (y-\mu)^2/(\mu(1-\mu))$

We have stressed throughout this book the importance of standard errors. Certainly, assessing the comparative worth of predictors to a model is crucial to the modeling process. It is one of the prime reasons we engage in data modeling, hence our emphasis on methods that use (3.67) as the basis of OIM standard errors. In the past, software designers and researchers used the EIM standard errors since most GLM software was implemented using Fisher scoring. Since Fisher scoring uses the expected Hessian, that matrix is available for standard error construction with no extra code or computer cycles.

The IRLS algorithm can be amended to produce an observed information matrix. Standard errors are then identical to those produced using maximum likelihood algorithms. They are implemented in the algorithm below.

(*Continued on next page*)

Listing 10.2: IRLS algorithm for binary probit regression using OIM

```
1   Dev = 0
2   μ = (y + 0.5)/2                    /* initialization of μ */
3   η = Φ^{-1}(μ)                      /* initialization of η */
4   WHILE (abs(ΔDev) > tolerance ) {
5       φ = exp(−η^2/2)/√(2π)
6       V = μ(1 − μ)
7       V' = 1 − 2μ
8       W = φ^2/V
9       g' = 1/φ
10      g'' = ηφ^2
11      W_o = W + (y − μ)(Vg'' + V'g')/(V^2(g')^3)
12      z = η + (y − μ)g'
13      β = (X^T W_o X)^{-1} X^T W_o z
14      η = Xβ
15      μ = Φ(η)
16      OldDev = Dev
17      Dev = 2∑ln(1/μ) if (y = 1)
18      Dev = 2∑ln(1/(1 − μ)) if (y = 0)
19      ΔDev = Dev - OldDev
20  }
21  χ^2 = ∑(y − μ)^2/(μ(1 − μ))
```

The algorithm is a bit more complicated, but not so much that we should avoid using it.

Finally, sometimes maximum likelihood programs are required for modeling endeavors. We list the link functions for both logit and probit parameterized in the requisite $\mathbf{x}\boldsymbol{\beta}$ format. The binomial log-likelihood function, in terms of μ, can be parameterized to the logit or probit model by substituting the inverse link functions in place of μ. That is, each μ value in the log-likelihood function is replaced by the inverse links below. With appropriate initial or starting values, maximum likelihood software should be able to use the resulting log-likelihood functions to derive parameter estimates and standard errors based on the observed Hessian. Below we continue our notation where $y = 0$ indicates a failure and $y = 1$ indicates a success.

$$
\begin{array}{rcl}
\text{Logit inverse link}: & 1/\{1+\exp(\mathbf{x}\boldsymbol{\beta})\} & \\
y=0 & & -\ln\{1+\exp(\mathbf{x}\boldsymbol{\beta})\} \\
y=1 & & \mathbf{x}\boldsymbol{\beta}-\ln\{1+\exp(\mathbf{x}\boldsymbol{\beta})\} \\
\text{Probit inverse link}: & \Phi(\mathbf{x}\boldsymbol{\beta}) & \\
y=0 & & \ln\{1-\Phi(\mathbf{x}\boldsymbol{\beta})\} \\
y=1 & & \ln\{\Phi(\mathbf{x}\boldsymbol{\beta})\}
\end{array}
$$

Binary logit and probit models assume that the response comprises fairly equal numbers of ones compared with zeros. When there is a significant disparity, clog-log or log-log models may provide better models (as indicated by smaller BIC statistics) because of their asymmetric nature. The asymmetry in these links will sometimes be a better fit to rare outcomes, though not always.

We model the `heart` dataset by using a probit model to see whether there is any significant reduction in deviance from the logit model. The binomial or proportional logistic model yielded a deviance of 1,273.251. Also the BIC statistic for the probit model is substantially less than that for the logistic model—suggesting a strong preference for the probit model. The probit model below produces a deviance of 1,268.626. If we are not interested in assessing odds ratios but are interested in the predictive and classification value of the model, then the probit model is preferable to that of logit.

```
. use http://www.stata-press.com/data/hh2/heart01
. glm death anterior hcabg kk2 kk3 kk4 age2-age4, family(bin) link(probit) nolog
Generalized linear models                          No. of obs      =       4483
Optimization     : ML                              Residual df     =       4474
                                                   Scale parameter =          1
Deviance         =  1268.626075                    (1/df) Deviance =   .2835552
Pearson          =  4469.044561                    (1/df) Pearson  =   .9988924
Variance function: V(u) = u*(1-u)                  [Bernoulli]
Link function    : g(u) = invnorm(u)               [Probit]
                                                   AIC             =   .2870011
Log likelihood   = -634.3130373                    BIC             =  -36348.98

------------------------------------------------------------------------------
             |                 OIM
       death |      Coef.   Std. Err.      z    P>|z|     [95% Conf. Interval]
-------------+----------------------------------------------------------------
    anterior |   .3002858   .0765854     3.92   0.000     .1501812    .4503903
       hcabg |   .3489019   .1770991     1.97   0.049     .0017941    .6960097
         kk2 |   .3757268   .0854744     4.40   0.000        .2082    .5432535
         kk3 |   .4104931    .130373     3.15   0.002     .1549667    .6660195
         kk4 |   1.428809   .1968568     7.26   0.000     1.042977    1.814642
        age2 |   .1833151   .1266684     1.45   0.148    -.0649504    .4315805
        age3 |   .6487667   .1102819     5.88   0.000     .4326182    .8649152
        age4 |   1.002231   .1164327     8.61   0.000     .7740269    1.230435
       _cons |   -2.56975   .1046419   -24.56   0.000    -2.774845   -2.364656
------------------------------------------------------------------------------
```

The GOF for a probit model may follow the same logic as many fit tests for logistic regression models. A Hosmer–Lemeshow table of observed versus predicted values for both ones and zeros can be constructed for fitted probit statistics. Graphs of standardized deviance or Anscombe residuals versus μ statistics can tell analysts whether the fit of the model is appropriate for the data. Residual analysis is vital to modeling.

In chapter 9, we examined the fit of logistic regression to the proportional Warsaw data. Here we compare that fit with specification of the probit link.

```
. glm menarche age, family(bin total) link(probit) nolog

Generalized linear models                          No. of obs      =        25
Optimization     : ML                              Residual df     =        23
                                                   Scale parameter =         1
Deviance         =  22.88743324                    (1/df) Deviance =  .9951058
Pearson          =  21.90102423                    (1/df) Pearson  =  .9522184

Variance function: V(u) = u*(1-u/total)            [Binomial]
Link function    : g(u) = invnorm(u/total)         [Probit]

                                                   AIC             =  4.437569
Log likelihood   = -53.46961796                    BIC             = -51.14671

------------------------------------------------------------------------------
             |                 OIM
    menarche |      Coef.   Std. Err.      z    P>|z|     [95% Conf. Interval]
-------------+----------------------------------------------------------------
         age |   .9078231   .0295303    30.74   0.000     .8499447    .9657015
       _cons |  -11.81894   .3873598   -30.51   0.000    -12.57815   -11.05973
------------------------------------------------------------------------------
```

Although a comparison of the predicted values in figure 10.2 does not indicate strong subjective evidence for distinguishing models on the basis of link, the AIC statistic is smaller for the probit link.

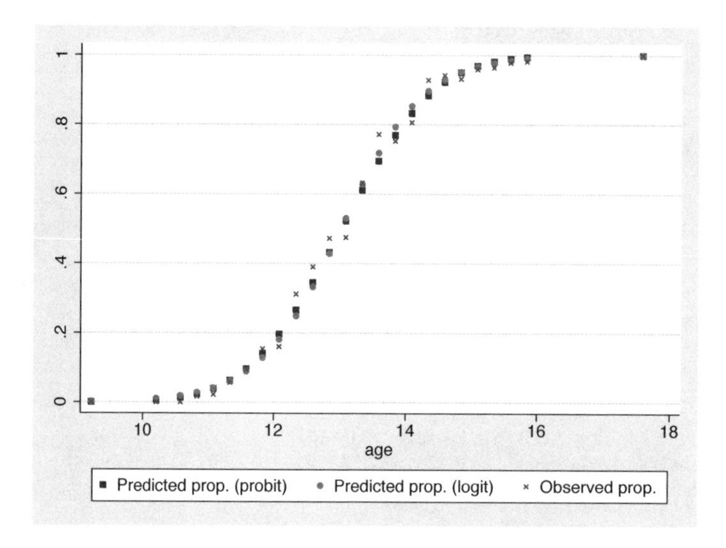

Figure 10.2: Predicted probabilities for probit and logit link function in proportional binary models. The observed (sample) proportions are included as well.

10.4 The clog-log and log-log models

Clog-log and log-log links are asymmetrically sigmoidal. For clog-log models, the upper part of the sigmoid is more elongated or stretched out than the logit or probit. Log-log models are based on the converse. The bottom of the sigmoid is elongated or skewed to the left.

We provide the link, inverse link, and derivative of the link for both clog-log and log-log in table 10.2. As we showed for the probit model, we can change logit or probit to clog-log or log-log by replacing the respective functions in the GLM–binomial algorithm.

Figure 10.3 shows the asymmetry of the clog-log and log-log links. We investigate the consequences of this asymmetry on the fitted values and discuss the relationship of the coefficient estimates from models using these links.

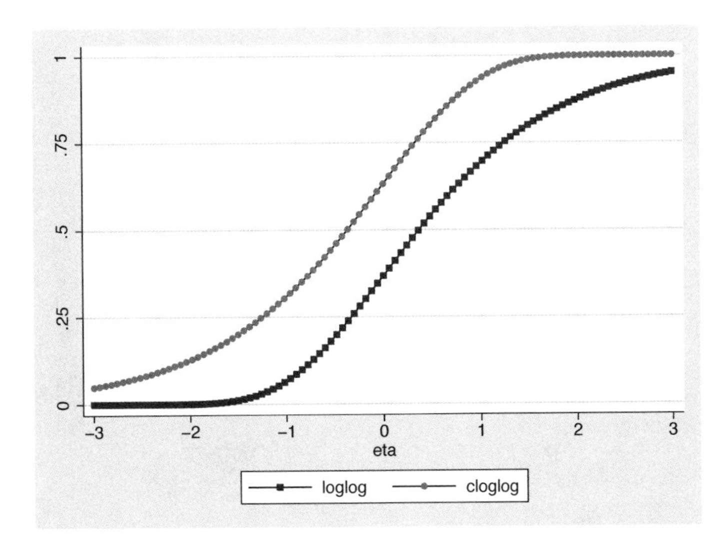

Figure 10.3: Complementary log-log and log-log functions

We may wish to change the focus of modeling from the positive to the negative outcomes. Assume that we have a binary outcome variable y and the reverse-outcome variable $z = 1 - y$. We have a set of covariates $\mathbf{X}$ for our model. If we use a symmetric link, then we can fit a binary model to either outcome. The resulting coefficients will differ in sign for the two models and the predicted probabilities (fitted values) will be complementary. They are complementary in that the fitted probabilities for the model on y plus the fitted probabilities for the model on z add to one. See our previous warning note on the sign change of the coefficients in section 9.2.

We say that probit(y) and probit(z) form a complementary pair. And logit(y) and logit(z) form a complementary pair. Although the equivalence of the link in the complementary pair is true with symmetric links, it is not true for the asymmetric links. The log-log and clog-log links form a complementary pair. This means that loglog(y) and cloglog(z) form a complementary pair and that loglog(z) and cloglog(y) form a complementary pair.

Algebraically, we can illustrate the change in coefficient signs for complementary pairs:

$$
\begin{aligned}
\text{probit}(z) &= \Phi^{-1}(z) = \Phi^{-1}(1-y) = -\Phi^{-1}(y) && (10.4)\\
&= -\text{probit}(y) && (10.5)\\
\text{logit}(z) &= \ln\left(\frac{z}{1-z}\right) = \ln\left\{\frac{1-y}{1-(1-y)}\right\} = \ln\left(\frac{1-y}{y}\right) = -\ln\left(\frac{y}{1-y}\right) && (10.6)\\
&= -\text{logit}(y) && (10.7)\\
\text{loglog}(z) &= -\ln\{-\ln(z)\} = -\ln\{-\ln(1-y)\} = -\left[\ln\{-\ln(1-y)\}\right] && (10.8)\\
&= -\text{cloglog}(y) && (10.9)\\
\text{cloglog}(z) &= \ln\{-\ln(1-z)\} = \ln[-\ln\{1-(1-y)\}] = -\left[-\ln\{-\ln(y)\}\right] && (10.10)\\
&= -\text{loglog}(y) && (10.11)
\end{aligned}
$$

Fitted values are functions of the inverse links given in table A.2. Using this information along with a superscript to identify the dependent variable and a subscript to denote the inverse link, we know that the linear predictors will differ in sign, $\eta^y_{\text{link}} = -\eta^z_{\text{complementary link}}$. We can algebraically illustrate the complementary fitted values for complementary pairs:

$$
\begin{aligned}
\mu^z_{\text{probit}} &= \Phi(\eta^z_{\text{probit}}) = \Phi(-\eta^y_{\text{probit}}) = 1 - \Phi(\eta^y_{\text{probit}}) && (10.12)\\
&= 1 - \mu^y_{\text{probit}} && (10.13)\\
\mu^z_{\text{logit}} &= \frac{\exp(\eta^z_{\text{probit}})}{1+\exp(\eta^z_{\text{probit}})} = \frac{\exp(-\eta^y_{\text{probit}})}{1+\exp(-\eta^y_{\text{probit}})} = 1 - \frac{\exp(\eta^y_{\text{probit}})}{1+\exp(\eta^y_{\text{probit}})} && (10.14)\\
&= 1 - \mu^y_{\text{logit}} && (10.15)\\
\mu^z_{\text{loglog}} &= \exp\{-\exp(-\eta^z_{\text{loglog}})\} = \exp\{-\exp(\eta^y_{\text{cloglog}})\} && (10.16)\\
&= 1 - \mu^y_{\text{cloglog}} && (10.17)\\
\mu^z_{\text{cloglog}} &= 1 - \exp\{-\exp(\eta^z_{\text{cloglog}})\} = 1 - \exp\{-\exp(-\eta^y_{\text{loglog}})\} && (10.18)\\
&= 1 - \mu^y_{\text{loglog}} && (10.19)
\end{aligned}
$$

Because of the asymmetric nature of the links,

$$
\begin{aligned}
\mu^y_{\text{loglog}} + \mu^z_{\text{loglog}} &\neq 1 && (10.20)\\
\mu^y_{\text{cloglog}} + \mu^z_{\text{cloglog}} &\neq 1 && (10.21)
\end{aligned}
$$

The proper relationship illustrates the *complementary* nature of the clog-log link.

$$
\begin{aligned}
\mu^y_{\text{loglog}} + \mu^z_{\text{cloglog}} &= 1 && (10.22)\\
\mu^z_{\text{loglog}} + \mu^y_{\text{cloglog}} &= 1 && (10.23)
\end{aligned}
$$

The clog-log inverse link function, which defines the fitted value for the model, is also used to predict the probability of death by following Cox proportional hazards modeling.

In a Cox model, given that death is the failure event, the probability of death given the various predictors in the model is specified by $\Pr(\text{death}) = 1 - \exp\{-\exp(\mathbf{x}\boldsymbol{\beta})\}$. One can then use the log-log inverse link to predict the probability of survival.

The binary clog-log algorithms for EIM and OIM variance calculations are displayed (respectively) below.

Listing 10.3: IRLS algorithm for binary clog-log regression

```
1   Dev = 0
2   μ = (y + 0.5)/2                  /* initialization of μ */
3   η = ln(−ln(1 − μ))               /* initialization of η */
4   WHILE (abs(ΔDev) > tolerance ) {
5         W = ((1 − μ) ln(1 − μ))^2/(μ(1 − μ))
6         z = η − (y − μ)/((1 − μ) ln(1 − μ))
7         β = (X^T W X)^{-1} X^T W z
8         η = Xβ
9         μ = 1 − exp(−exp(η))
10        OldDev = Dev
11        Dev = 2∑ln(1/μ) if (y = 1)
12        Dev = 2∑ln(1/(1 − μ)) if (y = 0)
13        ΔDev = Dev - OldDev
14  }
15  χ^2 = ∑(y − μ)^2/(μ(1 − μ))
```

Listing 10.4: IRLS algorithm for binary clog-log regression using OIM

```
1   Dev = 0
2   μ = (y + 0.5)/2                  /* initialization of μ */
3   η = ln(−ln(1 − μ))               /* initialization of η */
4   WHILE (abs(ΔDev) > tolerance ) {
5         V = μ(1 − μ)
6         V' = 1 − 2μ
7         W = ((1 − μ) ln(1 − μ))^2/(μ(1 − μ))
8         g' = −1/((1 − μ) ln(1 − μ))
9         g'' = (−1 − ln(1 − μ))/((μ − 1)^2 ln(1 − μ)^2)
10        W_o = w + (y − μ)(Vg'' + V'g')/(V^2(g')^3)
11        z = η + (y − μ)g'
12        β = (X^T W_o X)^{-1} X^T W_o z
13        η = Xβ
14        μ = 1 − exp(−exp(η))
15        OldDev = Dev
16        Dev = 2∑ln(1/μ) if (y = 1)
17        Dev = 2∑ln(1/(1 − μ)) if (y = 0)
18        ΔDev = Dev - OldDev
19  }
20  χ^2 = ∑(y − μ)^2/(μ(1 − μ))
```

If a maximum likelihood routine is being used to model clog-log data, then the likelihood function needs to be converted to $\mathbf{x}\boldsymbol{\beta}$ parameterization. The formulas for both the clog-log and log-log log likelihoods are

Inverse link	Formula	$y=0$	$y=1$
Clog-log	$1-\exp\{-\exp(\mathbf{x}\boldsymbol{\beta})\}$	$-\exp(\mathbf{x}\boldsymbol{\beta})$	$\ln[1-\exp\{-\exp(\mathbf{x}\boldsymbol{\beta})\}]$
Log-log	$\exp\{-\exp(-\mathbf{x}\boldsymbol{\beta})\}$	$\ln[1-\exp\{-\exp(-\mathbf{x}\boldsymbol{\beta})\}]$	$-\exp(-\mathbf{x}\boldsymbol{\beta})$

Finally, for completeness, we provide an algorithm for the simple log-log model.

Listing 10.5: IRLS algorithm for binary log-log regression

```
1   Dev = 0
2   μ = (y + 0.5)/2                    /* initialization of μ */
3   η = −ln(−ln(μ))                    /* initialization of η */
4   WHILE (abs(ΔDev) > tolerance ) {
5       W = (μ ln(μ))^2
6       z = η + (y − μ)/(μ ln(μ))
7       β = (X^T W X)^{-1} X^T W z
8       η = Xβ
9       μ = exp(−exp(−η))
10      OldDev = Dev
11      Dev = 2∑ln(1/μ) if (y = 1)
12      Dev = 2∑ln(1/(1 − μ)) if (y = 0)
13      ΔDev = Dev - OldDev
14  }
15  χ^2 = ∑(y − μ)^2/(μ(1 − μ))
```

Continuing with the **heart** dataset as our example, we now use a log-log model to fit the data. There is good reason to suspect that the log-log model is preferable to other binomial links. The following table gives the respective frequency of ones and zeros in the binary model:

```
. tab death if e(sample)

      death |      Freq.     Percent        Cum.
------------+-----------------------------------
          0 |      4,307       96.07       96.07
          1 |        176        3.93      100.00
------------+-----------------------------------
      Total |      4,483      100.00
```

We have used Stata's `e(sample)` function to restrict tabulation sample to be the same sample used in the previous estimation command.

Modeling the grouped data with the log-log link assumes that the data have significantly more zeros than ones in the binary form. Accordingly, the resulting log-log model should have a lower deviance statistic than that of other links.

```
. glm death anterior hcabg kk2 kk3 kk4 age2-age4, family(bin) link(loglog) nolog
Generalized linear models                    No. of obs      =       4483
Optimization     : ML                        Residual df     =       4474
                                             Scale parameter =          1
Deviance         =  1265.944904              (1/df) Deviance =   .2829559
Pearson          =  4719.701375              (1/df) Pearson  =   1.054918

Variance function: V(u) = u*(1-u)            [Bernoulli]
Link function    : g(u) = -ln(-ln(u))        [Log-log]

                                             AIC             =   .2864031
Log likelihood   = -632.972452               BIC             =  -36351.66

------------------------------------------------------------------------------
             |                 OIM
       death |      Coef.   Std. Err.      z    P>|z|     [95% Conf. Interval]
-------------+----------------------------------------------------------------
    anterior |   .2041431   .0519702     3.93   0.000     .1022834    .3060028
       hcabg |   .2318145   .1296487     1.79   0.074    -.0222923    .4859213
         kk2 |   .2523179   .0597734     4.22   0.000     .1351642    .3694715
         kk3 |   .3149235   .0952601     3.31   0.001     .1282171    .5016299
         kk4 |    1.18085   .1900758     6.21   0.000     .8083082    1.553392
        age2 |    .104686   .0782599     1.34   0.181    -.0487006    .2580726
        age3 |   .4162827   .0700609     5.94   0.000     .2789658    .5535995
        age4 |   .6921546   .0785702     8.81   0.000     .5381598    .8461494
       _cons |  -1.699495   .0640278   -26.54   0.000    -1.824987   -1.574003
------------------------------------------------------------------------------
```

We find the deviance to be 1,265.945, less than that produced by previously investigated links. Ranking of the respective deviance statistics (the clog-log model does not appear in the text) gives

Model	Deviance	BIC
Log-log	1265.945	−36351.66
Probit	1268.626	−36348.98
Logit	1273.251	−36344.35
Clog-log	1276.565	−36341.04

We do, in fact, find that the log-log model is preferable to the others on the basis of the minimum-deviance criterion for model selection. (We will not address here whether the difference in log-log and probit deviance values is significant.) However, before putting a final blessing on the model, we would need to evaluate the fit by using residual analysis. Particularly, we advise the use of Anscombe residuals versus μ plots. One can also use many other residuals to assess fit and goodness of link. See chapter 4. We may also construct a table analogous to that of the Hosmer–Lemeshow GOF table. The table, if grouped appropriately, helps to identify areas in the distribution of μ where fit may be lacking.

We can similarly assess the choice of link with the Warsaw data (described in chap. 9). Here we illustrate the header information for each of four candidate link functions.

```
. glm menarche age, family(bin total) link(logit) notable nolog
Generalized linear models                          No. of obs      =          25
Optimization     : ML                              Residual df     =          23
                                                   Scale parameter =           1
Deviance         =  26.70345269                    (1/df) Deviance =     1.16102
Pearson          =  21.86985435                    (1/df) Pearson  =    .9508632
Variance function: V(u) = u*(1-u/total)            [Binomial]
Link function    : g(u) = ln(u/(total-u))          [Logit]
                                                   AIC             =     4.59021
Log likelihood   = -55.37762768                    BIC             =   -47.33069

. glm menarche age, family(bin total) link(probit) notable nolog
Generalized linear models                          No. of obs      =          25
Optimization     : ML                              Residual df     =          23
                                                   Scale parameter =           1
Deviance         =  22.88743324                    (1/df) Deviance =    .9951058
Pearson          =  21.90102423                    (1/df) Pearson  =    .9522184
Variance function: V(u) = u*(1-u/total)            [Binomial]
Link function    : g(u) = invnorm(u/total)         [Probit]
                                                   AIC             =    4.437569
Log likelihood   = -53.46961796                    BIC             =   -51.14671

. glm menarche age, family(bin total) link(loglog) notable nolog
Generalized linear models                          No. of obs      =          25
Optimization     : ML                              Residual df     =          23
                                                   Scale parameter =           1
Deviance         =   34.6387326                    (1/df) Deviance =    1.506032
Pearson          =  33.90973603                    (1/df) Pearson  =    1.474336
Variance function: V(u) = u*(1-u/total)            [Binomial]
Link function    : g(u) = -ln(-ln(u/total))        [Log-log]
                                                   AIC             =    4.907621
Log likelihood   = -59.34526764                    BIC             =   -39.39541

. glm menarche age, family(bin total) link(cloglog) notable nolog
Generalized linear models                          No. of obs      =          25
Optimization     : ML                              Residual df     =          23
                                                   Scale parameter =           1
Deviance         =  118.8207752                    (1/df) Deviance =    5.166121
Pearson          =  228.6307867                    (1/df) Pearson  =    9.940469
Variance function: V(u) = u*(1-u/total)            [Binomial]
Link function    : g(u) = ln(-ln(1-u/total))       [Complementary log-log]
                                                   AIC             =    8.274903
Log likelihood   =  -101.436289                    BIC             =    44.78663
```

Again listing in order based on deviance values, we can assess the models with the following table.

Model	Deviance	BIC
Probit	22.887	−51.147
Logit	26.703	−47.331
Log-log	34.639	−39.395
Clog-log	118.821	44.787

Here we find evidence (minimum deviance and minimum BIC) that the probit model is the most desirable. The clog-log link function provides the least desirable fit to the data.

10.5 Other links

Various GLM implementations, including the `glm` command in Stata, will allow specifying other links with the binomial family. Wacholder (1986) describes using the `log`, `logc`, and `identity` links with binomial data to obtain measures of the risk ratio (`log`), health ratio (`logc`), and risk difference (`identity`).

Although this is possible on occasion, we cannot always use these links with the binomial family. Table A.1 specifies that the range of μ is limited to $(0,1)$ for the Bernoulli case and to $(0,k)$ for proportional data. The `glm` program may take steps to ensure that the values of μ are within the appropriate range (as Wacholder suggests), but μ is calculated from the linear predictor. The linear predictor is simply the product $\mathbf{x}\boldsymbol{\beta}$, and $\boldsymbol{\beta}$ is a vector of free parameters. There is no parameterization to ensure that the estimation of $\boldsymbol{\beta}$ limits the outcomes of the linear predictor and, thus, the fitted values μ. Our experience suggests that we should be grateful when a binomial model with these links converges to an answer and that we should not be surprised when the fitting algorithm fails. We can graphically illustrate why this is the case. Consider the identity link and compare it with the logit and probit functions depicted before:

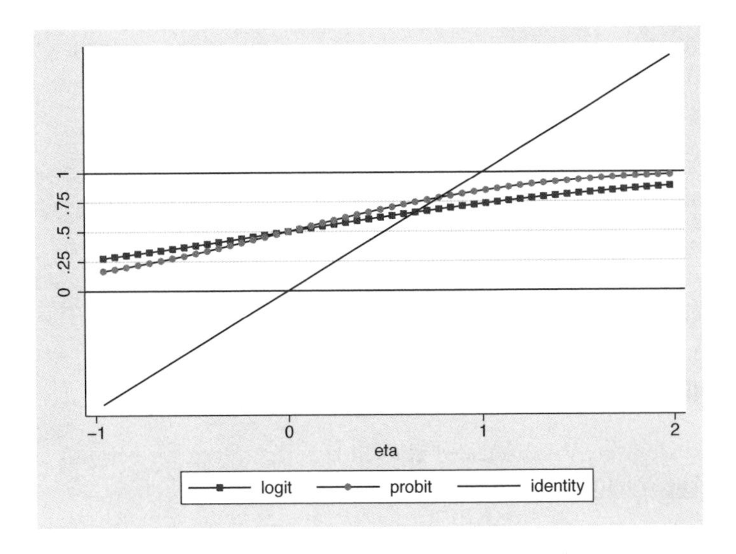

Figure 10.4: Probit, logit, and identity functions

The identity function is outside the required range for most of the domain. The same is true for the log and log complement links.

10.6 Interpretation of coefficients

In the previous chapter, we illustrated the interpretation of coefficients for logistic regression (binomial family with logit link). Here we expand that discussion and investigate the interpretation of coefficients and exponentiated coefficients for other links with the binomial family.

Interpreting coefficients is tenuous because of using link functions. Of course, interpretation is easier when the link is identity. Throughout the following, we assume that we have a binary outcome $\mathbf{y}$ and two covariates $\mathbf{x}_1$ and $\mathbf{x}_2$ (and a constant). The probability of a successful outcome ($\mathbf{y} = 1$) is given by

$$\Pr(y_i = 1) = p = g^{-1}(\beta_0 + x_{1i}\beta_1 + x_{2i}\beta_2) \tag{10.24}$$

where $g^{-1}()$ is the inverse link function. We will focus here on interpreting the β_1 coefficient.

In health studies, a successful outcome is generally negative (denotes death or onset of disease). The terminology used to refer to coefficients reflects this property.

10.6.1 Identity link

For the identity link, interpreting coefficients is straightforward. For a one-unit increase in $\mathbf{x}_1$, the β_1 coefficient represents the difference or change in the outcome. Since the outcome represents risk, the coefficients are sometimes called risk differences. We can see that the risk differences are constant and independent of the values of the covariates by looking at the change in the outcome when we increase $\mathbf{x}_1$ by one and hold the other covariates constant.

$$\Delta y_i = \{\beta_0 + (x_{1i}+1)\beta_1 + x_{2i}\beta_2\} - (\beta_0 + x_{1i}\beta_1 + x_{2i}\beta_2) \tag{10.25}$$

$$= \beta_1 \tag{10.26}$$

Therefore, for the identity link in a binomial family model, the coefficients are interpreted as risk differences.

10.6.2 Logit link

In the previous chapter, we discussed the odds ratio. Here we expand that discussion and illustrate the origin of the odds ratio term.

Since the inverse link is nonlinear, interpreting the coefficient is difficult.

$$\Delta y_i = \frac{\exp\{\beta_0 + (x_{1i}+1)\beta_1 + x_{2i}\beta_2\}}{1+\exp\{\beta_0 + (x_{1i}+1)\beta_1 + x_{2i}\beta_2\}} - \frac{\exp(\beta_0 + x_{1i}\beta_1 + x_{2i}\beta_2)}{1+\exp(\beta_0 + x_{1i}\beta_1 + x_{2i}\beta_2)} \tag{10.27}$$

The difficulty arises because the difference is not constant and depends on the values of the covariates.

Transforming the coefficients makes interpretation easier. We motivate this transformation by introducing an alternative metric.

The odds of a successful outcome is the ratio of the probability of success to the probability of failure. We can write the odds as

$$\text{Odds} = \frac{p}{1-p} \tag{10.28}$$

For the logit inverse link, we have for our specific model

$$\begin{aligned}
\text{Odds} = \frac{p}{1-p} &= \frac{\exp(\mathbf{x}\boldsymbol{\beta})/\{1+\exp(\mathbf{x}\boldsymbol{\beta})\}}{1/\{1+\exp(\mathbf{x}\boldsymbol{\beta})\}} && (10.29)\\
&= \exp(\mathbf{x}\boldsymbol{\beta}) && (10.30)\\
&= \exp(\beta_0 + x_{1i}\beta_1 + x_{2i}\beta_2) && (10.31)
\end{aligned}$$

The odds may be calculated for any collection of values of $\mathbf{x}$.

The odds ratio is the ratio of two odds. We calculate the odds ratio for each of the specific covariates in the model. The numerator odds is calculated such that the specific covariate is incremented by one relative to the value used in the denominator. The odds ratio for $\mathbf{x}_1$ is calculated as

$$\begin{aligned}
\text{Odds ratio for } \mathbf{x}_1 &= \frac{\exp\{\beta_0 + (x_{1i}+1)\beta_1 + x_{2i}\beta_2\}}{\exp(\beta_0 + x_{1i}\beta_1 + x_{2i}\beta_2)} && (10.32)\\
&= \exp(\beta_1) && (10.33)
\end{aligned}$$

The calculation of the odds ratio simplifies such that there is no dependence on a particular observation. The odds ratio is therefore constant—it is independent of the particular values of the covariates. This interpretation does not hold for other inverse links.

Interpretation for binomial family models using the logit link is straightforward for exponentiated coefficients. These exponentiated coefficients represent odds ratios. An odds ratio of 2 indicates that a successful outcome is twice as likely if the associated covariate is increased by one.

10.6.3 Log link

Since the inverse link is nonlinear, interpreting the coefficient is difficult.

$$\Delta y_i = \exp\{\beta_0 + (x_{1i}+1)\beta_1 + x_{2i}\beta_2\} - \exp(\beta_0 + x_{1i}\beta_1 + x_{2i}\beta_2) \tag{10.34}$$

The difficulty arises because the difference is not constant and depends on the values of the covariates.

Transforming the coefficients makes interpretation easier. We motivate this transformation by introducing an alternative metric.

The risk ratio is a ratio, for each covariate, of the probabilities of a successful outcome. A successful outcome under the log inverse link is characterized by

$$\begin{aligned} \Pr(y_i = 1) &= g^{-1}(\beta_0 + x_{1i}\beta_1 + x_{2i}\beta_2) && (10.35) \\ &= \exp(\beta_0 + x_{1i}\beta_1 + x_{2i}\beta_2) && (10.36) \end{aligned}$$

For the risk ratio, the numerator probability is calculated such that the specific covariate is incremented by one relative to the value used in the denominator probability. The risk ratio for $\mathbf{x}_1$ is calculated as

$$\begin{aligned} \text{Risk ratio for } \mathbf{x}_1 &= \frac{\exp\{\beta_0 + (x_{1i}+1)\beta_1 + x_{2i}\beta_2\}}{\exp(\beta_0 + x_{1i}\beta_1 + x_{2i}\beta_2)} && (10.37) \\ &= \exp(\beta_1) && (10.38) \end{aligned}$$

There is no dependence on a particular observation. The risk ratio is therefore constant—it is independent of the particular values of the covariates. This interpretation does not hold for other inverse links.

Interpretation for binomial family models using the log link is straightforward for exponentiated coefficients. These exponentiated coefficients represent risk ratios. A risk ratio of 2 indicates that a successful outcome is twice as likely (the risk is doubled) if the associated covariate is increased by one.

10.6.4 Log complement link

Since the inverse link is nonlinear, interpreting the coefficient is difficult.

$$\Delta y_i = 1 - \exp\{\beta_0 + (x_{1i}+1)\beta_1 + x_{2i}\beta_2\} - [1 - \exp(\beta_0 + x_{1i}\beta_1 + x_{2i}\beta_2)] \quad (10.39)$$

The difficulty arises because the difference is not constant and depends on the values of the covariates.

Transforming the coefficients makes interpretation easier. We motivate this transformation by introducing an alternative metric.

Instead of calculating the ratio of probabilities of success to get a measure of risk, we can calculate the ratio of probabilities of failure to get a measure of health. Under the log complement inverse link, an unsuccessful outcome is characterized by

$$\begin{aligned} \Pr(y_i = 0) &= 1 - g^{-1}(\beta_0 + x_{1i}\beta_1 + x_{2i}\beta_2) && (10.40) \\ &= 1 - \{1 - \exp(\beta_0 + x_{1i}\beta_1 + x_{2i}\beta_2)\} && (10.41) \\ &= \exp(\beta_0 + x_{1i}\beta_1 + x_{2i}\beta_2) && (10.42) \end{aligned}$$

For the health ratio, the numerator probability is calculated such that the specific covariate is incremented by one relative to the value used in the denominator probability. The health ratio for $\mathbf{x}_1$ is calculated as

$$\begin{aligned} \text{Health ratio for } \mathbf{x}_1 &= \frac{\exp\{\beta_0 + (x_{1i}+1)\beta_1 + x_{2i}\beta_2\}}{\exp(\beta_0 + x_{1i}\beta_1 + x_{2i}\beta_2)} && (10.43) \\ &= \exp(\beta_1) && (10.44) \end{aligned}$$

There is no dependence on a particular observation. The health ratio is therefore constant—it is independent of the particular values of the covariates. This interpretation does not hold for other inverse links.

Interpretation for binomial family models using the log complement link is straightforward for exponentiated coefficients. These exponentiated coefficients represent health ratios. A health ratio of 2 indicates that a successful outcome is half as likely (the risk is cut in half) if the associated covariate is increased by one.

10.6.5 Summary

As we have illustrated, exponentiated coefficients for models using the binomial family offer a much clearer interpretation for certain inverse link functions. Notwithstanding our desire for clearer interpretation, the calculation of risk differences, health ratios, and risk ratios are subject to data that support optimization using links that do not necessarily restrict fitted values to the supported range of the binomial family variance function.

10.7 Generalized binomial regression

Jain and Consul (1971) introduced the generalized negative binomial distribution that Famoye and Kaufman (2002) further investigated as a regression model. Later, Famoye (1995) found that by restricting the ancillary parameter to integer values, the regression model could be used for proportional binomial data.

The generalized negative binomial distribution is given by

$$\Pr(Y=y) = \frac{m}{m+\phi y}\binom{m+\phi y}{y}\theta^y(1-\theta)^{m+\phi y-y} \tag{10.45}$$

for $y = 0, 1, 2, \ldots$, $0 < \theta < 1$, $1 \leq \phi < \theta^{-1}$, for $m > 0$. The distribution reduces to the binomial distribution when $\phi = 0$ and reduces to the negative binomial distribution when $\phi = 1$.

A generalized binomial distribution results when we take the m parameter and treat it as the denominator (known values) in a proportional-data binomial model. Such a model may be specified as

$$\Pr(Y_i=y_i) = \frac{m_i}{m_i+\phi y_i}\binom{m_i+\phi y_i}{y_i}\left(\frac{pi_i}{1+\phi\pi_i}\right)^{y_i}\left(1-\frac{pi_i}{1+\phi\pi_i}\right)^{m_i+\phi y_i-y_i} \tag{10.46}$$

We see that the expected value and variance are related as

$$\mathrm{E}(Y_i) = \mu_i = m_i\theta(1-\theta\phi)^{-1} = m_i\pi_i \qquad \mathrm{V}(Y_i) = m_i\pi_i(1+\phi\pi_i)(1+\phi\pi_i-\pi_i) \tag{10.47}$$

In the usual GLM approach, we can introduce covariates into the model by substituting a link function of the linear predictor for the expected value. The probability mass function has a related log likelihood given by

$$\begin{aligned}\mathcal{L}(\boldsymbol{\beta},\phi) &= \sum_{i=1}^{n}\Big\{\log m_i + y_i\log\pi_i + (m_i+\phi y_i - y_i)\log(1+\phi\pi_i-\pi_i)\\ &\quad -(m_i+\phi y_i)\log(1+\phi\pi_i) + \log\Gamma(m_i+\phi y_i+1)\\ &\quad -\log\Gamma(y_i+1) - \log\Gamma(m_i+\phi y_i - y_i+1)\Big\} \qquad (10.48)\end{aligned}$$

Wheatley and Freeman (1982) discuss a model for the plant-to-plant distribution of carrot fly damage on carrots and parsnips in experiments (fields and microplots). They noted that the distribution of larvae was likely to require two descriptive parameters. Our use of these data is meant to illustrate the relationship to GLMs and the utility of Stata's `ml` programming capabilities. Since the authors used a clog-log link within a binomial model to describe the data, we will mimic that approach here. The experiment yielding the data includes three replicates (`R1`, `R2`, and `R3` indicator variables) and 11 different treatment combinations of insecticide and depth (`T1`, ..., `T11` indicator variables) given in table 10.3; the two rows for no insecticide constitute the last treatment, `T11`.

Table 10.3: 1964 microplot data of carrot fly damage

Treatment		Replicate 1		Replicate 2		Replicate 3	
Insecticide	Depth	No. damaged	No. examined	No. damaged	No. examined	No. damaged	No. examined
Diazinon	1.0	120	187	145	181	123	184
	2.5	60	184	85	191	113	171
	5.0	35	179	23	181	33	179
	10.0	5	178	40	184	18	171
	25.0	66	182	104	186	106	181
Disulfoton	1.0	97	187	128	173	112	179
	2.5	85	186	95	167	137	194
	5.0	72	190	89	184	92	192
	10.0	49	188	50	187	65	186
	25.0	132	179	122	182	120	179
Nil	I	138	187	169	184	163	175
	II	156	176	169	190	159	175

Reprinted from Wheatley and Freeman (1982).

Observed proportions of damage for each treatment are illustrated in figure 10.5.

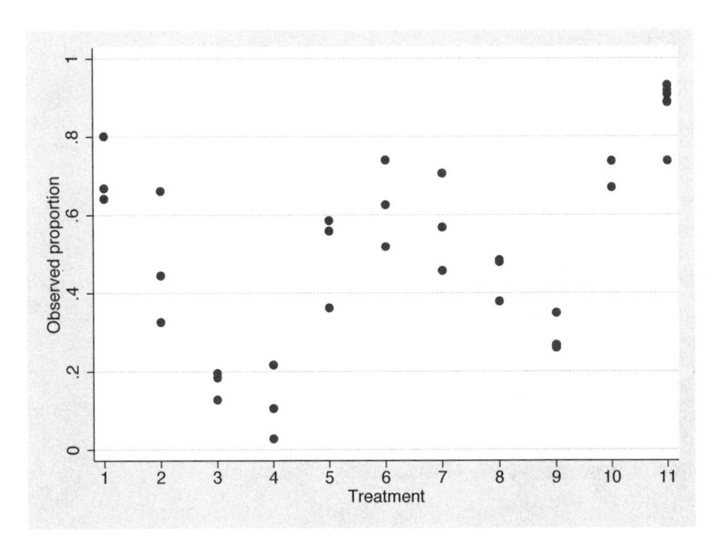

Figure 10.5: Observed proportion of carrot fly damage for each treatment (see table 10.3)

The clog-log binomial regression estimates are easily obtained using **glm**:

```
. glm damage R2 R3 T1 T2 T3 T4 T5 T6 T7 T8 T9 T10, family(bin examine) link(clog)
Iteration 0:   log likelihood = -158.43213
Iteration 1:   log likelihood = -156.76371
Iteration 2:   log likelihood = -156.75711
Iteration 3:   log likelihood = -156.75711

Generalized linear models                    No. of obs      =         36
Optimization     : ML                        Residual df     =         23
                                             Scale parameter =          1
Deviance         =  123.5631356              (1/df) Deviance =    5.37231
Pearson          =  121.3150965              (1/df) Pearson  =   5.274569

Variance function: V(u) = u*(1-u/examine)    [Binomial]
Link function    : g(u) = ln(-ln(1-u/examine))  [Complementary log-log]

                                             AIC             =    9.43095
Log likelihood   = -156.7571088              BIC             =    41.1422

------------------------------------------------------------------------------
             |                 OIM
      damage |      Coef.   Std. Err.      z    P>|z|     [95% Conf. Interval]
-------------+----------------------------------------------------------------
          R2 |   .3330675   .0453895     7.34   0.000     .2441058    .4220293
          R3 |   .3765869   .0452898     8.32   0.000     .2878204    .4653533
          T1 |  -.5672674   .0669523    -8.47   0.000    -.6984915   -.4360433
          T2 |   -1.20193   .0746685   -16.10   0.000    -1.348277   -1.055582
          T3 |  -2.462982   .1121606   -21.96   0.000    -2.682812   -2.243151
          T4 |  -2.840763   .1321047   -21.50   0.000    -3.099683   -2.581842
          T5 |  -1.117579   .0730595   -15.30   0.000    -1.260773   -.9743852
          T6 |  -.7731088   .0691728   -11.18   0.000     -.908685   -.6375326
          T7 |  -.8988995   .0701799   -12.81   0.000     -1.03645   -.7613494
          T8 |  -1.289745   .0750403   -17.19   0.000    -1.436822   -1.142669
          T9 |  -1.831744   .0878473   -20.85   0.000    -2.003922   -1.659567
         T10 |  -.6123426   .0675081    -9.07   0.000     -.744656   -.4800291
       _cons |   .5233536   .0466372    11.22   0.000     .4319464    .6147608
------------------------------------------------------------------------------
```

After fitting the binomial model, one can store the coefficient estimates for use as starting values in the fitting of the generalized binomial model.

```
. matrix b0 = e(b)
. local ll0 = `e(ll)'
. matrix z0 = (0)
. matrix b0 = b0, z0
```

If our goal were to provide full support for the generalized binomial model, we would program several support routines allowing user specification of useful options. However, our purpose here is to demonstrate how quickly one can obtain answers for specific problems. We will therefore specify the log likelihood only for our particular analysis. Stata's `ml` program allows such specification and will calculate and use numeric derivatives to optimize the objective function. This ability is enormously useful. Thus we write a rather simple program that accepts arguments with the current parameter values and the name of a scalar in which we will store the log likelihood.

```
. capture program drop doit_ll
. program define doit_ll
  1.        version 9.0
  2.        args todo b lnf
  3.        tempvar  eta pi
  4.        tempname phi
  5.        mleval `eta' = `b', eq(1)
  6.        mleval `phi' = `b', eq(2) scalar
  7.        local y  "$ML_y1"    /* Dep var name is in global macro by ml  */
  8.        local M  "examine"   /* Specify directly -- problem specific   */
  9.        gen double `pi'    = 1-exp(-exp(`eta'))  /* Cloglog link here */
 10.        mlsum `lnf' = log(`M') + `y'*log(`pi') +               /*
>               */ (`M'+`phi'*`y'-`y')*log(1+`phi'*`pi'-`pi') - /*
>               */ (`M'+`phi'*`y')*log(1+`phi'*`pi') -          /*
>               */ log(`M'+`phi'*`y') +                         /*
>               */ lngamma(`M'+`phi'*`y'+1) - lngamma(`y'+1) -  /*
>               */ lngamma(`M'+`phi'*`y'-`y'+1)
 11. end
```

After having specified the program that should be called by the `ml` program, we need only start the optimization process by using `ml model`, wherein we specify our starting values, the functionality of our program (we wrote a `d0` or log-likelihood-only function), the starting values, and the nature of the parameter subvectors (what Stata calls "equations"). After convergence is achieved, we use `ml display` to display our estimates. These results mimic those given in Famoye (1995).

```
. ml model d0 doit_ll (gbr: damage = R2 R3 T1 T2 T3 T4 T5 T6 T7 T8 T9 T10)
> /phi, search(off) init(b0, copy) max iter(50)
Iteration 0:   log likelihood = -156.75711
Iteration 1:   log likelihood = -142.40792
Iteration 2:   log likelihood = -142.22792
Iteration 3:   log likelihood = -142.22626
Iteration 4:   log likelihood = -142.22626
```

```
. ml display, title("Generalized binomial regression")
Generalized binomial regression                   Number of obs   =         36
                                                  Wald chi2(12)   =     475.22
Log likelihood = -142.22626                       Prob > chi2     =     0.0000

------------------------------------------------------------------------------
      damage |      Coef.   Std. Err.      z    P>|z|     [95% Conf. Interval]
-------------+----------------------------------------------------------------
gbr          |
          R2 |    .346261   .0780868     4.43   0.000     .1932136    .4993083
          R3 |   .4157763   .0794801     5.23   0.000     .2599982    .5715545
          T1 |    -.54981   .1677757    -3.28   0.001    -.8786443   -.2209757
          T2 |  -1.224663   .1573275    -7.78   0.000     -1.53302   -.9163072
          T3 |  -2.455202   .1735342   -14.15   0.000    -2.795323   -2.115081
          T4 |  -2.852996   .1868674   -15.27   0.000    -3.219249   -2.486743
          T5 |  -1.134379   .1574367    -7.21   0.000    -1.442949   -.8258085
          T6 |  -.7764593   .1618621    -4.80   0.000    -1.093703   -.4592154
          T7 |  -.9148208   .1593118    -5.74   0.000    -1.227066   -.6025754
          T8 |  -1.288835   .1562914    -8.25   0.000     -1.59516   -.9825092
          T9 |  -1.830209   .1599901   -11.44   0.000    -2.143784   -1.516634
         T10 |  -.5553261   .1696358    -3.27   0.001    -.8878062    -.222846
       _cons |   .5070029   .1279714     3.96   0.000     .2561835    .7578222
-------------+----------------------------------------------------------------
phi          |
       _cons |   .7330808   .2476194     2.96   0.003     .2477556    1.218406
------------------------------------------------------------------------------
```

This example demonstrates the tools available to Stata users to extend the binomial model by using the suite of `ml` commands. See the reference manual, or the excellent book by Gould, Pitblado, and Sribney (2006), for documentation and more examples.

11 The problem of overdispersion

Contents

11.1 Overdispersion

The problem of overdispersion is central to all GLM models having discrete responses. We discuss what overdispersion is, how to identify it, and how to deal with it.

Overdispersion in discrete-response models occurs when the variance of the response is greater than the nominal variance. It is generally caused by positive correlation between responses or by an excess variation between response probabilities or counts. The problem with overdispersion is that it may cause underestimation of standard errors of the estimated coefficient vector. A variable may appear to be a significant predictor when in fact it is not.

Overdispersion affects only the discrete models because continuous models fit the scale parameter ϕ, whereas discrete models do not have this extra parameter (it is theoretically equal to one). Thus continuous models still require the variance to be a function of the mean, but that relationship may be scaled. Discrete models, on the other hand, do not have the extra parameter to scale the relationship of the variance and mean.

We can recognize possible overdispersion by observing that the value of the Pearson χ^2, or deviance, divided by the degrees of freedom $(n-p)$ is larger than 1. The quotient of either is called the dispersion. Small amounts of overdispersion are usually of little concern; however, if the dispersion statistic is greater than 2.0, then an adjustment to the standard errors may be required.

There is a distinct difference between true overdispersion and apparent overdispersion. Outward indicators such as a large dispersion statistic may be only a sign of apparent overdispersion. How do we tell?

Apparent overdispersion may arise from any of the following:

1. The model omits important explanatory predictors.
2. The data contain outliers.
3. The model fails to include enough interaction terms.
4. A predictor needs to be transformed (to the log or some other scale).
5. We have assumed a linear relationship, e.g., between the logit transform and the response or predictors when the actual relationship is quadratic (or some higher order).

Any of the above situations may cause a model to appear inherently overdispersed. However, if we add the requisite predictor, discover the important interactions, adjust outliers, transform where necessary, and so forth, then the apparent overdispersion may be remedied. As a consequence, the model will be better fitted.

As an example of how omitting an important explanatory predictor affects the model such that it appears to be overdispersed when in fact it is not, let us create a simulated model demonstrating how to accommodate apparent overdispersion.

```
. set seed0 23988                   /* Set seed for genbinomial pgm */
. set seed 238947                   /* Set seed for reproducibility */
. set obs 10000                     /* Number of obs = 10,000       */
obs was 0, now 10000
. gen x1 = invnormal(uniform())     /* Creation of random variates  */
. gen x2 = invnormal(uniform())
. gen x3 = invnormal(uniform())
. gen d  = 50+5*int((_n-1)/1000) /* Creation of 10 different denominators */
. tab d                             /* Tabulation of the denominator         */

          d |      Freq.     Percent        Cum.
------------+-----------------------------------
         50 |      1,000       10.00       10.00
         55 |      1,000       10.00       20.00
         60 |      1,000       10.00       30.00
         65 |      1,000       10.00       40.00
         70 |      1,000       10.00       50.00
         75 |      1,000       10.00       60.00
         80 |      1,000       10.00       70.00
         85 |      1,000       10.00       80.00
         90 |      1,000       10.00       90.00
         95 |      1,000       10.00      100.00
------------+-----------------------------------
      Total |     10,000      100.00
```

Using these simulated denominators, we can now complete the generation of synthetic data and fit the binomial model. We use a handy utility program, `genbinomial`, by Roberto Gutierrez to help construct the data. Stata users can find and install this utility by using `net search genbinomial`. Or they can use the user-written `rnd` suite of random-number generators (Hilbe and Linde-Zwirble 1995) to create the same simulated set of binomial denominators.

```
. generate xb = 1 + .5*x1 - .75*x2 + .25*x3  /* Create linear predictor */
. genbinomial y, xbeta(xb) de(d)             /* Create simulated dataset */
. glm y x1 x2 x3, family(bin d) nolog        /* Model y on x1, x2, and x3 */
Generalized linear models                          No. of obs      =     10000
Optimization     : ML                              Residual df     =      9996
                                                   Scale parameter =         1
Deviance         =  10205.6227                     (1/df) Deviance =  1.020971
Pearson          =  10058.89719                    (1/df) Pearson  =  1.006292
Variance function: V(u) = u*(1-u/d)                [Binomial]
Link function    : g(u) = ln(u/(d-u))              [Logit]
                                                   AIC             =  5.309671
Log likelihood   = -26544.35733                    BIC             = -81860.94

------------------------------------------------------------------------------
             |                 OIM
           y |      Coef.   Std. Err.      z    P>|z|     [95% Conf. Interval]
-------------+----------------------------------------------------------------
          x1 |   .5033589   .0029086   173.06   0.000     .4976582    .5090597
          x2 |  -.7503928   .0030589  -245.31   0.000    -.7563882   -.7443974
          x3 |   .2520998   .0028398    88.77   0.000     .2465338    .2576657
       _cons |   .9954485   .0029139   341.62   0.000     .9897374     1.00116
------------------------------------------------------------------------------
```

Other simulated datasets will have slightly different values. If we ran several hundred simulated models, however, we would discover that the parameter estimates would equal the assigned values and that the Pearson χ^2 statistic divided by the model degrees of freedom would center about 1.0. The Pearson χ^2 dispersion statistic here is 1.01, and the parameter estimates are close to the values we specified. A Pearson statistic near one indicates that the model is not overdispersed; i.e., it is well specified. Values more than 1.0 indicate overdispersion. When the number of observations is high, as in this simulated dataset, values around 1.05 or more tell us that there may be greater correlation in the data than is allowed based on the distributional assumptions of the binomial probability distribution function (PDF). This situation biases the standard errors.

In the following, we consider what happens to the model fit when we omit the `x1` covariate.

(Continued on next page)

```
. glm y x2 x3, family(bin d) nolog          /* Here, we omit the x1 covariate */
Generalized linear models                    No. of obs        =      10000
Optimization     : ML                        Residual df       =       9997
                                             Scale parameter   =          1
Deviance         =  42352.92767              (1/df) Deviance   =   4.236564
Pearson          =  41802.16757              (1/df) Pearson    =   4.181471
Variance function: V(u) = u*(1-u/d)          [Binomial]
Link function    : g(u) = ln(u/(d-u))        [Logit]
                                             AIC               =   8.524202
Log likelihood   = -42618.00982              BIC               =  -49722.85

------------------------------------------------------------------------------
             |                 OIM
           y |      Coef.   Std. Err.      z    P>|z|     [95% Conf. Interval]
-------------+----------------------------------------------------------------
          x2 |  -.7178564   .0029679  -241.87   0.000    -.7236734   -.7120394
          x3 |   .2416217   .0027734    87.12   0.000     .2361859    .2470574
       _cons |   .9365708   .0027888   335.83   0.000     .9311049    .9420367
------------------------------------------------------------------------------
```

The remaining parameter estimates are fairly close to the original, or true, values. The Pearson dispersion is a very high 4.18. Such a value indicates an ill-fitted model. If this were our initial model, the value of the Pearson dispersion would tell us that something was wrong and that some type of adjustment must be made. We know in this situation that the overdispersion is only apparent and that including predictor `x1` eliminates the excess dispersion. But we usually do not know this when modeling data at the outset. Before assuming that the overdispersion is real, checking for apparent overdispersion is wise.

Note also the difference in the AIC and BIC statistics between the two models. The values generated by the "correct" model are respectively, 5.31 and −81,861; the values for the apparently overdispersed model are: 8.52 and −49,723. Because lower values of these GOF statistics indicate a better-fitted model, the latter values clearly represent a poorly fitted model. If we had not paid attention to the Pearson dispersion statistic, we may not have known that the model was poor, especially since the parameter estimate p-values all indicate that they contribute to the model.

Next we will consider an example of a model that indicates overdispersion due to not having a requisite interaction term.

The interaction term, `x23`, is generated by multiplying `x2` and `x3`. We give it a value of 0.2 and model the data by using the main-effects terms and interaction. The main effects are the same as in the previous example.

```
. gen x23 = x2*x3                                  /* create interaction         */
. gen xb1 = 1 + .5*x1 - .75*x2 + .25*x3 + .2*x23 /* create linear predictor    */
. genbinomial y1, xbeta(xb1) de(d)               /* create simulated dataset */
```

Here we model the main effects and interaction:

```
. glm y1 x1 x2 x3 x23, family(bin d) nolog
Generalized linear models                         No. of obs      =      10000
Optimization     : ML                             Residual df     =       9995
                                                  Scale parameter =          1
Deviance         =  10300.80471                   (1/df) Deviance =   1.030596
Pearson          =   10149.3606                   (1/df) Pearson  =   1.015444
Variance function: V(u) = u*(1-u/d)               [Binomial]
Link function    : g(u) = ln(u/(d-u))             [Logit]
                                                  AIC             =   5.305336
Log likelihood   = -26521.67958                   BIC             = -81756.55

------------------------------------------------------------------------------
             |                 OIM
          y1 |      Coef.   Std. Err.      z    P>|z|     [95% Conf. Interval]
-------------+----------------------------------------------------------------
          x1 |   .5030161   .0029273   171.84   0.000     .4972787    .5087535
          x2 |  -.7542533   .0031003  -243.29   0.000    -.7603298   -.7481769
          x3 |   .2487125   .0029374    84.67   0.000     .2429554    .2544696
         x23 |   .1963328   .0031953    61.44   0.000     .1900701    .2025955
       _cons |   1.000638   .0029252   342.08   0.000     .9949043    1.006371
------------------------------------------------------------------------------
```

Parameter estimates and the Pearson dispersion appear as we expect. We next model the data without the interaction term.

```
. glm y1 x1 x2 x3, family(bin d) nolog
Generalized linear models                         No. of obs      =      10000
Optimization     : ML                             Residual df     =       9996
                                                  Scale parameter =          1
Deviance         =  14157.55105                   (1/df) Deviance =   1.416322
Pearson          =  14092.45562                   (1/df) Pearson  =   1.409809
Variance function: V(u) = u*(1-u/d)               [Binomial]
Link function    : g(u) = ln(u/(d-u))             [Logit]
                                                  AIC             =   5.690811
Log likelihood   = -28450.05275                   BIC             = -77909.01

------------------------------------------------------------------------------
             |                 OIM
          y1 |      Coef.   Std. Err.      z    P>|z|     [95% Conf. Interval]
-------------+----------------------------------------------------------------
          x1 |   .5009411   .0029174   171.71   0.000     .4952231    .5066591
          x2 |  -.7579314   .0030744  -246.53   0.000    -.7639571   -.7519056
          x3 |   .2866747   .0028603   100.23   0.000     .2810686    .2922807
       _cons |   1.008843   .0029298   344.34   0.000       1.0031    1.014585
------------------------------------------------------------------------------
```

The overdispersion statistic rose by 38%, the AIC statistic is 0.39 higher, and the BIC is about 4,000 higher. The model may appear to be well fitted at first glance, but the Pearson dispersion statistic indicates that the model needs adjustment. If statistical output does not include a Pearson χ^2 statistic, or if the user ignores it, a model having parameter estimate p-values of all 0.000 means nothing. If the model is overdispersed, or for that matter underdispersed, it does not fit and we cannot trust the parameter estimates. On the other hand, if we observe overdispersion in the model by virtue of the Pearson χ^2 statistic, then we need to determine whether the overdispersion is real.

Here it is not. The model is only apparently overdispersed. Testing the criteria we listed will help ensure that we have genuine overdispersion. See Hilbe (2007) for an in-depth examination of overdispersion, as well as simulations for the remaining criteria.

Sometimes the overdispersion is indeed inherent in the dataset itself—no external remedy can be found to remove the overdispersion. Then we must use methods that are designed to deal with true overdispersion.

The following methods have been used with binomial models to adjust a model for real overdispersion:

1. Scaling of standard errors by the dispersion statistic
2. Using Williams' procedure
3. Using robust variance estimators as well as bootstrapping or jackknifing standard errors
4. Using GEE or random-effects models, mixed models, or some other variety of binomial model that accounts for extra correlation in the data

We look at the first three of these alternatives in this chapter. We will relegate the fourth method to part IV of the text.

11.2 Scaling of standard errors

Standard errors may be adjusted post hoc by scaling with the inverse square root of either the Pearson χ^2 or the deviance dispersion. Most major commercial software implementations of GLM allow such an option.

The method of scaling is simple. First, the algorithm fits the full model but does not display it on the screen. Second, depending upon which scale the user has chosen, the algorithm abstracts it from the results and calculates a new weight equal to the inverse of the square root of the dispersion.

Then the algorithm calculates one more iteration of the main weighted regression routine,

$$\boldsymbol{\beta} = (\mathbf{X}'\mathbf{W}\mathbf{X})^{-1}\mathbf{X}'\mathbf{W}\mathbf{z} \tag{11.1}$$

with the new values of the weight matrix, $\mathbf{W}$. The result is a scaling of the standard errors—a result that in effect adjusts the standard errors to a value that would have been obtained if the original dispersion value was 1. Of course, the greatest change in standard errors will be for those models having large dispersions.

Scaling is performed using the deviance-based dispersion. McCullagh and Nelder (1989) and other sources prefer the deviance dispersion for scaling binomial and count data; χ^2 dispersion is suggested for continuous-response models. The χ^2 dispersion statistic was used to fit the scale parameter for two-parameter continuous models. We noted then that the χ^2-based scale estimate was nearly identical to the scale or ancillary parameter produced using maximum likelihood methods. Except for the Gaussian model, the deviance dispersion was typically far from the true scale estimate. This is just the reverse for binary and count models, where scaling is based on the deviance instead.

The difference exists because continuous distributions do in fact have a scale parameter to be estimated. Now GLM theory traditionally ignores the scale, constraining it to the value of 1 during the estimation process. The scale is then point estimated by the χ^2 dispersion. However, the binomial and Poisson models have no scale parameter. Their scale is set to 1. Hence, estimating a true scale is inherently a mistake. As we shall see in the following chapter, the negative binomial does have a scale parameter, and as a consequence, many authors exclude it from the GLM family. We will show that this should not be the case.

The scaling of standard errors for binomial models is performed solely for adjusting standard errors for accommodating overdispersion. There is another catch. Overdispersion makes no sense for strictly binary models whose response is 0/1. We can, however, make sense of binary overdispersion by converting the dataset to a proportion format. This method is generally unavailable for models with one or more continuous predictors. Continuous predictors can be transformed to factors or leveled predictors, as for `age` in our example dataset. Certainly, information is lost to the model when continuous variables are factored, but the loss may be offset by gaining greater interpretability and more checks for model fit.

Recall the `heart` data table of estimates produced from modeling the binary logistic model in chapter 9:

death	Odds Ratio	EIM Std. Err.	z	P>|z|	[95% Conf.	Interval]
anterior	1.901333	.3185755	3.83	0.000	1.369103	2.640464
hcabg	2.105275	.7430692	2.11	0.035	1.054076	4.204801
kk2	2.251732	.4064421	4.50	0.000	1.580786	3.207453
kk3	2.172105	.5844269	2.88	0.004	1.281907	3.680486
kk4	14.29137	5.087652	7.47	0.000	7.112966	28.71423
age2	1.63726	.5078572	1.59	0.112	.8914272	3.007112
age3	4.532029	1.20653	5.68	0.000	2.689572	7.636636
age4	8.893222	2.417514	8.04	0.000	5.219999	15.15123

Next we observe the effect of scaling the standard errors.

The output below displays parameter results for the `heart` model with scaled standard errors. We have omitted the header portion of the output. The z-values have all increased, hence reducing p-values down the table. The odds ratios remain the same; only standard errors and their effects have changed. Again standard errors are displayed as if the deviance dispersion had originally been 1. Since this is a binary model, the scaling is not dealing with true overdispersion.

```
. use http://www.stata-press.com/data/hh2/heart01, clear
. glm death anterior hcabg kk2 kk3 kk4 age2-age4, family(bin) irls scale(dev)
> eform nolog noheader
```

death	Odds Ratio	EIM Std. Err.	z	P>|z|	[95% Conf.	Interval]
anterior	1.901333	.1699501	7.19	0.000	1.595783	2.265388
hcabg	2.105275	.3964042	3.95	0.000	1.455577	3.044965
kk2	2.251732	.2168242	8.43	0.000	1.864458	2.719448
kk3	2.172105	.3117735	5.40	0.000	1.63947	2.877784
kk4	14.29137	2.714104	14.00	0.000	9.849643	20.73611
age2	1.63726	.270926	2.98	0.003	1.183764	2.26449
age3	4.532029	.6436463	10.64	0.000	3.430865	5.98662
age4	8.893222	1.289668	15.07	0.000	6.692994	11.81674

(Standard errors scaled using square root of deviance-based dispersion)

Using the Stata programming language, one can manipulate binary-response models such as our example for true over- or underdispersion by converting the data to proportion format:

```
. egen grp=group(anterior-age4)       /* Covariate pattern is assigned a group # */
. drop if grp == .                    /* Discard missing values       */
. egen cases = count(grp), by(grp) /* Size of covariate pattern */
. egen die = total(death), by(grp) /* Deaths in pattern          */
. sort grp                            /* Sort by covariate pattern */
. qui by grp: keep if _n==1           /* Keep one case per pattern */
```

The new model will have `die` as the binomial numerator and `cases` as the binomial denominator. It is fitted using

```
. glm die anterior hcabg kk2 kk3 kk4 age2-age4, family(bin cases) irls eform
> nolog

Generalized linear models                      No. of obs      =         53
Optimization     : MQL Fisher scoring          Residual df     =         44
                   (IRLS EIM)                  Scale parameter =          1
Deviance         =  54.5713578                 (1/df) Deviance =   1.240258
Pearson          =  54.94033061                (1/df) Pearson  =   1.248644

Variance function: V(u) = u*(1-u/cases)        [Binomial]
Link function    : g(u) = ln(u/(cases-u))      [Logit]

                                               BIC             =  -120.1215
```

die	Odds Ratio	EIM Std. Err.	z	P>|z|	[95% Conf.	Interval]
anterior	1.901333	.3185757	3.83	0.000	1.369103	2.640464
hcabg	2.105275	.7430694	2.11	0.035	1.054076	4.204801
kk2	2.251732	.4064423	4.50	0.000	1.580786	3.207453
kk3	2.172105	.584427	2.88	0.004	1.281907	3.680487
kk4	14.29137	5.087654	7.47	0.000	7.112964	28.71423
age2	1.63726	.5078582	1.59	0.112	.8914261	3.007115
age3	4.532029	1.206534	5.68	0.000	2.689568	7.636647
age4	8.893222	2.41752	8.04	0.000	5.219991	15.15125

The odds ratios and standard errors are identical to the binary model. This is as it should be. However, there is a difference in value for the dispersion (the header for this model was included in chap. 9 where the deviance dispersion had a value of .285). The deviance dispersion is indeed a measure of overdispersion or of underdispersion. It is not an estimate of a scale or ancillary parameter as the Pearson χ^2 dispersion was for continuous models.

We scale the model to accommodate the overdispersion found in the binomial model.

```
. glm die anterior hcabg kk2 kk3 kk4 age2-age4, family(bin cases) irls
> scale(dev) eform nolog noheader
```

die	Odds Ratio	EIM Std. Err.	z	P>|z|	[95% Conf.	Interval]
anterior	1.901333	.3547878	3.44	0.001	1.318938	2.740892
hcabg	2.105275	.8275332	1.89	0.058	.9743652	4.548789
kk2	2.251732	.4526421	4.04	0.000	1.518478	3.339065
kk3	2.172105	.6508581	2.59	0.010	1.207323	3.907853
kk4	14.29137	5.665961	6.71	0.000	6.57062	31.08433
age2	1.63726	.5655858	1.43	0.154	.8319038	3.222273
age3	4.532029	1.343679	5.10	0.000	2.534686	8.103286
age4	8.893222	2.692317	7.22	0.000	4.913241	16.0972

(Standard errors scaled using square root of deviance-based dispersion)

Finally, we model the data by using the log-log link. We found in the last chapter that using the log-log link produced a smaller deviance statistic than using any of the other links. Scaling will also be applied. However, the coefficients are not exponentiated. Noncanonical binomial links cannot be interpreted using odds ratios. Hence, probit, log-log, clog-log, and other noncanonical links are usually left in basic coefficient form. Exponentiated coefficients for the log-log and clog-log links do not admit an odds ratio interpretation as explained in section 10.6.2, though some other noncanonical links do admit other interpretations for exponentiated coefficients as shown in section 10.6. Of course, this has no bearing on standard errors or p-values.

(*Continued on next page*)

```
. glm die anterior hcabg kk2 kk3 kk4 age2-age4, family(bin cases) l(loglog)
> scale(dev) eform irls nolog

Generalized linear models                     No. of obs      =          53
Optimization     : MQL Fisher scoring         Residual df     =          44
                   (IRLS EIM)                 Scale parameter =           1
Deviance         =  47.26519568               (1/df) Deviance =    1.074209
Pearson          =  42.22305114               (1/df) Pearson  =    .9596148

Variance function: V(u) = u*(1-u/cases)       [Binomial]
Link function    : g(u) = -ln(-ln(u/cases))   [Log-log]

                                              BIC             =   -127.4276

------------------------------------------------------------------------------
             |                 EIM
         die |     exp(b)   Std. Err.      z    P>|z|     [95% Conf. Interval]
-------------+----------------------------------------------------------------
    anterior |   1.226473   .0662428     3.78   0.000     1.103276    1.363428
       hcabg |   1.260886   .1687254     1.73   0.083     .9700012    1.639001
         kk2 |   1.287005   .0794218     4.09   0.000     1.140387    1.452474
         kk3 |   1.370155   .1374372     3.14   0.002     1.125609     1.66783
         kk4 |   3.257141   .6504386     5.91   0.000      2.20219    4.817461
        age2 |   1.110362   .0900216     1.29   0.197     .9472271    1.301592
        age3 |   1.516314   .1102095     5.73   0.000     1.314988    1.748464
        age4 |   1.998016     .16335     8.47   0.000      1.70219    2.345254
------------------------------------------------------------------------------
(Standard errors scaled using square root of deviance-based dispersion)
```

hcabg becomes a bit more suspect under this parameterization. Let us use a likelihood-ratio test to evaluate its worth to the model where the likelihood-ratio χ^2 statistic is given by

$$\chi^2_{(\nu)} = -2\left(\mathcal{L}_1 - \mathcal{L}_0\right) \tag{11.2}$$

where $\mathcal{L}_1$ is the log-likelihood value for fitting the full model and $\mathcal{L}_0$ is the log-likelihood value for the reduced model. Significance is based on a χ^2 distribution with the degrees of freedom ν equal to the number of predictors dropped from the full model. For our case we have

$$\chi^2_{(1)} = -2\left\{(-76.70504926) - (-78.27579023)\right\} = 3.14 \tag{11.3}$$

With 1 degree of freedom, we can calculate the significance by using

```
. display chiprob(1, 3.14)
.07639381
```

which shows that deleting **hcabg** from the model does not significantly reduce the log-likelihood function. We should remove it in favor of the more parsimonious model. The likelihood-ratio test provides more evidence to the results obtained by the Wald test (displayed in the output).

The method of scaling is a post hoc method of adjustment and is fairly easy to implement. In the previous model, the measure of deviance dispersion does not indicate a problem of overdispersion. We included the scaled output in this model only for continuity of the illustration. The other methods we examine are more sophisticated but usually do not provide a better adjustment. An example of this is Williams' procedure presented in the next section.

All binary models need to be evaluated for the possibility of extra binomial dispersion. When it is significant, means need to be taken to adjust standard errors to take account for the excess correlation in the data. Ascertaining the source of the dispersion is helpful, though this cannot always be done. Extra correlation is often a result of a clustering effect or of longitudinal data taken on a group of individuals over time. Scaling and implementing robust variance estimators often addresses this issue. Researchers may, in fact, need more sophisticated methods to deal with the extra correlation they have found in their data. The Stata `glm` command incorporates several sophisticated variance adjustment options, including jackknife, bootstrap, Newey–West, and others. The output below is obtained for the log-log model, assuming the data are in temporal order, where the variance matrix is a weighted sandwich estimate of variance of Newey–West type with Anderson quadratic weights for the first two lags.

```
. glm die anterior hcabg kk2 kk3 kk4 age2-age4, family(bin cases) l(loglog) irls
> nwest(anderson 2) nolog

Generalized linear models                          No. of obs      =        53
Optimization     : MQL Fisher scoring              Residual df     =        44
                   (IRLS EIM)                      Scale parameter =         1
Deviance         =  47.26519568                    (1/df) Deviance =  1.074209
Pearson          =  42.22305114                    (1/df) Pearson  =  .9596148

Variance function: V(u) = u*(1-u/cases)            [Binomial]
Link function    : g(u) = -ln(-ln(u/cases))        [Log-log]

HAC kernel (lags): Anderson (2)                    BIC             = -127.4276

------------------------------------------------------------------------------
             |                HAC
         die |      Coef.   Std. Err.      z    P>|z|     [95% Conf. Interval]
-------------+----------------------------------------------------------------
    anterior |    .204143    .051312     3.98   0.000     .1035733    .3047126
       hcabg |   .2318145   .0610595     3.80   0.000       .11214     .351489
         kk2 |   .2523179   .0386794     6.52   0.000     .1765076    .3281281
         kk3 |   .3149235   .1678259     1.88   0.061    -.0140091    .6438562
         kk4 |    1.18085   .1188533     9.94   0.000     .9479016    1.413798
        age2 |    .104686   .0545669     1.92   0.055    -.0022632    .2116352
        age3 |   .4162827     .07014     5.94   0.000     .2788107    .5537546
        age4 |   .6921545   .0626031    11.06   0.000     .5694546    .8148544
       _cons |  -1.699495   .0675317   -25.17   0.000    -1.831855   -1.567135
------------------------------------------------------------------------------
```

GEE or random-effects models may be required. These are not post hoc methods. GEE methods incorporate an adjustment into the variance–covariance matrix, which is iteratively updated with new adjustment values. Adjustments are calculated for each cluster group and are summed across groups before entering the weight matrix. Random-effects models can also be used but require specification of a distribution for the random element causing extra dispersion. The *Stata User's Guide* describes these possibilities.

11.3 Williams' procedure

Another method that has been widely used to deal with overdispersed grouped logit models is Williams' procedure. Williams' procedure, Williams (1982), iteratively reduces the χ^2-based dispersion to approximately 1.0. The procedure uses an extra parameter, called ϕ, to scale the variance or weight function. It changes with each iteration. The value of ϕ, which results in a χ^2-based dispersion of 1.0, is then used to weight a standard grouped or proportion logistic regression. The weighting formula is $1/\{1+(k-1)\phi\}$, where k is the proportional denominator.

Pearson's χ^2 is defined as

$$\chi^2 = \sum_{i=1}^{n} \frac{(y_i-\mu_i)^2}{\mathrm{V}(\mu_i)} = \sum_{i=1}^{n} \frac{(y_i-\mu_i)^2}{\mu_i(1-\mu_i/k_i)} \tag{11.4}$$

Listing 11.1: Williams' procedure algorithm

```
1      φ = user input
2      else {φ = 1}
3      sc = 1 + (k − 1)φ
4      sc = 1  if φ = 1
5      μ = k(y + 0.5)/(k + 1)
6      η = ln(μ/(k − μ))
7      WHILE (abs(ΔDev) > tolerance)  {
8             w = (sc(μ)(k − μ))/k
9             z = η + (k(y − μ))/(sc(μ)(k − μ))
10            β = (x^T wx)^(−1) x^T wz
11            η = xβ
12            μ = k/(1 + exp(−η))
13            OldDev = Dev
14            Dev_i = k ln((k − y)/(k − μ))     if (y = 0)
15            Dev_i = y ln(y/μ)                  if (y = k)
16            Dev_i = y ln(y/μ) + (k − y) ln((k − y)/(k − μ)) if (y ≠ 0, y ≠ k)
17            Dev = 2∑((p)Dev_i)/sc
18            ΔDev = Dev - OldDev
19     }
20     χ^2 = ∑(y − μ)^2/(sc(μ)(1 − μ/k))
21     dof = #observations - #predictors
22     dispersion = χ^2/dof
23     * Note: Adjust φ so dispersion=1, then with final φ value:
24     w = 1/(1 + (k − 1)φ)
25     Rerun algorithm with prior weight equal to w and φ = 1
```

The user must input a value of ϕ in the above algorithm. To avoid delay in finding the optimal ϕ value to begin, we can use the algorithm below that iterates through the Williams' procedure, searching for the optimal scale value. We refer to the above algorithm simply as `williams`.

Listing 11.2: IRLS algorithm for Williams' procedure with optimal scale

```
1   Let ks = scale (sc above)
2   wt = weight
3   williams y k predictors
4   χ² = Σ(y − μ)²/(μ(1 − μ/k))
5   dof = #observations - #predictors
6   Disp = χ²/dof
7   φ = 1/Disp
8   WHILE (abs(ΔDisp) > tolerance) {
9       OldDisp = Disp
10      williams y k predictors, ks=φ
11      χ² = Σ((y − μ)²/(μ(1 − μ/k)))
12      Disp = χ²/dof
13      φ = Dispφ
14      ΔDisp = Disp - OldDisp
15  }
16  wt = 1/(1 + (k − 1)φ)
17  williams y k predictors [w=wt]
```

You can obtain Stata code (which we use below) for this model by using Stata's powerful net installation features. You must be connected to the Internet to use these commands.

```
. net from http://www.stata.com/users/jhilbe
. net install williams
```

Running this version of the Williams' procedure yields the following output:

```
. williams die cases anterior hcabg kk2 kk3 kk4 age2-age4, eform nolog
Resid DF   =       44                          No obs.      =         53
Chi2       =  38.25804                         Deviance     =   41.05846
Dispersion =  .8695009                         Prob>chi2    =   .5984348
                                               Dispersion   =   .9331467
CHI2(8)    =  90.77051
Prob>CHI2  =  3.24e-16                         Pseudo R2    =   .1279448
logistic regression: Williams' procedure
```

__00000I	OR	Std. Err.	t	P>|t|	[95% Conf.	Interval]
anterior	1.633152	.3783214	2.12	0.034	1.037152	2.571644
hcabg	1.970145	.733856	1.82	0.069	.9493615	4.088506
kk2	2.115705	.5606647	2.83	0.005	1.258586	3.556538
kk3	2.23946	.7389409	2.44	0.015	1.172935	4.275751
kk4	13.53077	5.290082	6.66	0.000	6.28821	29.11509
age2	1.563912	.6983679	1.00	0.317	.6517818	3.752516
age3	3.717232	1.482288	3.29	0.001	1.701334	8.121754
age4	6.52112	2.579006	4.74	0.000	3.003865	14.15676

```
Log Likelihood = -309.34012
Phi            =   .0080142
```

The estimates are different from those produced in the standard model. An extra parameter was put into the estimating algorithm itself, rather than being attached only to the standard errors as in the scaling method. And the final weighted logistic regression displays p-values based on t statistics per the recommendation of Collett (1991) on page 196.

The procedure changed the results only slightly, but this is not always the case. Standard errors and coefficients will change little when the standard model has a Pearson χ^2 statistic near 1. Here the standard model had a Pearson statistic of 1.25.

11.4 Robust standard errors

Parameter standard errors are derived from the variance–covariance matrix produced in the estimation process. As discussed in chapter 3, model standard errors are calculated as the square root of the matrix diagonal. If the model is fitted using Fisher scoring, then the variance–covariance matrix from which standard errors are produced is the matrix of expected second derivatives. This is called the expected information matrix. If, on the other hand, standard errors are based on maximum likelihood using a variety of Newton–Raphson algorithms, then we have the matrix of observed second derivatives. As we discovered earlier, the more complex observed information matrix reduces to the expected information for canonical linked models.

In either case, the source of the variance–covariance matrix must not be forgotten. It is the matrix of observed second derivatives of the log-likelihood function, which itself is derived from an underlying probability distribution.

One of the prime requisites of a log-likelihood function is that the individual observations defined by it be independent of one another. We assume that there is no correlation between observations. When there is, the reliability of the model is in question.

We have discussed various types of overdispersion—some apparent, some genuine. One type of genuine overdispersion occurs when there is correlation between observations. Such correlation comes from a variety of sources. Commonly, correlation among response observations comes from the clustering effect due to grouping in the dataset itself. For instance, suppose that we are attempting to model the probability of death on the basis of patient characteristics as well as certain types of medical procedures. Data come from several different facilities. Perhaps there are differences in how facilities tend to treat patients. Facilities may even prefer treating certain types of patients over other types. If there are treatment differences between hospitals, then there may well be extra correlation of treatment within hospitals, thus violating the independence-of-observations criterion. Treatment across facilities may be independent, but not within facilities. When this type of situation occurs, as it does so often in practice, then standard errors need to be adjusted.

Methods have been designed to adjust standard errors by a means that does not depend directly on the underlying likelihood of the model. One such method is that of redefining the weight such that "robust" standard errors are produced. As we will see, robust standard errors still have a likelihood basis, but the assumption of independence is relaxed. The method we discuss was first detailed by Huber (1967) and White (1980). It was introduced into Stata software and discussed by (a former student of Huber) Rogers (1992) and others. It plays an important role in GEE and in the many survey-specific algorithms. Robust variance estimators have only recently found favor in the biological and health sciences, but they are now popular in the area. In fact, they are probably overused.

Robust standard errors, and there are several versions of them (see sec. 3.6), can be understood as a measurement of the standard error of the calculated parameter if we indefinitely resample and refit the model. For instance, if we use a 95% confidence interval, we can interpret robust standard errors as follows:

> If the sampling of data were continually repeated and estimated, we would expect the collection of robust constructed confidence intervals to contain the "true" coefficient 95% of the time.

We place the quotes around "true" for the same reason that we mentioned that the robust variance estimate is probably overused. Although the robust variance estimate is robust to several violations of assumptions in the linear regression model, the same cannot be said in other circumstances. For example, the robust variance estimate applied to a probit model does not have the same properties. If the probit model is correctly specified, then the issue of robustness is not relevant, since either the usual Hessian matrix or the sandwich estimate may be used. The probit estimator itself is not consistent in the presence of heteroskedasticity, for example, and the sandwich estimate of variance is appropriate when the estimator converges in probability. With our example, the probit model may not converge to anything interesting. Using the robust variance estimate does not change this, and so the utility of the estimate is limited to inferring coverage of "true" parameters (parameters of the probit model) for the infinite population from which we obtained our sample. As Bill Sribney, one of the authors of Stata's survey commands, likes to say, "There are no free lunches!"

Part IV

Count Response Models

12 The Poisson family

Contents

12.1 Count response regression models

In previous chapters, we discussed regression models having either binary or grouped response data. We now turn to models having counts as a response. Included are the Poisson, geometric, and negative binomial regression models. We will discuss multinomial and ordinal models, which entail multiple levels of a response, in later chapters.

Counts refer to a simple counting of events, whether they are classified in tables or found as raw data. The model is named for the study presented in Poisson (1837). Examples of count response data include hospital length of stay (LOS), number of rats that died in an drug experiment, or the classic count of Prussian soldiers who died as a result of being kicked by horses in von Bortkewitsch (1898).

Count data may also take the form of a rate. In a manner analogous to the grouped or proportional binomial model, we may think of rate data as counts per unit of population, for example, the comparative incidence rate of death for people exposed to a particular virus in Arizona. Incidence rates will be computed for each county and a model will be developed to help understand comparative risks.

We begin the discussion of both the count and rate data with the Poisson model. It is a comparatively simple one-parameter model and has been used extensively in epidemiological studies. We discover, however, that it is too simple for many real-life data situations.

12.2 Derivation of the Poisson algorithm

The Poisson PDF can be formulated as

$$f(y;\mu) = e^{-\mu}\mu^y/y! \tag{12.1}$$

or in exponential-family form as

$$f(y;\mu) = \exp\{y\ln(\mu) - \mu - \ln\Gamma(y+1)\} \tag{12.2}$$

The link and cumulant functions are then derived as

$$\theta = \ln(\mu) \tag{12.3}$$
$$b(\theta) = \mu \tag{12.4}$$

The canonical link is found to be the log link. Hence, the inverse link is $\exp(\eta)$ where η is the linear predictor. The mean and variance functions are calculated as the first and second derivatives with respect to θ.

$$b'(\theta) = \frac{\partial b}{\partial \mu}\frac{\partial \mu}{\partial \theta} \tag{12.5}$$
$$= (1)(\mu) \tag{12.6}$$
$$= \mu \tag{12.7}$$
$$b''(\theta) = \frac{\partial^2 b}{\partial \mu^2}\left(\frac{\partial \mu}{\partial \theta}\right)^2 + \frac{\partial b}{\partial \mu}\frac{\partial^2 \mu}{\partial \theta^2} \tag{12.8}$$
$$= (0)(1)^2 + (\mu)(1) \tag{12.9}$$
$$= \mu \tag{12.10}$$

The mean and variance functions of the Poisson distribution are identical. As a prelude to future discussion, models having greater values for the variance than the mean are said to be overdispersed.

The derivative of the link is essential to the Poisson algorithm. It is calculated as

$$\frac{\partial \theta}{\partial \mu} = \frac{\partial\{\ln(\mu)\}}{\partial \mu} = \frac{1}{\mu} \tag{12.11}$$

The log-likelihood function can be abstracted from the exponential form of the distribution by removing the initial exponential and associated braces.

$$\mathcal{L}(\mu;y) = \sum_{i=1}^{n}\{y_i\ln(\mu_i) - \mu_i - \ln\Gamma(y_i+1)\} \tag{12.12}$$

When the response has the value of zero, the individual log-likelihood functions reduce to

$$\mathcal{L}_i(\mu_i; 0) = -\mu_i \tag{12.13}$$

The log-likelihood function can also be parameterized in terms of $\mathbf{x}\boldsymbol{\beta}$. This is required for all maximum likelihood methods. Substituting the inverse link, $\exp(\mathbf{x}\boldsymbol{\beta})$, for each instance of μ in the GLM parameterization above,

$$\begin{aligned} \mathcal{L}(\mathbf{x}\boldsymbol{\beta}; y) &= \sum_{i=1}^{n} [y_i \ln\{\exp(\mathbf{x}_i\boldsymbol{\beta})\} - \exp(\mathbf{x}_i\boldsymbol{\beta}) - \ln\Gamma(y_i+1)] && (12.14) \\ &= \sum_{i=1}^{n} \{y_i(\mathbf{x}_i\boldsymbol{\beta}) - \exp(\mathbf{x}_i\boldsymbol{\beta}) - \ln\Gamma(y_i+1)\} && (12.15) \\ \mathcal{L}_i(\mathbf{x}\boldsymbol{\beta}; y_i = 0) &= -\exp(\mathbf{x}_i\boldsymbol{\beta}) && (12.16) \end{aligned}$$

The deviance function, required for the traditional GLM algorithm, is defined as $2\{\mathcal{L}(y; y) - \mathcal{L}(\mu; y)\}$. It is derived as

$$\begin{aligned} D &= 2\sum_{i=1}^{n} \{y_i \ln(y_i) - y_i - y_i \ln(\mu_i) + \mu_i\} && (12.17) \\ &= 2\sum_{i=1}^{n} \left\{y_i \ln\left(\frac{y_i}{\mu_i}\right) - (y_i - \mu_i)\right\} && (12.18) \end{aligned}$$

Again, when the response is zero, the deviance function reduces to

$$D_i(y_i = 0) = 2\mu_i \tag{12.19}$$

We can now enter the necessary functions into the basic IRLS GLM algorithm.

(*Continued on next page*)

Listing 12.1: IRLS algorithm for Poisson regression

```
1    Dev = 0
2    μ = (y − mean(y))/2            /* initialization of μ */
3    η = ln(μ)                      /* initialization of η */
4    WHILE (Abs(ΔDev) > tolerance ) {
5        W = μ
6        z = η + (y − μ)/μ − offset
7        β = (X^T W X)^{-1} X^T W z
8        η = Xβ + offset
9        μ = exp(η)
10       OldDev = Dev
11       Dev = 2∑(y ln(y/μ) − (y − μ)) if (y ≥ 1)
12       Dev = 2∑μ if (y = 0)
13       ΔDev = Dev - OldDev
14   }
15   χ^2 = ∑(y − μ)^2/μ
```

The log-likelihood function entails using the log-gamma function. Most commercial software packages have the log-gamma as a user-accessible function. However, the traditional GLM algorithm favors excluding the log likelihood except when used to calculate certain fit statistics. Some software packages have the algorithm iterate on the basis of the log-likelihood function, just as we do here on the deviance. For example, Stata's `poisson` command iterates on the basis of log-likelihood difference. The results of iteration based on the deviance or the log likelihood will be identical. However, some programs (not Stata), when using the log likelihood to iterate, drop the log-gamma term from the function. Usually, results vary only slightly, though the reported value of the log likelihood will differ.

When the deviance is used as the basis of iteration, the log-likelihood functions can be used to construct a (likelihood ratio) χ^2 GOF statistic and pseudo-R^2 given in (4.46) formula defined by

$$\chi^2 = -2\left({}_0 - {}_F \right) \tag{12.20}$$

$$R^2 = 1 - \frac{{}_F}{{}_0} \tag{12.21}$$

where ${}_0$ is the value of the log likelihood for a constant-only model and ${}_F$ is the final or full-value function. Different versions exist for both χ^2 and R^2 functions. One should always check the software manual to determine which formulas are being implemented. As we outlined in section 4.6.1, there are several definitions for R^2. From the list in that section, the Cameron–Windmeijer definition is the preferred one. Calculating

$$G^2 = 2\sum_{i=1}^{n} y_i \ln(y_i/\widehat{\lambda}_i) \tag{12.22}$$

as a GOF statistic is also common. This statistic is zero for a perfect fit.

The above algorithm varies from the standard algorithm we have been using thus far. We have added an offset to the algorithm. In fact, previous GLM algorithms could well have had offsets. We chose not to put them in until now because they are rarely used for anything but Poisson and negative binomial algorithms. Sometimes offsets are used with clog-log data, but this is not the ordinary case.

John Nelder first conceived of offsets as an afterthought to the IRLS algorithm that he and Wedderburn designed in 1972. The idea began as a method to put a constant term directly into the linear predictor without that term's being estimated. It affects the algorithm directly before and after regression estimation. Only later did Nelder discover that the notion of an offset could be useful for modeling rate data.

An offset must be put into the same metric as the linear predictor, η. Thus an offset enters into a (canonical link) Poisson model as the natural log of the variable. We must log-transform the variable to be used as an offset before inserting it into the estimation algorithm unless the software does this automatically or unless the software has a log option, like Stata (see documentation on `lnoffset`).

Epidemiologists interpret the notion of offset as exposure, for instance, the number of people who are exposed to a disease. The incidence, or number of people who suffer some consequence, enter as the response. Hence, we can think of the Poisson model, parameterized as a rate, as modeling incidence over exposure.

The Poisson mean and variance functions are identical. This means that there is no variation in how counts enter into the group by which they are defined. We demonstrate using a simple example illustrated in figure 12.1.

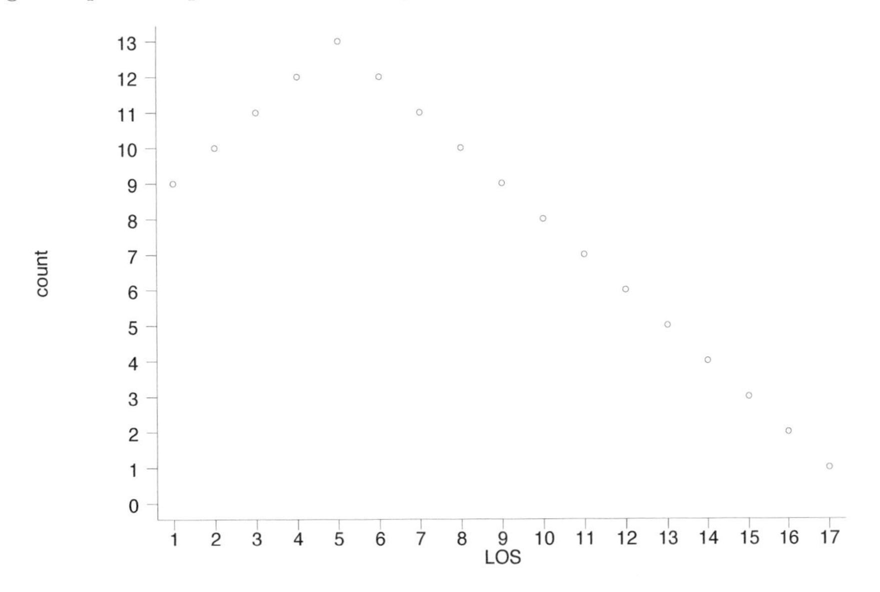

Figure 12.1: Frequency of occurrence versus LOS

Suppose that the circles or dots represent the number of patients that are discharged from the hospital after surgery. If a patient enters the facility, has an operation, and is discharged the same day, he or she is still credited with a length of stay of 1 day. We see that nine patients were discharged on the first day, 10 on the second, 11 on the third, and so on, until we have only one patient discharged on day 17.

Let us think of each day as a cell. Patients can be discharged at any time within the cell. The Poisson model assumes that they are uniformly discharged. There is no preference for the morning, afternoon, or evening. It makes no difference. The model is "blind" to any variation other than a uniform random time of discharge. Even though we are oversimplifying the notion with this example, this is essentially what it means to have an identity of mean and variance.

We know from experience that most patients are typically discharged in the morning. A few may be discharged at later times. If there is defined variability of discharge time within a cell or day, then we have Poisson overdispersion. The variance is greater than the mean. Moreover, if we can assume that the shape of discharge counts within each cell itself takes a gamma distribution, then we have a Poisson–gamma mixture model, a form of which is called the negative binomial distribution. We discuss it thoroughly in the next chapter.

Returning to the example data, we find that there is no possibility that 0 can be a value for the response hospital length of stay. Day 1, representing a count of 1, begins the day that a patient is admitted to the hospital. To be included in the dataset, the patient must have a count of 1 or greater.

The Poisson model assumes that there is the possibility of 0 counts even if there are none in the data. When the possibility of a 0 count is precluded, as in our example, then the Poisson model is not appropriate. A zero-truncated Poisson model should be used, which adjusts the Poisson probability appropriately to exclude 0 counts. To do this, subtract the probability of a Poisson 0 count, $\exp(-\mu)$, from 1, and condition this event out of the model. The resulting log likelihood is given by

$$\mathcal{L}(\mu; y|y>0) = \sum_{i=1}^{n} \left(y_i(\mathbf{x}_i\boldsymbol{\beta}) - \exp(\mathbf{x}_i\boldsymbol{\beta}) - \ln\Gamma(y_i+1) - \ln\left[1-\exp\{-\exp(\mathbf{x}_i\boldsymbol{\beta})\}\right]\right) \tag{12.23}$$

This expression amounts to apportioning the probability of the 0 count to the remaining counts (weighted by their original probability).

We mention this non-GLM model because such data are often inappropriately modeled using unadjusted Poisson regression. Even so, experience has shown us that estimates and standard errors do not differ greatly between the two models, particularly when there are many observations in the data. Stata has built-in commands for such models; see the documentation for `ztp` and `ztnb` commands. We discuss this model more and give examples in chapter 14.

12.3 Poisson regression: Examples

Let us use an example from the health field. We model length of stay (LOS), adjusted by several predictors, including

`hmo`	binary variable coding membership in an HMO
`white`	binary variable coding (self-reported) white race
`type1`	binary variable coding elective admission
`type2`	binary variable coding urgent admission
`type3`	binary variable coding emergency admission

`type` is a factor variable with three levels. It was separated into three factors before estimation. We will have `type1` serve as the reference.

All hospitals in this dataset are in Arizona and have the same diagnostic related group (DRG). Patient data have been randomly selected to be part of this dataset.

```
. use http://www.stata-press.com/data/hh2/medpar
. glm los hmo white type2 type3, family(poisson) eform nolog
Generalized linear models                          No. of obs      =       1495
Optimization     : ML                              Residual df     =       1490
                                                   Scale parameter =          1
Deviance         =  8142.666001                    (1/df) Deviance =   5.464877
Pearson          =  9327.983215                    (1/df) Pearson  =   6.260391

Variance function: V(u) = u                        [Poisson]
Link function    : g(u) = ln(u)                    [Log]

                                                   AIC             =   9.276131
Log likelihood   = -6928.907786                    BIC             =  -2749.057

------------------------------------------------------------------------------
             |                 OIM
         los |        IRR   Std. Err.      z    P>|z|     [95% Conf. Interval]
-------------+----------------------------------------------------------------
         hmo |   .9309504   .0222906    -2.99   0.003     .8882708    .9756806
       white |   .8573826   .0235032    -5.61   0.000     .8125327     .904708
       type2 |   1.248137   .0262756    10.53   0.000     1.197685    1.300713
       type3 |   2.032927   .0531325    27.15   0.000     1.931412    2.139778
------------------------------------------------------------------------------
```

We requested that the software provide incidence-rate ratios rather than coefficients. Incidence-rate ratios are interpreted similarly to odds ratios, except that the response is a count. We can interpret the IRR for `type2` and `type3` as the following:

Urgent admissions are expected to have a LOS that is 25% longer than elective admissions. Emergency admissions stay twice as long as elective admissions.

We can also see that HMO patients stay in the hospital for slightly less time than do non-HMO patients, 7% less time, and nonwhite patients stay in the hospital about 14% longer than do white patients.

At first glance, each predictor appears to have a statistically significant relationship to the response, adjusted by other predictors. However, by checking the deviance-based dispersion, we find substantial overdispersion in the data. The dispersion statistic is 5.46, much higher than it should be for a viable Poisson model.

There are two approaches to handing the data at this point. We can simply scale the standard errors by the dispersion, adjusting standard errors to a value that would be the case if there were no dispersion in the data, i.e., a dispersion of 1 . We may also seek to discover the source of the overdispersion and take more directed measures to adjust for it. First, we scale standard errors.

```
. glm los hmo white type2 type3, family(poisson) irls scale(dev) eform nolog
> noheader
```

los	IRR	EIM Std. Err.	z	P>|z|	[95% Conf.	Interval]
hmo	.9309504	.052109	-1.28	0.201	.8342215	1.038895
white	.8573826	.0549437	-2.40	0.016	.7561833	.9721253
type2	1.248137	.0614248	4.50	0.000	1.13337	1.374524
type3	2.032927	.1242082	11.61	0.000	1.803495	2.291547

```
(Standard errors scaled using square root of deviance-based dispersion)
```

The standard errors have markedly changed. `hmo` now appears to be noncontributory to the model. Scaling by the χ^2 dispersion yields similar results.

Provider type may have bearing on LOS. Some facilities may tend to keep patients longer than others. We now attempt to see whether overdispersion is caused by extra correlation due to provider effect.

```
. glm los hmo white type2 type3, family(poisson) cluster(provnum) eform nolog
> noheader
                          (Std. Err. adjusted for 54 clusters in provnum)
```

los	IRR	Robust Std. Err.	z	P>|z|	[95% Conf.	Interval]
hmo	.9309504	.0490889	-1.36	0.175	.8395427	1.03231
white	.8573826	.0625888	-2.11	0.035	.7430825	.9892642
type2	1.248137	.0760289	3.64	0.000	1.107674	1.406411
type3	2.032927	.4126821	3.49	0.000	1.365617	3.02632

Again we see somewhat similar results—`hmo` does not contribute. There indeed appears to be an excess correlation of responses within providers. That is, LOS is more highly correlated within providers, given particular patterns of covariates, than between providers. This correlation accounts for overdispersion, but we know nothing more about the nature of the overdispersion.

Finally, we show an example of a rate model. You can find this example in the *Stata Base Reference Manual* for Poisson regression, and it comes from Doll and Hill (1966). The data deal with coronary deaths among British physicians due to smoking background. Person years is entered as an offset. We let the software log `pyears` prior to estimation. The variables `a1`–`a5` represent 10-year-age categories, starting with age 35 and ending at age 84.

```
. use http://www.stata-press.com/data/hh2/doll, clear
. glm deaths smokes a2-a5, family(poisson) lnoffset(pyears) eform nolog
Generalized linear models                   No. of obs      =         10
Optimization     : ML                       Residual df     =          4
                                            Scale parameter =          1
Deviance         =  22.44834464             (1/df) Deviance =   5.612086
Pearson          =  19.12819936             (1/df) Pearson  =    4.78205
Variance function: V(u) = u                 [Poisson]
Link function    : g(u) = ln(u)             [Log]
                                            AIC             =   8.951629
Log likelihood   = -38.75814256             BIC             =     13.238

------------------------------------------------------------------------------
             |                 OIM
      deaths |        IRR   Std. Err.      z    P>|z|     [95% Conf. Interval]
-------------+----------------------------------------------------------------
      smokes |   1.126808   .1088536     1.24   0.217     .9324403    1.361693
          a2 |   4.470665   .8722725     7.68   0.000     3.049959    6.553153
          a3 |   13.56641   2.551449    13.86   0.000       9.3838    19.61333
          a4 |   29.07069   5.372727    18.23   0.000      20.2367      41.761
          a5 |     40.855   7.853267    19.30   0.000     28.03011    59.54779
      pyears | (exposure)
------------------------------------------------------------------------------
```

We model a main effects model only. The *Stata Base Reference Manual* explores interactions, finding several significant contributions to the model.

The age categories all appear to significantly contribute to the model. In fact, we see a rather abrupt increase in number of deaths as the age category increases. However, once again we find overdispersion. Note the deviance-based dispersion statistic of 5.61. We may then attempt to deal with the overdispersion by scaling standard errors.

```
. glm deaths smokes a2-a5, family(poisson) irls scale(dev) lnoffset(pyears)
> eform nolog noheader

------------------------------------------------------------------------------
             |                 EIM
      deaths |        IRR   Std. Err.      z    P>|z|     [95% Conf. Interval]
-------------+----------------------------------------------------------------
      smokes |   1.126808   .2578725     0.52   0.602     .7195331    1.764613
          a2 |   4.470665   2.066397     3.24   0.001     1.806909    11.06135
          a3 |   13.56641   6.044333     5.85   0.000      5.66531    32.48677
          a4 |   29.07069   12.72789     7.70   0.000     12.32473    68.56982
          a5 |     40.855   18.60424     8.15   0.000     16.73522    99.73761
      pyears | (exposure)
------------------------------------------------------------------------------
(Standard errors scaled using square root of deviance-based dispersion)
```

We find that smoking is not significant at a significance level of .05. A likelihood-ratio test provides the same result. Age categories remain significant.

In chapter 11, we detailed the various components to overdispersion. We pointed out the difference between apparent and inherent overdispersion. This is a case of the former. By creating various interaction terms based on smoking and age category, the model loses its apparent overdispersion. Smoking risk is associated with different levels of age. Understanding `a` as age-category and `s` as smoking, we can create the variables for the following interactions. The variable sa12 indicates smokers in age groups a1 and a2, sa34 indicates smokers in age groups a3 and a4, and sa5 indicates smokers in age group a5. We can model the data with these covariates by using

```
. gen sa12 = smokes*(a1+a2)
. gen sa34 = smokes*(a3+a4)
. gen sa5  = smokes*(a5)
. glm deaths sa12 sa34 sa5 a2-a5, family(poisson) irls scale(dev)
> lnoffset(pyears) eform nolog noheader
```

deaths	IRR	EIM Std. Err.	z	P>|z|	[95% Conf.	Interval]
sa12	2.636259	1.500935	1.70	0.089	.8636994	8.046623
sa34	.958517	.2456049	-0.17	0.869	.5800866	1.583824
sa5	.9047304	.3759254	-0.24	0.810	.4007151	2.042691
a2	4.294559	1.69886	3.68	0.000	1.977865	9.324822
a3	33.062	22.85543	5.06	0.000	8.529077	128.1611
a4	67.0972	44.28982	6.37	0.000	18.40091	244.6636
a5	97.87965	69.50933	6.45	0.000	24.33385	393.7078
pyears	(exposure)					

(Standard errors scaled using square root of deviance-based dispersion)

Scaling produces no noticeable effect on standard errors, thus demonstrating that the model is appropriate for the data and is not truly overdispersed.

The moral of our discussion is that we must be constantly aware of overdispersion when dealing with count data. If the model appears to be overdispersed, we must seek out the source of overdispersion. Maybe interactions are required or predictors need to be transformed. If the model still shows evidence of overdispersion, then scaling or applying robust variance estimators may be required. Perhaps other types of models need to be applied. The negative binomial would be a logical alternative.

12.4 Example: Testing overdispersion in the Poisson model

As we have mentioned in this chapter, a shortcoming of the Poisson model is its assumption that the variance is equal to the mean. We can use a regression-based test for this assumption by following the procedure shown below:

1. Obtain the fitted values $\widehat{\mu}_i$.

2. Calculate

$$z = \frac{(y_i - \widehat{\mu}_i)^2 - y_i}{\widehat{\mu}_i\sqrt{2}}$$

3. Regress z as a constant-only model.

The test of the hypotheses

$$\begin{aligned} H_0\colon \mathrm{V}(y) &= \mathrm{E}(y) \\ H_1\colon \mathrm{V}(y) &= \mathrm{E}(y) + \alpha g\{\mathrm{E}(y)\} \end{aligned}$$

is carried out by the t test of the constant in the regression.

Returning to our example, we may assess the overdispersion in our model by using the following procedure:

```
. use http://www.stata-press.com/data/hh2/medpar, clear
. glm los hmo white type2 type3, family(poisson) eform nolog noheader

------------------------------------------------------------------------------
             |                 OIM
         los |        IRR   Std. Err.      z    P>|z|     [95% Conf. Interval]
-------------+----------------------------------------------------------------
         hmo |   .9309504   .0222906    -2.99   0.003     .8882708    .9756806
       white |   .8573826   .0235032    -5.61   0.000     .8125327     .904708
       type2 |   1.248137   .0262756    10.53   0.000     1.197685    1.300713
       type3 |   2.032927   .0531325    27.15   0.000     1.931412    2.139778
------------------------------------------------------------------------------

. predict double mu, mu
. generate z = ((los-mu)^2-los) / (mu*sqrt(2))
. regress z
      Source |       SS       df       MS              Number of obs =    1495
-------------+------------------------------           F(  0,  1494) =    0.00
       Model |           0     0           .           Prob > F      =       .
    Residual |  348013.947  1494  232.941062           R-squared     =  0.0000
-------------+------------------------------           Adj R-squared =  0.0000
       Total |  348013.947  1494  232.941062           Root MSE      =  15.262

------------------------------------------------------------------------------
           z |      Coef.   Std. Err.      t    P>|t|     [95% Conf. Interval]
-------------+----------------------------------------------------------------
       _cons |   3.704561   .3947321     9.39   0.000     2.930273    4.478849
------------------------------------------------------------------------------
```

The hypothesis of no overdispersion is rejected. This is not surprising, since we have already seen evidence that the overdispersion exists.

There are other tests for overdispersion in the Poisson model, including a Lagrange multiplier test given by

$$\chi^2 = \frac{(\sum_i \widehat{\mu}_i^2 - n\overline{y})^2}{2\sum_i \widehat{\mu}_i^2} \tag{12.24}$$

This test is distributed χ^2 with 1 degree of freedom. We can continue our use of Stata to compute the test:

```
. quietly summarize los, meanonly
. quietly scalar nybar = r(sum)
. quietly generate double musq = mu*mu
. quietly summarize musq, meanonly
. quietly scalar mu2 = r(sum)
```

```
. quietly scalar chival = (mu2-nybar)^2/(2*mu2)
. display as txt "LM value = " as res chival _n as txt
> "P-value  = " as res %6.4f      chi2tail(1, chival)
LM value = 62987.844
P-value  = 0.0000
```

This approach yields the same conclusions.

12.5 Using the Poisson model for survival analysis

In section 6.6, we used the gamma model for survival analysis. Although the gamma model cannot incorporate censoring into its routine, a right censored exponential regression model may be duplicated with GLM by using the Poisson family. We simply log-transform the time response, incorporating it into the model as an offset. The new response is then the 0/1 censor variable. Using the Stata `cancer` dataset, we compare the two methods.

```
. sysuse cancer, clear
(Patient Survival in Drug Trial)
. tab drug, gen(drug)
  (output omitted)
. glm died age drug2 drug3, lnoffset(studytime) family(poisson) nolog
Generalized linear models                       No. of obs      =         48
Optimization     : ML                           Residual df     =         44
                                                Scale parameter =          1
Deviance         =  34.33673608                 (1/df) Deviance =   .7803804
Pearson          =  42.74908562                 (1/df) Pearson  =   .9715701
Variance function: V(u) = u                     [Poisson]
Link function    : g(u) = ln(u)                 [Log]
                                                AIC             =   2.173682
Log likelihood   = -48.16836804                 BIC             =  -135.9961

------------------------------------------------------------------------------
             |                 OIM
        died |      Coef.   Std. Err.      z    P>|z|     [95% Conf. Interval]
-------------+----------------------------------------------------------------
         age |   .0854187   .0334969     2.55   0.011     .0197659    .1510716
       drug2 |  -1.461711   .4761143    -3.07   0.002    -2.394878   -.5285437
       drug3 |  -1.857458   .4683077    -3.97   0.000    -2.775324   -.9395916
       _cons |  -6.962982   1.899393    -3.67   0.000    -10.68572   -3.240241
   studytime |  (exposure)
------------------------------------------------------------------------------
```

```
. stset studytime died

     failure event:  died != 0 & died < .
obs. time interval:  (0, studytime]
 exit on or before:  failure

------------------------------------------------------------------------------
       48  total obs.
        0  exclusions
------------------------------------------------------------------------------
       48  obs. remaining, representing
       31  failures in single record/single failure data
      744  total analysis time at risk, at risk from t =         0
                             earliest observed entry t =         0
                                  last observed exit t =        39

. streg age drug2 drug3, nohr dist(exp) nolog

         failure _d:  died
   analysis time _t:  studytime

Exponential regression -- log relative-hazard form

No. of subjects =           48                     Number of obs   =        48
No. of failures =           31
Time at risk    =          744
                                                   LR chi2(3)      =     26.35
Log likelihood  =   -48.168368                     Prob > chi2     =    0.0000

------------------------------------------------------------------------------
          _t |      Coef.   Std. Err.      z    P>|z|     [95% Conf. Interval]
-------------+----------------------------------------------------------------
         age |   .0854187   .0334969     2.55   0.011     .0197659    .1510716
       drug2 |  -1.461711   .4761143    -3.07   0.002    -2.394878   -.5285437
       drug3 |  -1.857458   .4683077    -3.97   0.000    -2.775324   -.9395916
       _cons |  -6.962982   1.899393    -3.67   0.000    -10.68572   -3.240241
------------------------------------------------------------------------------
```

The scale value produced by the Poisson model is nearly 1, which is to be expected. Theoretically, the exponential distribution has a scale of 1. In that, it is a subset of the gamma distribution. The `streg` model displayed above is parameterized in terms of a log hazard rate (as opposed to log-time parameterization), hence the `nohr` option. Log-time rates would reverse the signs of the parameter estimates.

12.6 Using offsets to compare models

If we wish to test a subset of coefficients, we can fit a model and obtain a Wald test. Stata provides this functionality through the `test` command. For example, if we wish to test that H_0: $\beta_{\text{age}} = 0$ and $\beta_{\text{drug2}} = -1.5$, for the survival model presented in section 12.5, we first run the model

(*Continued on next page*)

```
. glm died age drug2 drug3, lnoffset(studytime) family(poisson) noheader nolog
```

```
------------------------------------------------------------------------------
             |                 OIM
        died |      Coef.   Std. Err.      z    P>|z|     [95% Conf. Interval]
-------------+----------------------------------------------------------------
         age |   .0854187   .0334969     2.55   0.011     .0197659    .1510716
       drug2 |  -1.461711   .4761143    -3.07   0.002    -2.394878   -.5285437
       drug3 |  -1.857458   .4683077    -3.97   0.000    -2.775324   -.9395916
       _cons |  -6.962982   1.899393    -3.67   0.000    -10.68572   -3.240241
   studytime | (exposure)
------------------------------------------------------------------------------
```

To test our hypothesis using a Wald test, we use the `test` command.

```
. test age=0, notest
 ( 1)  [died]age = 0
. test drug2=-1.5, accumulate
 ( 1)  [died]age = 0
 ( 2)  [died]drug2 = -1.5
           chi2(  2) =    6.80
         Prob > chi2 =    0.0333
```

If we prefer a likelihood-ratio test, we can generate an offset that reflects our constraints and fit the model. Since our model already includes an offset on the log scale, we must combine all our constraints into one variable. First, we tell Stata to save the log likelihood from the full model.

```
. estimates store Unconstrained
```

Next we create an offset reflecting the constraints that we wish to impose on the model.

```
. generate double offvar = 0*age - 1.5*drug2 + log(studytime)
```

Finally, we can fit the constrained model by specifying our new variable as the offset and then asking for the likelihood-ratio test of our constraints.

```
. glm died drug3, offset(offvar) family(poisson) nolog
Generalized linear models                          No. of obs      =        48
Optimization     : ML                              Residual df     =        46
                                                   Scale parameter =         1
Deviance         =  40.83400585                    (1/df) Deviance =  .8876958
Pearson          =  57.71533028                    (1/df) Pearson  =  1.254681
Variance function: V(u) = u                        [Poisson]
Link function    : g(u) = ln(u)                    [Log]
                                                   AIC             =  2.225708
Log likelihood   = -51.41700292                    BIC             = -137.2412

------------------------------------------------------------------------------
             |                 OIM
        died |      Coef.   Std. Err.      z    P>|z|     [95% Conf. Interval]
-------------+----------------------------------------------------------------
       drug3 |  -1.875897   .4546061    -4.13   0.000    -2.766908   -.9848854
       _cons |  -2.204461         .2   -11.02   0.000    -2.596454   -1.812469
      offvar |   (offset)
------------------------------------------------------------------------------
```

```
. lrtest Unconstrained
Likelihood-ratio test                               LR chi2(2)  =       6.50
(Assumption: . nested in Unconstrained)             Prob > chi2 =     0.0388
```

Both tests yield the conclusion that the full model is significantly different from the constrained model. The likelihood-ratio test can be verified, by hand, using the listed values of the log likelihood from each model $[6.5 = -2\{(-51.417) - (-48.168)\}]$.

12.7 Interpretation of coefficients

Interpreting the coefficients is hard because of the nonlinearity of the link function. We motivate an alternative metric that admits a transformation of the coefficients for easier interpretation for the canonical log link.

For illustration, we assume a model with two covariates $\mathbf{x}_1$ and $\mathbf{x}_2$ along with a constant. We focus on the interpretation of β_1.

Since the inverse canonical link is nonlinear, interpreting the coefficient is difficult.

$$\Delta y_i = \exp\{\beta_0 + (x_{1i}+1)\beta_1 + x_{2i}\beta_2\} - \exp(\beta_0 + x_{1i}\beta_1 + x_{2i}\beta_2) \quad (12.25)$$

The difficulty arises because the difference is not constant and depends on the values of the covariates.

Instead of focusing on the difference in the outcome, we define a measure of the incidence-rate ratio as the rate of change in the outcome (incidence)

$$\begin{aligned} \text{Incidence} - \text{rate ratio for } \mathbf{x}_1 &= \frac{\exp\{\beta_0 + (x_{1i}+1)\beta_1 + x_{2i}\beta_2\}}{\exp(\beta_0 + x_{1i}\beta_1 + x_{2i}\beta_2)} && (12.26) \\ &= \exp(\beta_1) && (12.27) \end{aligned}$$

The calculation of the incidence-rate ratio simplifies such that there is no dependence on a particular observation. The incidence-rate ratio is therefore constant—it is independent of the particular values of the covariates. This interpretation does not hold for other inverse links.

Interpretation for canonical Poisson-family models is straightforward for exponentiated coefficients. These exponentiated coefficients represent incidence-rate ratios. An incidence-rate ratio of 2 indicates that there is twice the incidence (the outcome is twice as prevalent) if the associated covariate is increased by one.

13 The negative binomial family

Contents

The negative binomial is the only GLM distribution, supported by Stata's `glm` command, with an ancillary parameter, τ. Unlike the scale parameters associated with Gaussian, gamma, and inverse Gaussian distributions, τ is not conceived as a dispersion statistic. When τ is equal to 1, the negative binomial reduces to the geometric distribution. The geometric is the discrete analog to the continuous negative exponential distribution, and the negative binomial is the discrete analog to the continuous gamma distribution. The exponential model can be fitted using log-gamma regression with the scale constrained to 1. Likewise, the geometric model can be fitted using negative binomial regression with the ancillary scale parameter τ set to 1.

In the previous chapter we illustrated the Poisson distribution in a graph with counts entering cells at a uniform rate. We mentioned the situation where counts may enter each cell with a predefined gamma shape. This was the Poisson–gamma mixture or contagion distribution model, or negative binomial regression. Figure 13.1 shows counts from 6–8. Instead of counts entering uniformly, as with the Poisson assumption, we see counts entering with a specific gamma-distributed shape. This, then, is how we may simplify the concept of a negative binomial process.

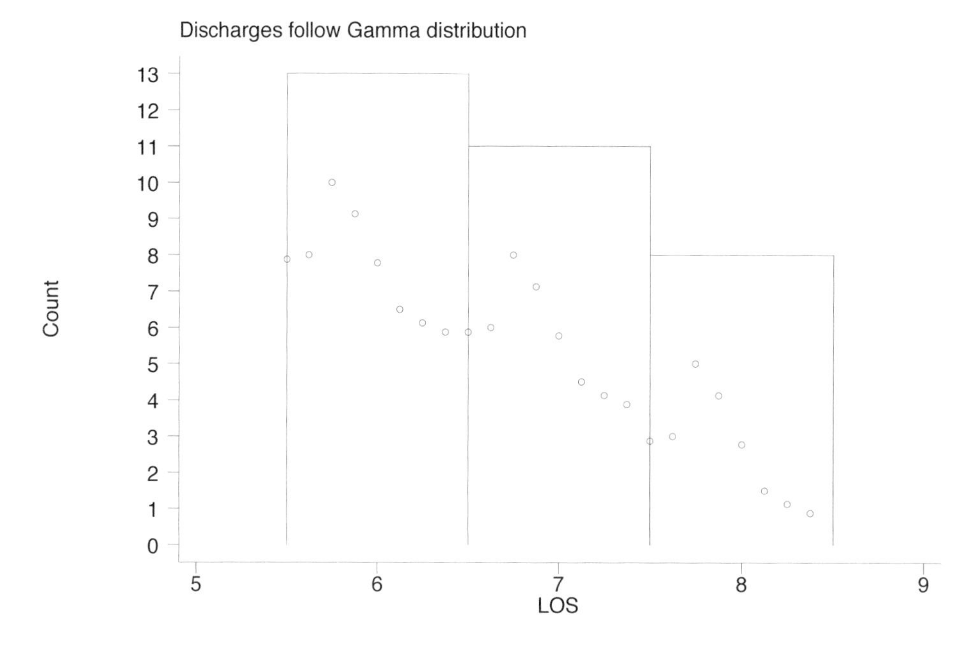

Figure 13.1: Frequency of occurrence versus LOS (with gamma mixing)

The negative binomial has long been recognized as a full member of the exponential family of distributions. Likewise, the geometric distribution is listed in nearly every source as an exponential family member. However, until recently neither the geometric nor the negative binomial was thought to be suitable for entry into the mainstream family of GLMs because of the complexity of the ancillary parameter. Strictly speaking, the negative binomial distribution is a member of the two-parameter exponential family and we allow it in this discussion by assuming that the ancillary parameter is fixed rather than stochastic.

The negative binomial regression model seems to have been first discussed in 1949 by Anscombe. Others have mentioned it, pointing out its usefulness for dealing with overdispersed count data. Lawless (1987) detailed the mixture model parameterization of the negative binomial, providing formulas for its log likelihood, mean, variance, and moments. Later, Breslow (1990) cited Lawless' work while manipulating the Poisson model to fit negative binomial parameters. From its inception to the late 1980s, the negative binomial regression model has been construed as a mixture model that is useful for accommodating otherwise overdispersed Poisson data. The models were fitted either using maximum likelihood or as an adjusted Poisson model.

McCullagh and Nelder (1989) gave brief attention to the negative binomial but did recognize that, with a fixed value for τ, it could be considered a GLM. The authors mentioned the existence of a canonical link, breaking away from the strict mixture model concept, but they did not proceed beyond a paragraph explanation.

Not until the mid 1990s was the negative binomial construed to be a full-fledged member of the GLM family. Hilbe (1993b) detailed how both the GLM-based negative binomial and geometric regression models could be derived from their respective probability distributions. The idea behind the arguments is that the model that is traditionally called the negative binomial is, within the GLM framework, a log-linked negative binomial. The canonical link, although usable, has properties that often result in nonconvergence. This is particularly the case when the data are highly unbalanced and occurs because the ancillary parameter is a term in both the link and inverse link functions. This problem, as well as how the canonical model is to be interpreted, spurred the creation of a SAS log-negative binomial and geometric regression macro and the inclusion of the negative binomial within the Stata `glm` command. It was also made part of the XploRe statistical package. The negative binomial is now widely available in GLM programs and is used to model overdispersed count data in many disciplines.

The negative binomial, when fitted using maximum likelihood, is almost always based on a Poisson–gamma mixture model. This is the log-linked version of the GLM negative binomial and is the default parameterization for the `nbreg` command. The negative binomial, as mentioned above, may have other links, including the canonical and the identity. Except for demonstrating how these links are constructed, though, we spend little time with them here.

There are two methods for motivating the negative binomial regression model, and we present them both for completeness. Only one derivation, however, fits within the GLM framework. Both methods are parameterizations of a Poisson–gamma mixture and yield what the literature has begun to distinguish as the NB-1 (constant overdispersion) and the NB-2 (variable overdispersion) regression models.

13.1 Constant overdispersion

In the first derivation, we consider a Poisson–gamma mixture in which

$$y_i = e^{-\lambda_i}\lambda_i^{y_i}/y_i! \tag{13.1}$$

where the Poisson parameter λ_i is a random variable. The λ_i are distributed $\Gamma(\delta, \mu_i)$, where

$$\mu_i = e^{\mathbf{x}_i\boldsymbol{\beta}+\text{offset}_i} \tag{13.2}$$

As such, the expected value of the Poisson parameter is

$$\text{E}[\lambda_i] = e^{\mathbf{x}_i\boldsymbol{\beta}+\text{offset}_i}/\delta \tag{13.3}$$

and the variance is

$$\text{V}[\lambda_i] = e^{\mathbf{x}_i\boldsymbol{\beta}+\text{offset}_i}/\delta^2 \tag{13.4}$$

The resulting mixture distribution is then derived as

$$
\begin{aligned}
f(y_i|x_i) &= \int_0^\infty \frac{e^{-(\lambda_i)}(\lambda_i)^{y_i}}{y_i!} \frac{\delta^{\mu_i}}{\Gamma(\mu_i)} \lambda_i^{\mu_i-1} e^{-\lambda_i \delta} d\lambda_i & (13.5)\\
&= \frac{\delta^{\mu_i}}{\Gamma(y_i+1)\Gamma(\mu_i)} \int_0^\infty \lambda_i^{(y_i+\mu_i)-1} e^{-\lambda_i(\delta+1)} d\lambda_i & (13.6)\\
&= \frac{\delta^{\mu_i}}{\Gamma(y_i+1)\Gamma(\mu_i)} \frac{\Gamma(y_i+\mu_i)}{(\delta+1)^{y_i+\mu_i}} \\
&\quad \int_0^\infty \frac{(\delta+1)^{y_i+\mu_i}}{\Gamma(y_i+\mu_i)} \lambda_i^{(y_i+\mu_i)-1} e^{-\lambda_i(\delta+1)} d\lambda_i & (13.7)\\
&= \frac{\delta^{\mu_i}}{\Gamma(y_i+1)\Gamma(\mu_i)} \frac{\Gamma(y_i+\mu_i)}{(\delta+1)^{y_i+\mu_i}} & (13.8)\\
&= \frac{\Gamma(y_i+\mu_i)}{\Gamma(\mu_i)\Gamma(y_i+1)} \left(\frac{\delta}{1+\delta}\right)^{\mu_i} \left(\frac{1}{1+\delta}\right)^{y_i} & (13.9)
\end{aligned}
$$

The moments of this distribution are given by

$$
\begin{aligned}
\mathrm{E}[y_i] &= e^{\mathbf{x}_i\boldsymbol{\beta}+\mathrm{offset}_i}/\delta & (13.10)\\
\mathrm{V}[y_i] &= e^{\mathbf{x}_i\boldsymbol{\beta}+\mathrm{offset}_i}(1+\delta)/\delta^2 & (13.11)
\end{aligned}
$$

such that the variance-to-mean ratio is given by $(1+\delta)/\delta$ (constant for all observations). The variance is a scalar multiple of the mean and is thus referred to as NB-1 in the literature. For a complete treatment of this and other models, see Hilbe (2007).

We may rewrite the model as

$$
\frac{\Gamma(y_i+\mu_i)}{\Gamma(\mu_i)\Gamma(y_i+1)} \left(\frac{1}{1+\alpha}\right)^{\mu_i} \left(\frac{\alpha}{1+\alpha}\right)^{y_i} \qquad (13.12)
$$

where $\alpha = 1/\delta$. We then have that the Poisson is the limiting case when $\alpha = 0$

$$
\begin{aligned}
\alpha &= \exp(\tau) \qquad \mu_i = \exp\{\mathbf{x}_i\boldsymbol{\beta} + \mathrm{offset}_i - \ln(\alpha)\} & (13.13)\\
\mathcal{L} &= \sum_{i=1}^n \Big[\ln\{\Gamma(\mu_i+y_i)\} - \ln\{\Gamma(y_i+1)\} - \ln\{\Gamma(\mu_i)\} + \\
&\quad y_i \ln(\alpha) - (y_i+\mu_i)\ln(1+\alpha) \Big] & (13.14)
\end{aligned}
$$

In the above, we allowed that

$$
\mu_i = \exp\{\mathbf{x}_i\boldsymbol{\beta} + \mathrm{offset}_i - \ln(\alpha)\} \qquad (13.15)
$$

where the inclusion of the $-\ln(\alpha)$ term standardizes the $\boldsymbol{\beta}$ coefficients such that the expected value of the outcome is then equal to the exponentiated linear predictor $\exp(\mathbf{X}\boldsymbol{\beta})$; to obtain the expected outcome without this term, we would have to multiply the exponentiated linear predictor by α. This specification parameterizes $\alpha > 0$ as $\exp(\tau)$ to enforce the boundary condition that $\alpha > 0$. The model is fitted using Stata's `nbreg` command along with specification of the `dispersion(constant)` option. However, this distribution cannot be written as a member of the exponential family of distributions.

13.2 Variable overdispersion

One can illustrate the derivation of the negative binomial, within the GLM framework, in two ways. The first is to conceive the negative binomial distribution in the traditional sense—as a Poisson model with gamma heterogeneity where the gamma noise has a mean of 1. The second way of looking at the negative binomial is as a probability function in its own right (independent of the Poisson). In this sense, one can think of the negative binomial probability function as the probability of observing y failures before the rth success in a series of Bernoulli trials. Each of these two ways of looking at the negative binomial converges to the same log-likelihood function. Though the approaches are different, reviewing them is educational. Regardless of the approach, the variance is a quadratic function of the mean and is thus referred to as NB-2 in the literature. For a complete treatment of the various forms of these models, see Hilbe (2007).

13.2.1 Derivation in terms of a Poisson–gamma mixture

The derivation of the negative binomial (NB-2) is usually illustrated as a Poisson model with gamma heterogeneity where the gamma noise has a mean of 1. First, we introduce an individual unobserved effect into the conditional Poisson mean:

$$\begin{aligned} \ln \mu_i &= \mathbf{x}_i\boldsymbol{\beta} + \epsilon_i && (13.16) \\ &= \ln \lambda_i + \ln u_i && (13.17) \end{aligned}$$

The distribution of y conditioned on x_i and u_i remains Poisson with the conditional mean and variance given by μ_i.

$$f(y_i|u_i) = \frac{e^{-\lambda_i u_i}(\lambda_i u_i)^{y_i}}{y_i!} \tag{13.18}$$

The conditional mean of y_i under the gamma heterogeneity is then given by $\lambda_i u_i$ instead of just λ_i. The unconditional distribution of y is then derived as

$$f(y_i|x_i) = \int_0^\infty \frac{e^{-(\lambda_i u_i)}(\lambda_i u_i)^{y_i}}{y_i!} g(u_i) du_i \tag{13.19}$$

where the choice of $g()$ defines the unconditional distribution. Here we specify a gamma distribution for $u_i = \exp(\epsilon_i)$. Like other models of heterogeneity, the mean of the distribution is unidentified if the gamma specification contains a constant term (because the disturbance enters multiplicatively). Therefore, the gamma mean is assumed to be one. Using this normalization, we have that

$$f(y_i|x_i) = \int_0^\infty \frac{e^{-(\lambda_i u_i)}(\lambda_i u_i)^{y_i}}{y_i!} \frac{\nu^\nu}{\Gamma(\nu)} u_i^{\nu-1} e^{-\nu u_i} du_i \tag{13.20}$$

$$= \frac{\lambda_i^{y_i}}{\Gamma(y_i+1)} \frac{\nu^\nu}{\Gamma(\nu)} \int_0^\infty e^{-(\lambda_i+\nu)u_i} u_i^{(y_i+\nu)-1} du_i \tag{13.21}$$

$$= \frac{\lambda_i^{y_i}}{\Gamma(y_i+1)} \frac{\nu^\nu}{\Gamma(\nu)} \frac{\Gamma(y_i+\nu)}{(\lambda_i+\nu)^{y_i+\nu}} \int_0^\infty \frac{(\lambda_i+\nu)^{y_i+\nu}}{\Gamma(y_i+\nu)} e^{-(\lambda_i+\nu)u_i} u_i^{(y_i+\nu)-1} du_i \tag{13.22}$$

$$= \frac{\lambda_i^{y_i}}{\Gamma(y_i+1)} \frac{\nu^\nu}{\Gamma(\nu)} \frac{\Gamma(y_i+\nu)}{(\lambda_i+\nu)^{y_i+\nu}} \tag{13.23}$$

$$= \frac{\lambda_i^{y_i}}{\Gamma(y_i+1)} \frac{\nu^\nu}{\Gamma(\nu)} \Gamma(y_i+\nu) \left(\frac{\nu}{\lambda_i+\nu}\right)^\nu \frac{1}{\nu^\nu} \left(\frac{\lambda_i}{\lambda_i+\nu}\right)^{y_i} \frac{1}{\lambda_i^{y_i}} \tag{13.24}$$

$$= \frac{\Gamma(y_i+\nu)}{\Gamma(y_i+1)\Gamma(\nu)} \left(\frac{\nu}{\lambda_i+\nu}\right)^\nu \left(\frac{\lambda_i}{\lambda_i+\nu}\right)^{y_i} \tag{13.25}$$

$$= \frac{\Gamma(y_i+\nu)}{\Gamma(y_i+1)\Gamma(\nu)} \left(\frac{1}{1+\lambda_i/\nu}\right)^\nu \left(1-\frac{1}{1+\lambda_i/\nu}\right)^{y_i} \tag{13.26}$$

We may then rewrite the result as

$$\frac{\Gamma(y_i+1/\alpha)}{\Gamma(y_i+1)\Gamma(1/\alpha)} \left(\frac{1}{1+\alpha\mu_i}\right)^{1/\alpha} \left(1-\frac{1}{1+\alpha\mu_i}\right)^{y_i} \tag{13.27}$$

such that

$$\mu_i = \lambda_i = \exp(\mathbf{x}_i\boldsymbol{\beta} + \text{offset}_i) \tag{13.28}$$

which allows that the mean of the negative binomial is given by $\exp(\mathbf{x}_i\boldsymbol{\beta} + \text{offset}_i)$ (parameterized using the exponential function to restrict the outcomes to positive counts) and the overdispersion (variance divided by the mean) is given by $1 + \alpha\mu_i$.

In this parameterization, α is positive. The boundary case, $\alpha = 0$, corresponds to the nested Poisson model, and as α increases, so does the assumed overdispersion of the model. The likelihood is sometimes coded in terms of $\tau = \ln(\alpha)$, so the interesting boundary condition for the nested Poisson model corresponds to $\tau = -\infty$.

Inverting the scale parameter from that appearing in McCullagh and Nelder (1989) allows a direct relationship between the mean and scale. Estimation results will not differ between parameterizations, as long as each is used consistently for all related formulas. However, we think that the direct relationship approach allows for enhanced interpretation of excess model dispersion. John Nelder first introduced this approach to one of the authors of this book. Moreover, Nelder used this parameterization in his GENSTAT macro, called k-system, and now prefers this specification.

In exponential-family notation, we have that

$$
\begin{aligned}
f(y;\mu,\alpha) &= \exp\left\{y\ln\left(\frac{\alpha\mu}{1+\alpha\mu}\right)+\frac{1}{\alpha}\ln\left(\frac{1}{1+\alpha\mu}\right)+\ln\Gamma\left(y+\frac{1}{\alpha}\right)-\right. \\
&\qquad \left.\ln\Gamma(y+1)-\ln\Gamma\left(\frac{1}{\alpha}\right)\right\} \qquad (13.29)
\end{aligned}
$$

so that we now have the major formulas that are required for the GLM negative binomial algorithm.

$$
\begin{aligned}
\theta &= \ln\left(\frac{\alpha\mu}{1+\alpha\mu}\right) && (13.30)\\
b(\theta) &= -\frac{1}{\alpha}\ln\left(\frac{1}{1+\alpha\mu}\right) && (13.31)\\
b'(\theta) &= \frac{\partial b}{\partial\mu}\frac{\partial\mu}{\partial\theta} && (13.32)\\
&= \left(\frac{1}{1+\alpha\mu}\right)\{\mu(1+\alpha\mu)\} && (13.33)\\
&= \mu && (13.34)\\
b''(\theta) &= \frac{\partial^2 b}{\partial\mu^2}\left(\frac{\partial\mu}{\partial\theta}\right)^2+\frac{\partial b}{\partial\mu}\frac{\partial^2\mu}{\partial\theta^2} && (13.35)\\
&= \left\{-\frac{\alpha}{(1+\alpha\mu)^2}\right\}\{\mu^2(1+\alpha\mu)^2\}+\left(\frac{1}{1+\alpha\mu}\right)(1+\alpha\mu)(\mu+2\alpha\mu^2) && (13.36)\\
&= -\alpha\mu^2+\mu+2\alpha\mu^2 && (13.37)\\
&= \mu+\alpha\mu^2 && (13.38)\\
\mathrm{V}(\mu) &= \mu+\alpha\mu^2 && (13.39)\\
\frac{\partial \mathrm{V}(\mu)}{\partial\mu} &= 1+2\alpha\mu && (13.40)
\end{aligned}
$$

The full log likelihood and deviance functions are given by

$$
\begin{aligned}
\mathcal{L}(\mu;y,\alpha) &= \sum_{i=1}^{n}\left\{y_i\ln\left(\frac{\alpha\mu_i}{1+\alpha\mu_i}\right)-\frac{1}{\alpha}\ln(1+\alpha\mu_i)+\ln\Gamma\left(y_i+\frac{1}{\alpha}\right)-\right. \\
&\qquad \left.\ln\Gamma(y_i+1)-\ln\Gamma\left(\frac{1}{\alpha}\right)\right\} \qquad (13.41)
\end{aligned}
$$

The deviance function, defined as usual, $2\{\mathcal{L}(y;y) - \mathcal{L}(\mu;y)\}$, is calculated as

$$D = 2\sum_{i=1}^{n}\left\{y_i \ln\left(\frac{y_i}{\mu_i}\right) - \left(y_i + \frac{1}{\alpha}\right)\ln\left(\frac{1+\alpha y_i}{1+\alpha\mu_i}\right)\right\} \tag{13.42}$$

13.2.2 Derivation in terms of the negative binomial probability function

The second approach to understanding the negative binomial within the framework of GLM is to begin with the negative binomial probability function. As stated earlier, we can conceive of the distribution as the probability of observing y failures before the rth success in a series of Bernoulli trials. The negative binomial probability mass function is written

$$f(y;r,p) = \binom{y+r-1}{r-1} p^r (1-p)^y \tag{13.43}$$

In exponential-family form, the above translates to

$$f(y;r,p) = \exp\left\{y\ln(1-p) + r\ln(p) + \ln\binom{y+r-1}{r-1}\right\} \tag{13.44}$$

Thus

$$\theta = \ln(1-p) \tag{13.45}$$
$$e^\theta = 1-p \tag{13.46}$$
$$p = 1-e^\theta \tag{13.47}$$
$$b(\theta) = -r\ln(p) = -r\ln\left(1-e^\theta\right) \tag{13.48}$$

where the scale parameter $a(\phi) = 1$. The mean and variance may be obtained by taking the derivatives of $b(\theta)$ with respect to θ:

$$b'(\theta) = \frac{\partial b}{\partial p}\frac{\partial p}{\partial \theta} \tag{13.49}$$
$$= \left(-\frac{r}{p}\right)\{-(1-p)\} = \frac{r(1-p)}{p} \tag{13.50}$$
$$= \mu \tag{13.51}$$
$$b''(\theta) = \frac{\partial^2 b}{\partial p^2}\left(\frac{\partial p}{\partial \theta}\right)^2 + \frac{\partial b}{\partial p}\frac{\partial^2 p}{\partial \theta^2} \tag{13.52}$$
$$= \left(\frac{r}{p^2}\right)(1-p)^2 + \frac{-r}{p}(1-p) \tag{13.53}$$
$$= \frac{r(1-p)^2 + rp(1-p)}{p^2} = \frac{r(1-p)}{p^2} \tag{13.54}$$

In terms of the mean, the variance may be written

$$\mathrm{V}(\mu) = b''(\theta) = \mu + \frac{\mu^2}{r} \tag{13.55}$$

Reparameterizing $\alpha = 1/r$ so that the variance is directly (instead of inversely) proportional to the mean yields

$$p = 1 - e^{\theta} = \frac{1}{1+\alpha\mu} \tag{13.56}$$

$$\theta = \ln(1-p) = \ln\left(\frac{\alpha\mu}{1+\alpha\mu}\right) \tag{13.57}$$

$$b(\theta) = -\frac{1}{\alpha}\ln(p) = \frac{1}{\alpha}\ln(1+\alpha\mu) \tag{13.58}$$

$$b'(\theta) = \frac{1-p}{\alpha p} = \mu = \frac{1}{\alpha(e^{-\theta}-1)} \tag{13.59}$$

$$b''(\theta) = \frac{1-p}{\alpha p^2} = \mu + \alpha\mu^2 \tag{13.60}$$

$$g'(\theta) = \frac{\partial}{\partial\mu}\ln\left(\frac{\alpha\mu}{1+\alpha\mu}\right) = \frac{1}{\mu+\alpha\mu^2} \tag{13.61}$$

where the variance in terms of the mean is given by

$$\mathrm{V}(\mu) = b''(\theta) = \mu + \alpha\mu^2 \tag{13.62}$$

The full log likelihood and deviance functions are then found by substituting (13.56) and the reparameterization, $\alpha = 1/r$, into (13.44).

$$\mathcal{L}(\mu; y, \alpha) = \sum_{i=1}^{n}\left\{y_i \ln\left(\frac{\alpha\mu_i}{1+\alpha\mu_i}\right) - \frac{1}{\alpha}\ln(1+\alpha\mu_i) + \ln\Gamma\left(y_i + \frac{1}{\alpha}\right) - \ln\Gamma(y_i+1) - \ln\Gamma\left(\frac{1}{\alpha}\right)\right\} \tag{13.63}$$

The deviance function, defined as usual, $2(\mathcal{L}(y;y) - \mathcal{L}(\mu;y))$, is calculated as

$$D = 2\sum_{i=1}^{n}\left\{y_i \ln\left(\frac{y_i}{\mu_i}\right) - \left(y_i + \frac{1}{\alpha}\right)\ln\left(\frac{1+\alpha y_i}{1+\alpha\mu_i}\right)\right\} \tag{13.64}$$

where these results coincide with the results of the previous subsection. So, we may view the negative binomial as a mixture distribution of the Poisson and gamma, or we may view it as a probability function on its own.

13.2.3 The canonical link negative binomial parameterization

We construct a canonical link negative binomial algorithm by inserting all the required formulas into the standard GLM algorithm. This algorithm is the result of either derivation given in the last two subsections.

As in previous algorithm construction, we obtain the relevant terms for our algorithm by substituting the inverse link of the linear predictor for μ. For the canonical case, we have that $\eta = \theta$, and we can derive the relationships between μ and η by using the results from either of the previous two subsections.

$$\begin{aligned}
\eta &= \theta & (13.65)\\
&= \ln\left(\frac{\alpha\mu}{1+\alpha\mu}\right) & (13.66)\\
&= -\ln\left(\frac{1}{\alpha\mu}+1\right) & (13.67)\\
\mu &= \frac{1}{\alpha(e^{-\eta}-1)} & (13.68)
\end{aligned}$$

In algorithm construction, we also use the derivative of the link function that was given before as

$$\begin{aligned}
g'(\theta) &= \frac{\partial}{\partial\mu}\ln\left(\frac{\alpha\mu}{1+\alpha\mu}\right) & (13.69)\\
&= \frac{1}{\mu+\alpha\mu^2} & (13.70)
\end{aligned}$$

The relevant algorithm is then given by

Listing 13.1: IRLS algorithm for negative binomial regression

```
1    μ = (y + mean(y))/2
2    η = −ln(1/(αμ) + 1)
3    WHILE (abs(ΔDev) > tolerance) {
4          W = μ + αμ^2
5          z = (η + (y − μ)/W) − offset
6          β = (X^T W X)^{-1} X^T W z
7          η = Xβ + offset
8          μ = 1/(α(e^{−η} − 1)
9          OldDev = Dev
10         Dev = 2∑(y ln(y/μ) − (y + 1/α) ln((1 + αy)/(1 + αμ)))
11         Dev = 2∑(ln(1/(1 + αμ))/α)   if y = 0
12         ΔDev = Dev - OldDev
13   }
14   χ^2 = ∑((y − μ)^2/(μ + αμ^2))
```

In addition to the `glm` command, this model is also available through Stata's `nbreg` command by specifying the `dispersion(mean)` option. The option is not required, since this is the default model. Our emphasis on the option, in this case, serves to differentiate this case from the `dispersion(constant)` option associated with the model in section 13.1.

13.3 The log-negative binomial parameterization

The negative binomial is rarely used in canonical form. Its primary use is to serve as an overdispersed Poisson regression model. That is, since Poisson models are mostly overdispersed in real data situations, the negative binomial is used to model such data in their place. To facilitate the use of the Poisson to overdispersion, the negative binomial is nearly always parameterized using a log link given in (13.28).

To convert the canonical form to log form, we substitute the log and inverse link into the algorithm. We then amend the weight function, w, and the working response, z. Recalling that $w = 1/(\mathrm{V}g')$ and $z = \eta + (y-\mu)g'$, we see that using the log link results in

$$w = \frac{1}{(\mu+\alpha\mu^2)/\mu^2} \tag{13.71}$$

$$= \frac{\mu}{1+\alpha\mu} \tag{13.72}$$

$$z = \eta + \frac{y-\mu}{\mu} \tag{13.73}$$

so we may specify the log-negative binomial algorithm as

Listing 13.2: IRLS algorithm for log-negative binomial regression

```
1   μ = (y + mean(y))/2
2   η = ln(μ)
3   WHILE (abs(ΔDev) > tolerance) {
4       W = μ/(1 + αμ)
5       z = (η + (y − μ)/μ) − offset
6       β = (X^T W X)^{-1} X^T W z
7       η = Xβ + offset
8       μ = exp(η)
9       OldDev = Dev
10      Dev = 2∑(ln(y/μ) − (y + 1/α) ln((1 + αy)/(1 + αμ)))
11      Dev = 2∑(ln(1/(1 + αμ))/α)   if y = 0
12      ΔDev = Dev - OldDev
13  }
14  χ^2 = ∑((y − μ)^2/(μ + αμ^2))
```

We have emphasized throughout this book that, when possible, GLMs should be recast to use the observed rather than the traditional expected information matrix to abstract standard errors. The above algorithm uses the expected information matrix. We can change the above algorithm to use the observed information matrix by converting the weight and working response.

Formulas common to using the observed information matrix and expected information matrix are the following:

$$\begin{aligned}
\eta &= g(\mu) = \ln(\mu) && (13.74)\\
g'(\mu) &= g' = 1/\mu && (13.75)\\
g''(\mu) &= g'' = -1/\mu^2 && (13.76)\\
\mathrm{V}(\mu) &= \mathrm{V} = \mu + \alpha\mu^2 && (13.77)\\
\mathrm{V}^2 &= (\mu + \alpha\mu^2)^2 && (13.78)\\
\mathrm{V}'(\mu) &= \mathrm{V}' = 1 + 2\alpha\mu && (13.79)\\
w &= \mu/(1+\alpha\mu) && (13.80)
\end{aligned}$$

The observed information matrix adjusts the weights such that

$$\begin{aligned}
w_o &= w + (y-\mu)(\mathrm{V}g'' + \mathrm{V}'g')/\left\{\mathrm{V}^2(g')^3\right\} && (13.81)\\
z &= \eta + (y-\mu)/\{w_o(1+\alpha\mu)\} && (13.82)
\end{aligned}$$

Substituting these results into the GLM negative binomial algorithm:

Listing 13.3: IRLS algorithm for log-negative binomial regression using OIM

```
1   μ = (y + mean(y))/2
2   η = ln(μ)
3   WHILE (abs(ΔDev) > tolerance) {
4       W = μ/(1 + αμ) + (y − μ)(αμ/(1 + 2αμ + α²μ²))
5       z = (η + (y − μ)/(Wαμ)) − offset
6       β = (XᵀWX)⁻¹XᵀWz
7       η = Xβ + offset
8       μ = exp(η)
9       OldDev = Dev
10      Dev = 2∑ln(y/μ) − (y + 1/α) ln((1 + αy)/(1 + αμ))
11      Dev = 2∑ln(1/(1 + αμ))/α   if y = 0
12      ΔDev = Dev - OldDev
13  }
14  χ² = ∑((y − μ)²/(μ + αμ²))
```

Maximum likelihood methods require the log likelihood to be parameterized in terms of $\mathbf{x}\boldsymbol{\beta}$. We provide them here, letting C_i represent the normalization terms.

$$\begin{aligned}
C_i &= \ln\Gamma(y_i + 1/\alpha) - \ln\Gamma(y_i+1) - \ln\Gamma(1/\alpha) && (13.83)\\
\mathcal{L}(\mathbf{x}\boldsymbol{\beta}; y, \alpha) &= \sum_{i=1}^{n}\left\{y_i\ln(\alpha\exp(\mathbf{x}_i\boldsymbol{\beta})) - \left(y_i + \frac{1}{\alpha}\right)\ln(1+\alpha\exp(\mathbf{x}_i\boldsymbol{\beta})) + C_i\right\} && (13.84)\\
&= \sum_{i=1}^{n}\left\{y_i\ln\left(\frac{\alpha\exp(\mathbf{x}_i\boldsymbol{\beta})}{1+\alpha\exp(\mathbf{x}_i\boldsymbol{\beta})}\right) - \frac{\ln(1+\alpha\exp(\mathbf{x}_i\boldsymbol{\beta})}{\alpha} + C_i\right\} && (13.85)
\end{aligned}$$

These log-likelihood functions are the ones usually found in maximum likelihood programs that provide negative binomial modeling. The canonical link, as well as identity and other possible links, are usually unavailable.

One can fit a negative binomial together with an estimate of the scale parameter by iteratively forcing the χ^2 dispersion to unity. The following algorithm can be used for this purpose. It is the basis for the original SAS log-negative binomial macro. Hilbe's **negbin** program, available for download using Stata's Internet capabilities, implements this approach:

```
. net search negbin
```

We show results from this command in the next section.

Listing 13.4: IRLS algorithm for log-negative binomial regression with estimation of α

```
1    Poisson estimation: y <predictors>
2    $\chi^2 = \sum(y-\mu)^2/\mu$
3    Disp  = $\chi^2$/df
4    $\alpha$ = 1/disp
5    WHILE (abs($\Delta$Disp) > tolerance) {
6         OldDisp = Disp
7         estimation: y <predictors>, alpha$\approx\alpha$
8         $\chi^2 = \sum((y-\mu)^2/(\mu+\alpha\mu^2))$
9         Disp = $\chi^2$/df
10        $\alpha$ = disp($\alpha$)
11        $\Delta$Disp = Disp - OldDisp
12   }
```

Again the above algorithm provides accurate estimates of α and of negative binomial estimates. One can also amend the above algorithm to use the deviance-based dispersion rather than the χ^2. In any case, one can obtain both parameter and scale estimates by using the above method if maximum likelihood software is unavailable. We will give an example of its use in the following section.

13.4 Negative binomial examples

Examples of the negative binomial come from overdispersed Poisson data. We see how the negative binomial fits the example dataset we used in the last chapter.

The scale of the ancillary parameter is not estimated from the GLM model; rather, it must be supplied prior to estimation. The best method is to obtain the value of the ancillary parameter by maximum likelihood and then to use that value in the GLM algorithm. The value of doing this in practical application is that all the GLM diagnostics are available for evaluating the model fit.

```
. use http://www.stata-press.com/data/hh2/medpar
. nbreg los hmo white type2 type3, irr nolog
Negative binomial regression                    Number of obs     =       1495
                                                LR chi2(4)        =     118.03
Dispersion     = mean                           Prob > chi2       =     0.0000
Log likelihood = -4797.4766                     Pseudo R2         =     0.0122

------------------------------------------------------------------------------
         los |        IRR   Std. Err.      z    P>|z|     [95% Conf. Interval]
-------------+----------------------------------------------------------------
         hmo |   .9343023   .0497622    -1.28   0.202     .8416883    1.037107
       white |   .8789164   .0602425    -1.88   0.060     .7684307    1.005288
       type2 |   1.247634    .063121     4.37   0.000     1.129855     1.37769
       type3 |   2.026193   .1542563     9.28   0.000     1.745332    2.352252
-------------+----------------------------------------------------------------
    /lnalpha |   -.807982   .0444542                     -.8951107   -.7208533
-------------+----------------------------------------------------------------
       alpha |   .4457567   .0198158                      .4085624    .4863371
------------------------------------------------------------------------------
Likelihood-ratio test of alpha=0:  chibar2(01) = 4262.86 Prob>=chibar2 = 0.000
. glm los hmo white type2 type3, family(nb .4457567) eform nolog
Generalized linear models                          No. of obs      =      1495
Optimization     : ML                              Residual df     =      1490
                                                   Scale parameter =         1
Deviance         =  1568.142891                    (1/df) Deviance =  1.052445
Pearson          =  1624.538286                    (1/df) Pearson  =  1.090294
Variance function: V(u) = u+(.4457567)u^2          [Neg. Binomial]
Link function    : g(u) = ln(u)                    [Log]
                                                   AIC             =  6.424718
Log likelihood   = -4797.476603                    BIC             = -9323.581

------------------------------------------------------------------------------
             |                 OIM
         los |        IRR   Std. Err.      z    P>|z|     [95% Conf. Interval]
-------------+----------------------------------------------------------------
         hmo |   .9343023   .0497622    -1.28   0.202     .8416883    1.037107
       white |   .8789164   .0602424    -1.88   0.060      .768431    1.005288
       type2 |   1.247634    .063121     4.37   0.000     1.129855     1.37769
       type3 |   2.026193   .1542563     9.28   0.000     1.745332    2.352252
------------------------------------------------------------------------------
```

Note the default log-link parameterization for the GLM model. When using Stata, the canonical link, or others, must be specifically requested. The default value of α is 1. When using the provided software, the value of α is the # specified in the `family(nb #)` option. It is not estimated but is rather specified by the user.

The model results demonstrate that most of the overdispersion in the Poisson model has been accommodated in the negative binomial. Given its deviance scale of near unity, we would probably prefer it to the Poisson. Of course, we need to evaluate the model by using residual analysis. We always recommend plotting Anscombe residuals or standardized residuals versus predictors. We also observe here that the mean of μ, the predicted fit, and LOS are 9.85.

We mentioned that the scale can be estimated within the GLM approach independent of maximum likelihood techniques. We iterate a negative binomial in a loop that updates α at each iteration. Using this method, we can obtain an estimate of τ or α.

```
. negbin los hmo white type2 type3, eform
1:     Deviance:        3110  Alpha (k):      .1597  Disp:         2.298
2:     Deviance:        1812  Alpha (k):       .367  Disp:         1.274
3:     Deviance:        1512  Alpha (k):      .4677  Disp:         1.048
4:     Deviance:        1458  Alpha (k):      .4902  Disp:         1.008
5:     Deviance:        1449  Alpha (k):      .4942  Disp:         1.001
6:     Deviance:        1447  Alpha (k):      .4949  Disp:             1
7:     Deviance:        1447  Alpha (k):       .495  Disp:             1
8:     Deviance:        1447  Alpha (k):       .495  Disp:             1
                                                      No obs.      =        1495
Poisson Dp =      6.26                                Poisson Dev =        8143
Alpha (k)  =      .495                                Neg Bin Dev =        1447
                                                      Prob > Chi2 =           0

Chi2       =   1490.01                                Deviance    =  1447.139
Prob>chi2  =  .4950578                                Prob>chi2   =  .7824694
Dispersion =  1.000006                                Dispersion  =  .9712345

Log negative binomial regression
```

los	IRR	Std. Err.	z	P>|z|	[95% Conf.	Interval]
hmo	.9343301	.0519053	-1.22	0.221	.8379405	1.041808
white	.8792076	.0629791	-1.80	0.072	.7640442	1.011729
type2	1.247625	.0659398	4.19	0.000	1.124854	1.383795
type3	2.026132	.1615912	8.85	0.000	1.732932	2.36894

```
Loglikelihood = -4800.2524
```

The estimation of α would be closer to the maximum likelihood value if updating were based on the deviance rather than on the Pearson χ^2 dispersion. Still, the method provides a close approximation to the scale by iteratively reducing the χ^2 dispersion to 1.

We next look at the `smoke` dataset. After modeling the data by using maximum likelihood, we found that α, the value of the ancillary parameter, was near 0. Placing this value into the GLM program produces identical results.

(*Continued on next page*)

```
. use http://www.stata-press.com/data/hh2/doll, clear
. glm deaths smokes a2-a5, family(nb 0.000001) lnoffset(pyears) eform nolog
Generalized linear models                         No. of obs      =         10
Optimization     : ML                             Residual df     =          4
                                                  Scale parameter =          1
Deviance         =  22.44756894                   (1/df) Deviance =   5.611892
Pearson          =  19.12754753                   (1/df) Pearson  =   4.781887
Variance function: V(u) = u+(0.000001)u^2         [Neg. Binomial]
Link function    : g(u) = ln(u)                   [Log]
                                                  AIC             =   8.951624
Log likelihood   = -38.75812018                   BIC             =   13.23723

------------------------------------------------------------------------------
             |                 OIM
      deaths |        IRR   Std. Err.      z    P>|z|     [95% Conf. Interval]
-------------+----------------------------------------------------------------
      smokes |   1.126815   .1088576     1.24   0.216     .9324401    1.361708
          a2 |   4.470652   .8722852     7.68   0.000      3.04993    6.553177
          a3 |   13.56629   2.551483    13.86   0.000     9.383641    19.61332
          a4 |   29.07066   5.372829    18.23   0.000     20.23654    41.76127
          a5 |   40.85524   7.853452    19.30   0.000     28.03008    59.54853
      pyears | (exposure)
------------------------------------------------------------------------------
```

What does this tell us? The model still appears to be overdispersed. We could try scaling, but what happened?

A negative binomial having a scale parameter of 0 is a Poisson model. Recall that the overdispersion we found in the last chapter was due to apparent overdispersion, which could be corrected by appropriate interaction terms. The overdispersion was not inherent in the data.

We learned that we can make a first round test of apparent versus inherent overdispersion by modeling the data using both Poisson and negative binomial. If the negative binomial does not clear up the problem, and has a scale near 0, then the model is likely Poisson. You must look for interactions, and so forth. Adding the interactions to the negative binomial model does not help. We can scale if we wish, but it is better to revert to the Poisson model.

The log-link negative binomial exponentiated parameter estimates are to be understood as incidence-rate ratios, just like Poisson IRR estimates. This is not the case for other links.

A final point should be made prior to leaving our discussion of the GLM-based negative binomial. In the previous chapter we mentioned that the zero-truncated Poisson model may be preferable when modeling count data situations in which there is no possibility of 0 counts. That is, when the response is such that 0 counts are excluded, we should consider the zero-truncated Poisson. We used the hospital length of stay example as an example. The same situation may exist with an otherwise negative binomial model. If the response is not capable of having a 0 count, then one should consider the zero-truncated negative binomial.

The zero-truncated negative binomial log-likelihood function is a modification of the standard form. The modification is based on an adjustment to the probability function. The adjustment excludes 0 in the probability range, and adjusts the probability function so that it still sums to 1. If we label the log-link negative binomial likelihood as $L_{\rm NB}$ and the log-link negative binomial log likelihood as $_{\rm NB}$, then the zero-truncated log likelihood for a log-link negative binomial is given by

$$= \sum_{i=1}^{n} \ln\left\{\frac{L_{{\rm NB}\,i}}{1 - P_{\rm NB}(y_i = 0)}\right\} \tag{13.86}$$

$$= \sum_{i=1}^{n} \left({}_{{\rm NB}\,i} - \ln\left[1 - \quad 1 + \exp(\mathbf{x}\boldsymbol{\beta}) \quad ^{-1/\alpha}\right]\right) \tag{13.87}$$

If we find that there is little difference in parameter and scale estimates between the standard and the zero-truncated form, then using the standard form is wiser. We can interpret this situation to mean that the lack of 0 counts had little impact on the probability of the response, or μ. On the other hand, if large differences do exist, then we must maintain the zero-truncated model. Convergence can be a problem for some types of data situations.

Hilbe initially developed zero-truncated Poisson and negative binomial programs, which are now available as part of the Stata program; see the `ztp` and `ztnb` manual entries in the *Stata Base Reference Manual*. The original files can still be found using

```
. net from http://www.stata.com/users/jhilbe
```

We cover these and other models in chapter 14.

13.5 The geometric family

We mentioned at the beginning of this chapter that the geometric regression model is simply the negative binomial with the ancillary or scale parameter set at 1. It is the discrete correlate to the negative exponential distribution, taking a smooth downward slope from an initial high point at 0 counts. If the data are in fact distributed in such a manner, then the geometric model may be appropriate. The log-link form has the same gradual downward smooth slope as the canonical link.

(*Continued on next page*)

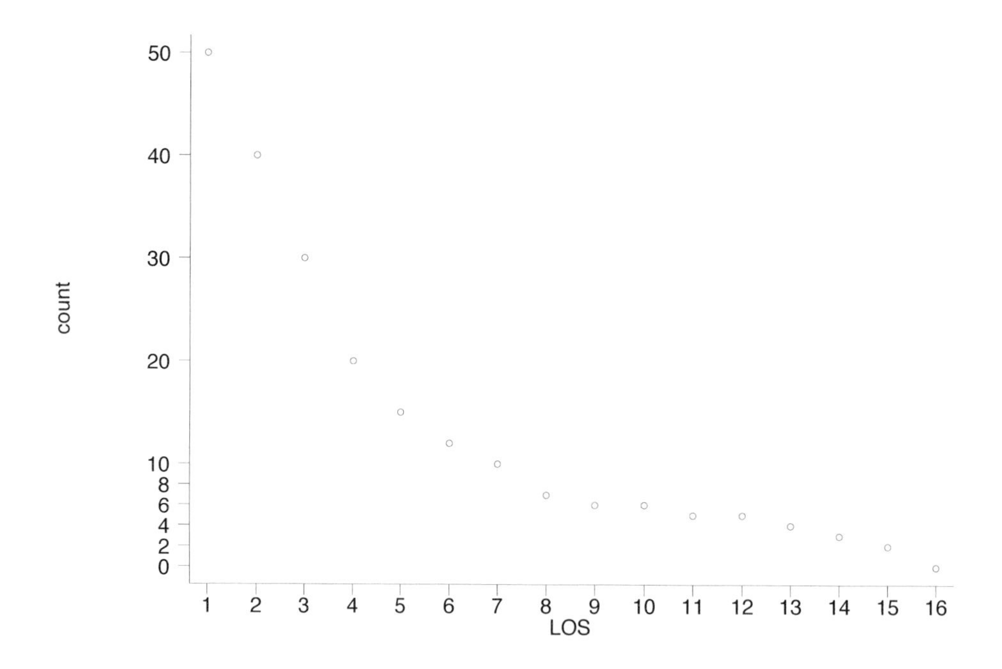

Figure 13.2: Frequency of occurrence versus LOS

The canonical geometric regression algorithm may be specified as

Listing 13.5: IRLS algorithm for geometric regression

```
1   μ = (y + mean(y))/2
2   η = −ln(1 + 1/μ)
3   WHILE (abs(ΔDev) > tolerance) {
4       W = μ + μ^2
5       z = (η + (y − μ)/W) − offset
6       β = (X^T W X)^{-1} X^T W z
7       η = Xβ + offset
8       μ = 1/((exp(−η) − 1)
9       OldDev = Dev
10      Dev = 2∑(ln(y/μ) − (1 + y) ln((1 + y)/(1 + μ)))
11      Dev = 2∑ − ln(1 + μ)   if y = 0
12      ΔDev = Dev - OldDev
13  }
14  χ^2 = ∑((y − μ)^2/(μ + μ^2))
```

and the log-geometric parameterized so that standard errors are produced from the observed information matrix, and may be specified as

Listing 13.6: IRLS algorithm for log-geometric regression

```
1   μ = (y + mean(y))/2
2   η = ln(μ)
3   WHILE (abs(ΔDev) > tolerance) {
4       W = μ/(1 + μ) + (y − μ)(μ/(1 + 2μ + μ^2))
5       z = (η + (y − μ)/(W(μ))) − offset
6       β = (X^T W X)^{-1} X^T W z
7       η = Xβ + offset
8       μ = exp(η)
9       OldDev = Dev
10      Dev = 2∑(ln(y/μ) − (1 + y) ln((1 + y)/(1 + μ)))
11      Dev = 2∑ln(1/(1 + μ)  if y = 0
12      ΔDev = Dev - OldDev
13  }
14  χ^2 = ∑((y − μ)^2/(μ + μ^2))
```

Sometimes the geometric distribution fits the data better than the negative binomial. For instance, we spent considerable time with the LOS data example in the previous section. We found a good negative binomial model for the data. Now let us use the same data with a canonical linked geometric model. We know from graphing the response, LOS, that it shows a high initial start and a steady monotonic decline. This is the appearance of a geometric distribution.

```
. use http://www.stata-press.com/data/hh2/medpar, clear
. glm los white type2 type3, family(nb 1) link(nb) eform nolog
Generalized linear models                         No. of obs      =       1495
Optimization     : ML                             Residual df     =       1491
                                                  Scale parameter =          1
Deviance         =  812.0438886                   (1/df) Deviance =   .5446304
Pearson          =  806.6148763                   (1/df) Pearson  =   .5409892
Variance function: V(u) = u+(1)u^2                [Neg. Binomial]
Link function    : g(u) = ln(u/(u+(1/1)))         [Neg. Binomial]
                                                  AIC             =   6.639057
Log likelihood   = -4958.695274                   BIC             =  -10086.99

                              OIM
         los      exp(b)   Std. Err.      z    P>|z|     [95% Conf. Interval]

       white    .9875833   .0073704    -1.67   0.094     .9732427    1.002135
       type2    1.020941   .0063076     3.35   0.001     1.008653    1.033379
       type3    1.054697   .0066635     8.43   0.000     1.041718    1.067839
```

The deviance and BIC statistics are considerably lower than the negative binomial model, and the incidence-rate ratios are nearly the same. In fact, the fitted values are a closer fit to the actual responses than to the fitted values from any other model.

We can also observe that the model is rather underdispersed. Scaling may help adjust standard errors, or we may adjust standard errors by clustering on the provider, as we did earlier when considering the negative binomial.

```
. glm los white type2 type3, family(nb 1) link(nb) irls scale(dev) eform
> nolog noheader
```

los	exp(b)	EIM Std. Err.	z	P>|z|	[95% Conf. Interval]	
white	.9875833	.0054393	-2.27	0.023	.9769798	.9983019
type2	1.020941	.0046549	4.55	0.000	1.011858	1.030105
type3	1.054697	.0049176	11.42	0.000	1.045103	1.06438

(Standard errors scaled using square root of deviance-based dispersion)

```
. glm los white type2 type3, family(nb 1) link(nb) cluster(provnum) eform
> nolog noheader
```

(Std. Err. adjusted for 54 clusters in provnum)

los	exp(b)	Robust Std. Err.	z	P>|z|	[95% Conf. Interval]	
white	.9875833	.0049677	-2.48	0.013	.9778946	.997368
type2	1.020941	.0052132	4.06	0.000	1.010774	1.03121
type3	1.054697	.0113407	4.95	0.000	1.032702	1.07716

In adjusting the standard errors, scaling provides results similar to those of the modified sandwich estimate of variance. This tells us that the underdispersion in the original model is accounted for by the correlation effect of each provider in the data. Since this model has significantly lower deviance and BIC statistics than the negative binomial or log-linked geometric, and since its residuals fit better than any of the GLM-based alternative models, we prefer it over the previously fitted models.

13.6 Interpretation of coefficients

Interpretating the coefficients is hard because of the nonlinearity of the link function. We motivate an alternative metric that admits a transformation of the coefficients for easier interpretation for the log link.

For illustration, we assume a model with two covariates $\mathbf{x}_1$ and $\mathbf{x}_2$ along with a constant. We focus on the interpretation of β_1.

Since the inverse canonical link is nonlinear, interpretation of the coefficient is difficult.

$$\Delta y_i = \exp\{\beta_0 + (x_{1i}+1)\beta_1 + x_{2i}\beta_2\} - \exp(\beta_0 + x_{1i}\beta_1 + x_{2i}\beta_2) \quad (13.88)$$

The difficulty arises because the difference is not constant and depends on the values of the covariates.

Instead of focusing on the difference in the outcome, we define a measure of the incidence-rate ratio as the ratio of change in the outcome (incidence)

$$\begin{aligned}\text{Incidence} - \text{rate ratio for } \mathbf{x}_1 &= \frac{\exp\ \beta_0 + (x_{1i}+1)\beta_1 + x_{2i}\beta_2}{\exp(\beta_0 + x_{1i}\beta_1 + x_{2i}\beta_2)} &(13.89)\\ &= \exp(\beta_1) &(13.90)\end{aligned}$$

The calculation of the incidence-rate ratio simplifies such that there is no dependence on a particular observation. The incidence-rate ratio is therefore constant—it is independent of the particular values of the covariates. This interpretation does not hold for other inverse links.

Interpretation for log-linked negative binomial or log-linked geometric family models is straightforward for exponentiated coefficients. These exponentiated coefficients represent incidence-rate ratios. An incidence-rate ratio of 2 indicates that there is twice the incidence (the outcome is twice as prevalent) if the associated covariate is increased by one.

14 Other count data models

Contents

14.1 Count response regression models

We observed in the last several chapters that both Poisson and negative binomial models, as count response regression models, can succumb to overdispersion. Overdispersion exists in discrete-response models when the observed model variance exceeds the expected model variance. A fundamental assumption of the Poisson model is that the mean and variance functions are equal. When the variance exceeds the mean, the model is overdispersed.

The negative binomial model can also be overdispersed. When the variance exceeds the calculated value of $\mu + a\mu^2$, the negative binomial model is overdispersed. For both the Poisson and negative binomial, overdispersion is indicated by the value of the Pearson χ^2 dispersion statistic. Values more than 1.0 indicate overdispersion.

We have also learned that indicated overdispersion is not always real—it may be only apparent overdispersion. Apparent overdispersion occurs when we can externally adjust the model to reduce the dispersion statistic to near 1.0. We indicated how to do so in chapter 11. Apparent overdispersion may occur because of a missing explanatory predictor(s), because the data contain outliers, because the model requires an interaction term, a predictor needs to be transformed to another scale, or the link function is misspecified. We demonstrated how apparent overdispersion disappears when the appropriate remedy is provided.

Once we determine that the model overdispersion is real, we next should, if possible, attempt to identify the cause or source for the overdispersion. What causes the extra correlation observed in the model? Scaling and applying robust variance estimators to adjust model standard errors may provide a post hoc remedy to accommodate bias in the standard errors, but it does not address the underlying cause of the overdispersion.

There are two general causes for real overdispersion in binomial and count response models, as outlined in Hilbe (2007). The first is a violation of the distributional assumptions upon which the model is based. For instance, with count models, the distributional mean specifies an expected number of 0 counts in the response. If there are far more zeros than expected under distributional assumptions, or if the response has no 0 counts, the resulting count model will be overdispersed. Zero-inflated Poisson (ZIP) and negative binomial (ZINB) models adjust for excessive zeros in the response. One can also use hurdle models to deal with such a situation. On the other hand, when there are no 0 counts, one can use a zero-truncated Poisson (ZTP) or negative binomial (ZTNB) to model the data. Usually these types of models fit better than standard models, but not always. We list these models in table 14.1, together with several others that have also been used to overcome distributional problems with the respective base model.

Table 14.1: Other count data models

Zero-truncated Poisson and zero-truncated negative binomial (ZTP, ZTNB)
Truncated Poisson and truncated negative binomial
Zero-inflated Poisson and zero-inflated negative binomial (ZIP, ZINB)
Censored Poisson and censored negative binomial
Poisson and negative binomial selection models
Negative binomial with endogenous stratification
Generalized Poisson regression
Generalized binomial regression

Other similar types of models have also been developed, but biases and other difficulties have been found for them. As such, these models have not yet found support in commercial applications, for example, negative multinomial regression and generalized negative binomial regression. We will not consider these models further in this text.

Another cause of overdispersion relates to misspecified variance. The binomial (actually Bernoulli) variance may be reparameterized to take the quasilikelihood form, $\mu^2(1-\mu)^2$, rather than the basic $\mu(1-\mu)$. For count models, we produce table 14.2, which is similar to information summarized in Hilbe (2007), detailing the various count variance functions.

Table 14.2: Variance functions V for count data models; ϕ and α above are constants

Poisson	μ
QL	$\phi\mu$
NB-1	$\mu(1+\alpha) = \mu + \alpha\mu$
Geometric	$\mu(1+\mu) = \mu + \mu^2$
NB-2	$\mu(1+\alpha\mu) = \mu + \alpha\mu^2$
NB-H	$\mu + (\alpha_i)\mu^2$ (where $\alpha_i = z_i\gamma$)
NB-P	$\mu + \alpha\mu^p$
GEE	$V_i R_i V_i$ where R_i is a correlation matrix

NB-H represents the heterogeneous negative binomial model, called generalized negative binomial in Stata. The term "generalized negative binomial" has a history in the literature of the area and refers to an entirely different model. We prefer to use the term heterogeneous for this model, following the usage of Greene (2002) and LIMDEP software.

NB-P refers to the power negative binomial, that is, the NB-2 variance function for which the exponent parameter is to be estimated rather than specified. The NB-P model is therefore a three-parameter model, with μ or $\boldsymbol{\beta}$ estimated, as well as the heterogeneity parameter, α, and the exponent, p. This model is not available in Stata but has been added in the newest release of LIMDEP. Use Stata's `net search` utility to check on availability, since development among the Stata community tends to be fast.

Panel models constitute another model type that relates to the distributional aspects of the Poisson and negative binomial PDFs. A collection of standard approaches is listed in table 14.3.

Table 14.3: Poisson and negative binomial panel data models

- Fixed-effects models
  - (Unconditional) fixed effects
  - Conditional fixed effects
- Random-effects models
  - Poisson with gamma-distributed random effects
  - Poisson with Gaussian-distributed random effects
  - Negative binomial with beta-distributed random effects
  - Negative binomial with Gaussian-distributed random effects
- GEE
- Mixed linear models
  - Random intercepts (multiple levels)
  - Random coefficients

Each of the models in table 14.3 was developed to adjust an overdispersed Poisson or negative binomial model by enhancing the variance function itself, or by addressing the distributional assumptions of the Poisson and negative binomial models. The statistician has at his or her disposal a variety of count response models that, if selected properly, will yield a well-fitted model to the data.

In this chapter we will look at only a few of the above count models. For a detailed examination of each of the above listed models, see Hilbe (2007).

14.2 Zero-truncated models

We often find that our count response model fails to have 0 counts. That is, we discover not only that the count has no zeros but also that it structurally excludes having 0 counts. Hospital length of stay data, common in the health analysis industry, are a prime example of this type of data situation. When a patient enters the hospital, the count begins with one. There is no zero length of hospital stay. Zero counts are structurally excluded from lengths of stay. A patient stays in the hospital 1, 2, 3, or more days before being discharged.

The problem with modeling zero-excluded counts relates to the distributional assumptions of both Poisson and negative binomial PDFs. Both probability functions require the possibility of zeros. In fact, for a specified value of the mean of the distribution, there is a given expected percentage of zeros, ones, twos, and so forth in the data. Figure 14.1 displays the probabilities outcomes given a mean of 4. The three distributions include the Poisson, negative binomial with $\alpha = 0.5$, and a negative binomial with $\alpha = 1.5$. Since a negative binomial with $\alpha = 0$ is a Poisson distribution, we have labeled the curve for Poisson as NB0. The predicted probability of 0 is about 2% for a Poisson distribution with a mean of 4, about 11% for a negative binomial with $\alpha = 0.5$, and about 2% for a negative binomial with an $\alpha = 1.5$. For a Poisson model having 100 observations with a count response mean of 4, we would expect that there will be two zeros. If the data in fact have 30 zeros, it should be obvious that distributional assumptions have been violated, and the parameter estimates and standard errors will therefore be biased.

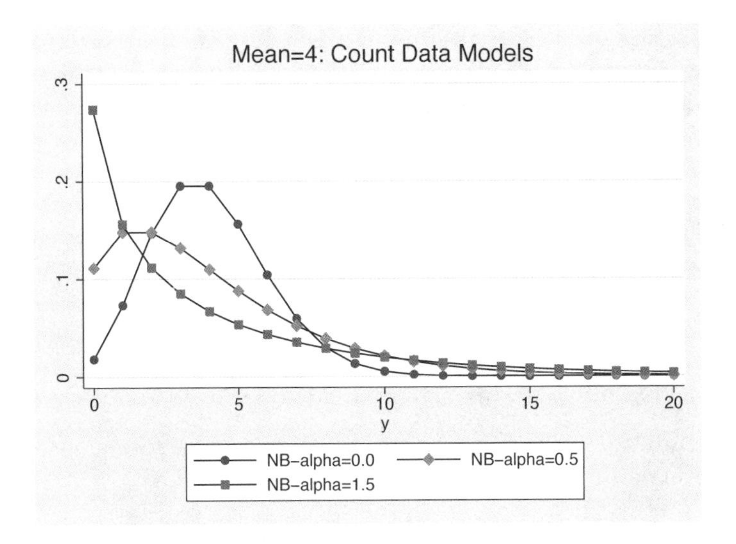

Figure 14.1: Probability mass functions for negative binomial models

A standard method of dealing with Poisson and negative binomial models that exclude zero counts is to use a model typically referred to as a zero-truncated model. The model is not adjusted by using some type of scaling or robust variance estimator since the parameter estimates are biased as well as standard errors. The solution is to adjust the probability function itself so that zeros are excluded. We do this by determining the formula to calculate 0 counts, subtract it from one, and then divide the Poisson or negative binomial probability function by the resultant value.

The Poisson PDF can be expressed as

$$f(y_i;\mu) = e^{-\mu}\mu^{y_i}/y_i! \tag{14.1}$$

The probability of a 0 count is given as $\exp(-\mu)$. The probability of not having a 0 count is then $1 - \exp(-\mu)$. Dividing the Poisson PDF for positive outcomes by the total probability of a positive outcome rescales each term so that the result will sum to unity.

$$f(y_i;\mu|y_i>0) = \frac{\exp(-\mu)\mu^{y_i}}{\{1-\exp(-\mu)\}y_i!} \tag{14.2}$$

With $\mu = \exp(x\boldsymbol{\beta})$, the log-likelihood form of the above is expressed as

$$\mathcal{L}(\mu;y|y>0) = \sum\Big(y\,(x\boldsymbol{\beta}) - \exp\,(x\boldsymbol{\beta}) - \ln\Gamma(y+1) - \ln\big[1-\exp\,\{-\exp(x\boldsymbol{\beta})\}\,\big]\Big) \tag{14.3}$$

We use the same logic to determine the log-likelihood function of the zero-truncated negative binomial. The probability of a zero negative binomial count is $(1+\alpha\mu)-1/\alpha$ or, in terms of the linear predictor $x\boldsymbol{\beta}$, $\{1+\alpha\exp(x\boldsymbol{\beta})\}-1/\alpha$.

After subtracting from 1, we use the resultant formula to rescale probabilities for strictly positive outcomes for the negative binomial log-likelihood function. The rescaled function is then given by

$$\mathcal{L}(\mu; y|y>0) = \sum \left[\mathcal{L}_{NB} - \ln\{-\alpha\exp(x\boldsymbol{\beta}) + 1/\alpha\}\right] \tag{14.4}$$

where $\mathcal{L}_{NB}$ is the (unscaled) negative binomial log likelihood.

The IRLS estimating algorithm used for standard GLM models is not appropriate for this type of model. The rescaled log-likelihood functions are not members of the exponential family of distributions and therefore cannot take advantage of the simpler Fisher Scoring method of estimation. We must use a Newton–Raphson type of maximum likelihood method to estimate parameters, standard errors, and other model statistics.

To demonstrate the difference between models, we will create a 50,000-observation synthetic Poisson dataset with the following format: `y = -1.0 + 0.5*x1 + 1.5*x2`.

```
. set seed 234987
. set obs 50000
obs was 0, now 50000
. generate x1 = invnormal(uniform())
. generate x2 = invnormal(uniform())
. generate xb = -1.0 + .5*x1 + 1.5*x2
. genpoisson yp, xbeta(xb)
```

We next model the data by using a Poisson model. We expect that the parameter estimates will be close to the values we specified in the construction of the dataset.

```
. poisson yp x1 x2
Iteration 0:   log likelihood = -46068.695
Iteration 1:   log likelihood = -46068.689
Iteration 2:   log likelihood = -46068.689
Poisson regression                                Number of obs   =      50000
                                                  LR chi2(2)      =  153824.38
                                                  Prob > chi2     =     0.0000
Log likelihood = -46068.689                       Pseudo R2       =     0.6254

------------------------------------------------------------------------------
          yp |      Coef.   Std. Err.      z    P>|z|     [95% Conf. Interval]
-------------+----------------------------------------------------------------
          x1 |   .4977036    .003926   126.77   0.000     .4900089    .5053984
          x2 |   1.495008    .004141   361.03   0.000     1.486892    1.503124
       _cons |  -.9965804   .0074978  -132.92   0.000    -1.011276   -.9818851
------------------------------------------------------------------------------
```

We find our expectations met. Next, for illustrative purposes, we model only the positive outcomes; we treat the data as if there are no zero outcomes.

```
. poisson yp x1 x2 if y>0, nolog
Poisson regression                                Number of obs   =      19925
                                                  LR chi2(2)      =   63221.14
                                                  Prob > chi2     =     0.0000
Log likelihood = -33286.771                       Pseudo R2       =     0.4871
```

yp	Coef.	Std. Err.	z	P>|z|	[95% Conf. Interval]	
x1	.3990055	.0040112	99.47	0.000	.3911437	.4068672
x2	1.17244	.0048163	243.43	0.000	1.163	1.18188
_cons	-.270428	.0085124	-31.77	0.000	-.2871121	-.2537439

The parameter estimates differ substantially from those we specified in constructing the data. We now investigate a zero-truncated model on the positive-outcome subset of data.

```
. ztp yp x1 x2 if y>0, nolog
Zero-truncated Poisson regression                 Number of obs   =      19925
                                                  LR chi2(2)      =   78141.93
                                                  Prob > chi2     =     0.0000
Log likelihood = -24969.996                       Pseudo R2       =     0.6101
```

yp	Coef.	Std. Err.	z	P>|z|	[95% Conf. Interval]	
x1	.4970893	.0045528	109.18	0.000	.488166	.5060125
x2	1.492862	.0056942	262.17	0.000	1.481702	1.504022
_cons	-.9923822	.0116607	-85.10	0.000	-1.015237	-.9695276

We again obtain parameter estimates that closely match those values used in the construction of the data. If we do in fact have data not including zero outcomes, we see that using a zero-truncated model provides a more superior fit than simply using the standard count model and ignoring the distributional violation involved. The next two sections addresses the occasion that there is an excess frequency of zero outcomes in the data, that the number of zero outcomes exceeds what would otherwise be expected for the given distribution.

14.3 Zero-inflated models

Using the same logic as we did with the zero-truncated models, count response models having far more zeros than expected by the distributional assumptions of the Poisson and negative binomial models result in incorrect parameter estimates as well as biased standard errors. We will briefly discuss the nature of the zero-inflated count models and how they adjust the likelihood function to produce appropriate model statistics. Of course, this statement hinges on excess zeros being the only problem in the data. There are often multiple sources resulting in correlated data and therefore overdispersion.

Zero-inflated models consider two distinct sources of zero outcomes. One source is generated from individuals who do not enter into the counting process, the other from those who do enter the count process but result in a zero outcome. Suppose that we have data related to physician visits. We intend to model number of visits made to the office of the patient's physician over a given period. However, suppose that our data also have information on individuals who do not have a physician during the specified period. Such patients are given a count of zero.

When there are far more 0 counts (individuals not visiting a physician) than allowed by the standard Poisson or negative binomial distribution, upon which the respective models are based, we may consider breaking up the model into a part indicating whether an individual has a physician ($y = 1$ or $y = 0$); that is, one part of the model focuses on whether the individual entered into the counting process, and another part models (if the individual enters the counting process) the number of visits made to the physician ($y \geq 0$).

Lambert (1992) first described this type of mixture model in the context of process control in manufacturing. It has since been used in many applications and is now found discussed in nearly every book or article dealing with count response models.

For the zero-inflated model, the probability of observing a zero outcome equals the probability that an individual is in the always-zero group plus the probability that the individual is not in that group times the probability that the counting process produces a zero; see Hilbe and Greene (2007). If we define $B(0)$ as the probability that the binary process results in a zero outcome and $\Pr(0)$ as the count process probability of a zero outcome, the probability of a zero outcome for the system is then given by

$$\Pr(y = 0) = B(0) + (1 - Z)\Pr(0) \tag{14.5}$$

and the probability of a nonzero count is

$$\Pr(y = k; k > 0) = [1 - B(0)]\Pr(k) \tag{14.6}$$

where $\Pr(k)$ is the count process probability that the outcome is k.

Logistic and probit are the most commonly used link functions for the Bernoulli part of the log-likelihood function. Mixing the binary logistic distribution for 0 counts with the negative binomial results in the following zero-inflated log-likelihood function in terms of the linear predictor $x\boldsymbol{\beta}$. In the parameterization below, we indicate the Bernoulli linear predictor as $x\boldsymbol{\beta}_b$ and the negative binomial as $x\boldsymbol{\beta}$ as in Hilbe (2007).

In the following, $F(z\gamma)$ may be in terms of probit or logit. This model has also been called the "with-zeros" or WZ model (for example, the "with zeros Poisson model"). Another estimator is available by allowing that $v_i = z_i\gamma = \tau(x_i\boldsymbol{\beta})$, which introduces only one new parameter, τ, rather than a new covariate list. In the formulas below, we denote the set of zero outcomes as S. We include both weights and offsets in the presentation of the log likelihood and estimating equations for the Poisson version.

$$
\begin{aligned}
\xi_i^{\beta} &= \mathbf{x}_i\boldsymbol{\beta} + \text{offset}_i^{\beta} & (14.7)\\
\xi_i^{\gamma} &= \mathbf{z}_i\boldsymbol{\gamma} + \text{offset}_i^{\gamma} & (14.8)\\
\lambda_i &= \exp(\xi_i^{\beta}) & (14.9)\\
\Pr(Y_i = 0) &= F(\xi_i^{\gamma}) + \ 1 - F(\xi_i^{\gamma}) \ \frac{e^{-\lambda_i}\lambda_i^{y_i}}{y_i!} & (14.10)\\
\Pr(Y_i > 0) &= \ 1 - F(\xi_i^{\gamma}) \ \frac{e^{-\lambda}\lambda^{y_i}}{y_i!} & (14.11)\\
&= \sum_{i\in S} w_i \ln\left[F(\xi_i^{\gamma}) + \ 1 - F(\xi_i^{\gamma}) \ \exp(-\lambda_i)\right]\\
&\quad + \sum_{i\notin S} w_i \left[\ln \ 1 - F(\xi_i^{\gamma}) \ - \lambda_i + \xi_i^{\beta} y_i - \ln(y_i!)\right] & (14.12)
\end{aligned}
$$

with associated derivatives

$$
\begin{aligned}
\frac{\partial}{\partial\beta_j} &= -\sum_{i\in S} w_i x_{ji} \frac{\ 1 - F(\xi_i^{\gamma}) \ \lambda_i \exp(-\lambda_i)}{F(\xi_i^{\gamma}) + \ 1 - F(\xi_i^{\gamma}) \ \exp(-\lambda_i)} + \sum_{i\notin S} w_i x_{ji}(y_i - \lambda_i) & (14.13)\\
\frac{\partial}{\partial\gamma_j} &= \sum_{i\in S} w_i z_{ji} \frac{f(\xi_i^{\gamma}) \ 1 - \exp(-\lambda_i)}{F(\xi_i^{\gamma}) + \ 1 - F(\xi_i^{\gamma}) \ \exp(-\lambda_i)} - \sum_{i\notin S} w_i z_{ji} \frac{f(\xi_i^{\gamma})}{1 - F(\xi_i^{\gamma})} & (14.14)
\end{aligned}
$$

Similarly, for the negative binomial the $F(z\gamma)$ may be in terms of probit or logit. Another estimator is available through assuming that $v_i = z_i\gamma = \tau(x_i\boldsymbol{\beta})$ (introducing only one new parameter τ). We include both weights and offsets in the presentation of the log likelihood and estimating equations for the negative binomial version of zero-inflated models.

$$
\begin{aligned}
\psi(z) &= \text{digamma function evaluated at } z & (14.15)\\
m &= 1/\alpha & (14.16)\\
p_i &= 1/(1 + \alpha\mu_i) & (14.17)\\
\mu_i &= \exp(x_i\boldsymbol{\beta} + \text{offset}_i) & (14.18)\\
\Pr(Y = 0) &= F(z_i\gamma) + \ 1 - F(z_i\gamma) \ p_i^m & (14.19)\\
\Pr(Y > 0) &= \ 1 - F(z_i\gamma) \ \frac{\Gamma(m + y_i)}{\Gamma(y_i + 1)\Gamma(m)} p_i^m (1 - p_i)^{y_i} & (14.20)\\
&= \sum_{i\in S} w_i \ln\left[F(z_i\gamma) + \ 1 - F(z_i\gamma) \ p_i^m\right] + \sum_{i\notin S} w_i \ln \ 1 - F(z_i\gamma)\\
&\quad + w_i \ln\Gamma(m + y_i) - w_i\Gamma(y_i + 1) - w_i \ln\Gamma(m) + w_i m \ln p_i\\
&\quad + w_i y_i \ln(1 - p_i) & (14.21)
\end{aligned}
$$

with associated estimating equations

$$\frac{\partial}{\partial\beta_j} = \sum_{i\in S} w_i x_{ji} \frac{-\ 1 - F(z_i\gamma)\ \mu_i p_i^{m+1}}{F(z_i\gamma) +\ 1 - F(z_i\gamma)\ p_i^m} + \sum_{i\notin S} w_i x_{ji} p_i (y_i - \mu_i) \tag{14.22}$$

$$\frac{\partial}{\partial\gamma_j} = \sum_{i\in S} w_i z_{ji} \frac{f(z_i\gamma)(1-p_i^m)}{F(z_i\gamma) +\ 1 - F(z_i\gamma)\ p_i^m} + \sum_{i\notin S} w_i z_{ji} \frac{-f(z_i\gamma)}{1 - F(z_i\gamma)} \tag{14.23}$$

$$\frac{\partial}{\partial\alpha} = -\sum_{i\in S} w_i \frac{m^2 p_i^m \ln p_i - m\mu_i p_i^{m-1}}{F(z_i\gamma) +\ 1 - F(z_i\gamma)\ p_i^m} - \sum_{i\notin S} w_i \alpha^{-2} \left\{ \frac{\alpha(\mu_i - y_i)}{1+\alpha\mu_i} - \ln(1+\alpha\mu_i) + \psi(y_i + 1/\alpha) - \psi(1/\alpha) \right\} \tag{14.24}$$

In the following examples, we use the same data for estimating zero-inflated models as we did for estimating the zero-truncated models, except that we will include the zero outcomes. A tabulation of the smallest 15 outcomes is given by

```
. tab yp
```

yp	Freq.	Percent	Cum.
0	30,075	60.15	60.15
1	9,488	18.98	79.13
2	3,995	7.99	87.12
3	2,064	4.13	91.24
4	1,176	2.35	93.60
5	758	1.52	95.11
6	466	0.93	96.04
7	373	0.75	96.79
8	287	0.57	97.36
9	201	0.40	97.77
10	166	0.33	98.10
11	142	0.28	98.38
12	113	0.23	98.61
13	93	0.19	98.79
14	67	0.13	98.93
15	67	0.13	99.06
16	56	0.11	99.17
17	36	0.07	99.25
18	30	0.06	99.31
19	27	0.05	99.36
20	29	0.06	99.42
21	30	0.06	99.48
22	23	0.05	99.52
23	11	0.02	99.55
24	12	0.02	99.57
25	15	0.03	99.60
26	19	0.04	99.64
27	16	0.03	99.67
28	14	0.03	99.70
29	10	0.02	99.72
30	13	0.03	99.74
31	4	0.01	99.75
32	6	0.01	99.76
33	12	0.02	99.79

34	2	0.00	99.79
35	8	0.02	99.81
36	4	0.01	99.82
37	7	0.01	99.83
38	6	0.01	99.84
39	5	0.01	99.85
40	6	0.01	99.86
41	5	0.01	99.87
42	1	0.00	99.88
43	6	0.01	99.89
44	6	0.01	99.90
45	2	0.00	99.90
46	3	0.01	99.91
47	4	0.01	99.92
48	2	0.00	99.92
50	1	0.00	99.92
51	1	0.00	99.93
52	2	0.00	99.93
53	2	0.00	99.93
54	1	0.00	99.94
56	1	0.00	99.94
57	1	0.00	99.94
59	2	0.00	99.94
60	1	0.00	99.95
61	1	0.00	99.95
62	2	0.00	99.95
64	1	0.00	99.95
66	2	0.00	99.96
69	1	0.00	99.96
73	2	0.00	99.96
74	2	0.00	99.97
75	2	0.00	99.97
77	1	0.00	99.97
78	1	0.00	99.98
79	2	0.00	99.98
80	1	0.00	99.98
87	1	0.00	99.98
94	1	0.00	99.99
95	1	0.00	99.99
101	1	0.00	99.99
113	1	0.00	99.99
114	1	0.00	99.99
131	1	0.00	100.00
200	1	0.00	100.00
249	1	0.00	100.00
Total	50,000	100.00	

(Continued on next page)

and the corresponding model is fitted by

```
. zip yp x1 x2, inflate(x1 x2) nolog
Zero-inflated Poisson regression                  Number of obs   =      50000
                                                  Nonzero obs     =      19925
                                                  Zero obs        =      30075
Inflation model = logit                           LR chi2(2)      =   78804.72
Log likelihood  = -46064.49                       Prob > chi2     =     0.0000

------------------------------------------------------------------------------
             |      Coef.   Std. Err.      z    P>|z|     [95% Conf. Interval]
-------------+----------------------------------------------------------------
yp           |
          x1 |   .4964941   .0040664   122.10   0.000     .4885242    .5044641
          x2 |   1.488969   .0048179   309.05   0.000     1.479526    1.498411
       _cons |  -.9834876   .0093508  -105.18   0.000    -1.001815   -.9651603
-------------+----------------------------------------------------------------
inflate      |
          x1 |  -.3477685   .2415469    -1.44   0.150    -.8211917    .1256547
          x2 |  -1.653593   .3627721    -4.56   0.000    -2.364613   -.9425725
       _cons |  -4.386779   .5364296    -8.18   0.000    -5.438162   -3.335396
------------------------------------------------------------------------------
```

The inflated frame provides parameter estimates for the binary part of the mixture model. We find here that x1 does not contribute to having zero outcomes, whereas x2 does provide significant information about the binary process. The upper frame is interpreted just like a standard Poisson model. See Long and Freese (2006) and Hilbe (2007) for more discussion, examples, and interpretation of the zero-inflated models.

14.4 Hurdle models

Like zero-inflated models, hurdle models were developed, for example, in Mullahy (1986), to deal with count response having more zeros than allowed by the distributional assumptions of Poisson and negative binomial regression. Unlike zero-inflated models, however, hurdle models, which have at times been referred to as zero-altered models (ZAP and ZANB), clearly partition the data into two groupings.

The first part consists of a binary process that generates only two counts, ones and zeros. This process continues until an unspecified event in the data occurs that begins the generation of a nonzero counting process. In effect, the hurdle to be crossed is the generation of zeros.

The primary difference that the hurdle models have from the zero-inflated models is that the count aspect of the model does not incorporate zero counts in its structure; the counting process is modeled using a zero-truncated model. The binary aspect is modeled using a Bernoulli model with the logit, probit, or clog-log link.

The binary and count aspects of the Poisson hurdle log-likelihood function are represented by

$$\xi_i^\beta = \mathbf{x}_i\boldsymbol{\beta} + \text{offset}_i^\beta \quad (14.25)$$
$$\xi_i^\gamma = \mathbf{z}_i\boldsymbol{\gamma} + \text{offset}_i^\gamma \quad (14.26)$$
$$\lambda_i = \exp(\xi_i^\beta) \quad (14.27)$$
$$\Pr(Y_i = 0) = F(\xi_i^\gamma) \quad (14.28)$$
$$\Pr(Y_i > 0) = 1 - F(\xi_i^\gamma) \frac{e^{-\lambda_i}\lambda_i^{y_i}}{y_i!\; 1 - \exp(\lambda_i)} \quad (14.29)$$
$$= \sum_{i \in S} w_i \ln\left[F(\xi_i^\gamma)\right] + \sum_{i \notin S} w_i \Big[\ln 1 - F(\xi_i^\gamma) - \lambda_i + \xi_i^\beta y_i - \ln(y_i!) - \ln 1 - \exp(\lambda_i)\Big] \quad (14.30)$$

Given the structure of the model, the estimating equations for $\boldsymbol{\beta}$ and $\boldsymbol{\gamma}$ from the hurdle model are going to be the same as those obtained separately from the binary and zero-truncated count models.

Here we use the Poisson-logit model on the same data we used with the zero-inflated Poisson (with logit inflation). The Stata command used here is available on this book's web site.

```
. hplogit yp x1 x2, nolog
Poisson-Logit Hurdle Regression                    Number of obs   =      50000
                                                   Wald chi2(2)    =   12069.08
Log likelihood = -46259.401                        Prob > chi2     =     0.0000

------------------------------------------------------------------------------
             |      Coef.   Std. Err.      z    P>|z|     [95% Conf. Interval]
-------------+----------------------------------------------------------------
logit        |
          x1 |    .689183   .0130306    52.89   0.000     .6636435    .7147225
          x2 |   2.089292   .0193773   107.82   0.000     2.051313    2.127271
       _cons |  -.7030531   .0126769   -55.46   0.000    -.7278994   -.6782068
-------------+----------------------------------------------------------------
poisson      |
          x1 |   .4970893   .0045528   109.18   0.000      .488166    .5060125
          x2 |   1.492862   .0056942   262.17   0.000     1.481702    1.504022
       _cons |  -.9923822   .0116607   -85.10   0.000    -1.015237   -.9695276
------------------------------------------------------------------------------
AIC Statistic =      1.850
```

To check the viability of the above hurdle model, we model the two parts separately. First we must create a new binary response variable for the logit model. We generate a variable called yp0 that takes the value of 1 if yp is greater than zero and 0 if yp is in fact 0. The results are

```
. gen yp0 = yp > 0

. logit yp0 x1 x2, nolog

Logistic regression                               Number of obs   =      50000
                                                  LR chi2(2)      =   24661.07
                                                  Prob > chi2     =     0.0000
Log likelihood = -21289.406                       Pseudo R2       =     0.3668

------------------------------------------------------------------------------
         yp0 |      Coef.   Std. Err.      z    P>|z|     [95% Conf. Interval]
-------------+----------------------------------------------------------------
          x1 |    .689183   .0130306    52.89   0.000     .6636435    .7147225
          x2 |   2.089292   .0193773   107.82   0.000     2.051313     2.12727
       _cons |  -.7030531   .0126769   -55.46   0.000    -.7278994   -.6782068
------------------------------------------------------------------------------
```

The independent logit model is identical to the hurdle model logit. We next model the positive outcomes in the data by using a zero-truncated Poisson.

```
. ztp yp x1 x2 if yp>0, nolog

Zero-truncated Poisson regression                 Number of obs   =      19925
                                                  LR chi2(2)      =   78141.93
                                                  Prob > chi2     =     0.0000
Log likelihood = -24969.996                       Pseudo R2       =     0.6101

------------------------------------------------------------------------------
          yp |      Coef.   Std. Err.      z    P>|z|     [95% Conf. Interval]
-------------+----------------------------------------------------------------
          x1 |   .4970893   .0045528   109.18   0.000      .488166    .5060125
          x2 |   1.492862   .0056942   262.17   0.000     1.481702    1.504022
       _cons |  -.9923822   .0116607   -85.10   0.000    -1.015237   -.9695276
------------------------------------------------------------------------------
```

You may obtain predicted values for each equation in the Hurdle model by specifying (you must run these commands immediately after fitting the hurdle model)

```
. predict xb1, eq(#1)         /* logit prediction                     */
. predict xb2, eq(#2)         /* zero-truncated Poisson prediction */
. gen mu1 = 1/(1+exp(-xb1)) /* predicted Pr(yp=0)                   */
. gen mu2 = exp(xb2)          /* predicted ztp count                  */
```

Taking 20 observations from well inside the dataset, we observe that the values of `mu1` and `mu2` are as we would expect; `mu1` values range from 0 to 1. They were based on `yp` being converted to `yp0` having a 0/1 format. As expected, though, values of `mu1` tend to be closer to 1 with the greater values of `yp`. Ideally, values of `mu2` should fairly match the values of `yp`.

```
. list yp0 mu1 yp mu2 in 1000/1020
```

	yp0	mu1	yp	mu2
1000.	0	.2842635	0	.3164035
1001.	0	.3500637	0	.391126
1002.	0	.3485616	0	.3907986
1003.	0	.0235478	0	.0427647
1004.	0	.0707165	0	.0968218
1005.	1	.9420996	6	4.500294
1006.	0	.2472051	0	.2751199
1007.	1	.1182253	1	.1462684
1008.	0	.0154965	0	.0318236
1009.	1	.6394172	2	.9197059
1010.	0	.4987893	0	.6157939
1011.	0	.0279743	0	.0484527
1012.	0	.0617883	0	.0874684
1013.	0	.2765903	0	.3076101
1014.	0	.1457443	0	.1733399
1015.	1	.6938534	1	1.096809
1016.	0	.9472021	0	4.830457
1017.	1	.6927031	1	1.095622
1018.	1	.9986951	80	69.96786
1019.	1	.5578175	1	.7262897
1020.	0	.1970691	0	.2255661

14.5 Heterogeneous negative binomial models

We next consider the heterogeneous negative binomial model. Stata refers to this model as a generalized negative binomial, which is modeled using the `gnbreg` command, but *generalized* negative binomial has long been used in the literature to refer to an entirely different model. The term *heterogeneous*, however, has been used to refer to this model, and by referring to it as such we follow common usage.

The heterogeneous negative binomial model not only estimates the standard list of parameters and their associated standard errors but also allows parameterization of the ancillary or overdispersion parameter α. Both the NB-2α and NB-1δ parameters may be parameterized. The log-likelihood function of the heterogeneous NB-2 model can be derived, allowing for an offset in each covariate list, as

$$\alpha_i = \exp(\mathbf{z}_i\boldsymbol{\gamma} + \text{offset}_i^{\gamma}) \tag{14.31}$$

$$m_i = 1/\alpha_i \qquad p_i = 1/(1+\alpha_i\mu_i) \qquad \mu_i = \exp(\mathbf{x}_i\boldsymbol{\beta} + \text{offset}_i^{\beta}) \tag{14.32}$$

$$= \sum_{i=1}^{n}\Big[\ln\,\Gamma(m_i+y_i) \;-\ln\,\Gamma(y_i+1) \;-\ln\,\Gamma(m_i) + m_i\ln(p_i) + y_i\ln(1-p_i)\Big] \tag{14.33}$$

For an example of this model, we return to the previously used MedPar data. We model a negative binomial (NB-2) before the `gnbreg` for comparison.

```
. use http://www.stata-press.com/data/hh2/medpar, clear
. nbreg los hmo white type2 type3 died, nolog
Negative binomial regression                      Number of obs   =       1495
                                                  LR chi2(5)      =     151.08
Dispersion     = mean                             Prob > chi2     =     0.0000
Log likelihood = -4780.9543                       Pseudo R2       =     0.0156

------------------------------------------------------------------------------
         los |      Coef.   Std. Err.      z    P>|z|     [95% Conf. Interval]
-------------+----------------------------------------------------------------
         hmo |   -.065878   .0527626    -1.25   0.212    -.1692908    .0375348
       white |  -.1218214   .0678367    -1.80   0.073    -.2547789     .011136
       type2 |   .2418841   .0502247     4.82   0.000     .1434454    .3403227
       type3 |   .7286303   .0754246     9.66   0.000     .5808008    .8764597
        died |  -.2355563    .040457    -5.82   0.000    -.3148506   -.1562619
       _cons |   2.372991   .0682086    34.79   0.000     2.239305    2.506678
-------------+----------------------------------------------------------------
    /lnalpha |  -.8347201   .0448234                     -.9225724   -.7468679
-------------+----------------------------------------------------------------
       alpha |   .4339959   .0194532                      .3974952    .4738484
------------------------------------------------------------------------------
Likelihood-ratio test of alpha=0:  chibar2(01) = 4108.03 Prob>=chibar2 = 0.000
```

Having determined that `hmo` and `white` do not contribute to the parameterization of α, we model the data by using `gnbreg`, with `type` and `died` as predictors of α.

```
. gnbreg los hmo white type2 type3 died, lnalpha(type2 type3 died) nolog
Generalized negative binomial regression          Number of obs   =       1495
                                                  LR chi2(5)      =     107.03
                                                  Prob > chi2     =     0.0000
Log likelihood = -4731.6185                       Pseudo R2       =     0.0112

------------------------------------------------------------------------------
             |      Coef.   Std. Err.      z    P>|z|     [95% Conf. Interval]
-------------+----------------------------------------------------------------
los          |
         hmo |  -.0630361   .0492268    -1.28   0.200    -.1595188    .0334466
       white |  -.0756874   .0642429    -1.18   0.239    -.2016013    .0502264
       type2 |   .2349034    .049132     4.78   0.000     .1386065    .3312003
       type3 |   .7701078   .0971028     7.93   0.000     .5797897    .9604258
        died |  -.2283633   .0444074    -5.14   0.000    -.3154003   -.1413263
       _cons |   2.327675    .063542    36.63   0.000     2.203135    2.452215
-------------+----------------------------------------------------------------
lnalpha      |
       type2 |   .1660303   .1165805     1.42   0.154    -.0624631    .3945238
       type3 |     .74406   .1517917     4.90   0.000     .4465538    1.041566
        died |     .77198   .0932257     8.28   0.000     .5892611    .9546989
       _cons |  -1.258687   .0657166   -19.15   0.000    -1.387489   -1.129885
------------------------------------------------------------------------------
```

The AIC statistic is lower for the heterogeneous model than for the basic NB-2. We also find that `type3`, that is, emergency admissions, and `died` significantly contribute to α. That is, both emergency admissions in reference to elective admissions and urgent admission, and the death of a patient, contribute to the overdispersion in the data.

Displaying only the header output from the **glm** command negative binomial output, with the ancillary or heterogeneity parameter given as the constant of the full-maximum likelihood model, we see

```
. glm los hmo white type2 type3 died, family(nb .4339959) notable nolog
Generalized linear models                    No. of obs      =       1495
Optimization     : ML                        Residual df     =       1489
                                             Scale parameter =          1
Deviance         =  1566.760836              (1/df) Deviance =   1.052224
Pearson          =  1687.716966              (1/df) Pearson  =   1.133457
Variance function: V(u) = u+(.4339959)u^2    [Neg. Binomial]
Link function    : g(u) = ln(u)              [Log]
                                             AIC             =   6.403952
Log likelihood   = -4780.954342              BIC             =  -9317.653
```

This output shows that there is some negative binomial overdispersion in the data, 13%. This overdispersion is derived largely from serious patients, including those that died. Other patients do not seem to contribute to the extra correlation of the data. This information is considerably superior to simply knowing that the Poisson model of the same data is overdispersed (as seen in the header output of a Poisson model where you should especially note the Pearson dispersion statistic).

```
. glm los hmo white type2 type3 died, family(poisson) notable nolog
Generalized linear models                    No. of obs      =       1495
Optimization     : ML                        Residual df     =       1489
                                             Scale parameter =          1
Deviance         =  7954.791152              (1/df) Deviance =   5.342371
Pearson          =  9251.713475              (1/df) Pearson  =   6.213374
Variance function: V(u) = u                  [Poisson]
Link function    : g(u) = ln(u)              [Log]
                                             AIC             =     9.1518
Log likelihood   = -6834.970362              BIC             =  -2929.622
```

We will show another method of interpreting the ancillary or heterogeneity parameter. We use Stata's **gnbreg** command to fit a heterogeneous (or Stata's generalized) negative binomial regression model on our data, but this time allowing only the scale parameter α to depend on whether the patient died.

```
. gnbreg los hmo white type2 type3, lnalpha(died) irr nolog
Generalized negative binomial regression          Number of obs   =       1495
                                                  LR chi2(4)      =     131.39
                                                  Prob > chi2     =     0.0000
Log likelihood = -4757.8933                       Pseudo R2       =     0.0136

------------------------------------------------------------------------------
         los |        IRR   Std. Err.      z    P>|z|     [95% Conf. Interval]
-------------+----------------------------------------------------------------
         hmo |   .9350806   .0479923    -1.31   0.191     .8455937    1.034038
       white |   .9107966   .0592616    -1.44   0.151     .8017471    1.034678
       type2 |   1.245232   .0612273     4.46   0.000      1.13083    1.371208
       type3 |    2.13208   .1611669    10.02   0.000     1.838485    2.472561
-------------+----------------------------------------------------------------
     lnalpha |   (type gnbreg to see ln(alpha) coefficient estimates)
------------------------------------------------------------------------------
```

The outcome shows similar incidence-rate ratios to what we obtained with previous models. To see the fitted values for the coefficients in the parameterization of α, we type

```
. gnbreg
Generalized negative binomial regression          Number of obs   =       1495
                                                  LR chi2(4)      =     131.39
                                                  Prob > chi2     =     0.0000
Log likelihood = -4757.8933                       Pseudo R2       =     0.0136

------------------------------------------------------------------------------
             |      Coef.   Std. Err.      z    P>|z|     [95% Conf. Interval]
-------------+----------------------------------------------------------------
los          |
         hmo |  -.0671226   .0513243    -1.31   0.191    -.1677163    .0334712
       white |  -.0934357   .0650656    -1.44   0.151    -.2209621    .0340906
       type2 |    .219322   .0491694     4.46   0.000     .1229518    .3156923
       type3 |   .7570981   .0755914    10.02   0.000     .6089417    .9052545
       _cons |   2.304254   .0642043    35.89   0.000     2.178416    2.430092
-------------+----------------------------------------------------------------
lnalpha      |
        died |   .8171296   .0917266     8.91   0.000     .6373487    .9969105
       _cons |  -1.135773   .0575386   -19.74   0.000    -1.248547      -1.023
------------------------------------------------------------------------------
```

From the output we can determine that

$$\alpha_{\text{died}} = \exp\left(-1.135773 + 1.0(.8171296)\right) \quad (14.34)$$

$$= \exp(-.3186434) \quad (14.35)$$

$$= .7271348 \quad (14.36)$$

$$\alpha_{\text{survived}} = \exp\left(-1.135773 + 0.0(.8171296)\right) \quad (14.37)$$

$$= \exp(-1.135773) \quad (14.38)$$

$$= .32117376 \quad (14.39)$$

compared with $\alpha = .4457567$ when we do not account for the survival of the patient. The coefficient of the `died` variable is significant, and we conclude that this is a better model for the data than the previous models. There is clearly even greater overdispersion for the subset of data in which patients died, and accounting for that property results in a better model than specifying a geometric model.

The final model we will consider in this chapter is the generalized Poisson. This model is similar to the negative binomial model but is not based on the negative binomial PDF. Moreover, as will be discussed, one can use it to model both overdispersion and underdispersion.

14.6 Generalized Poisson regression models

The most commonly used regression model for count data is the Poisson regression model with a log link function described by the probability mass function:

$$f(y_i;\theta_i,\mathbf{x}_i) = \frac{\theta_i^{y_i}e^{-\theta_i}}{y_i!}, \quad y_i = 0,1,2,\ldots, \quad \theta_i > 0 \tag{14.40}$$

Covariates and regression parameters are then introduced via the expected outcome by the inverse link function $\theta_i = \exp(\mathbf{x}_i\boldsymbol{\beta})$.

We see that $\mathbf{x}_i$ is a covariate vector and $\boldsymbol{\beta}$ is a vector of unknown coefficients to be estimated. When there is overdispersion due to latent heterogeneity or overdispersion, people often assume a gamma mixture of Poisson variables. Suppose that ϑ_i is an unobserved individual heterogeneity factor, with $\exp(\vartheta_i)$ following a gamma distribution with mean 1 and variance κ. We now assume the response vector to follow a modification of the Poisson model with mean $\theta_i^\star = \exp(\mathbf{x}_i\boldsymbol{\beta} + \vartheta_i)$, which describes the well-known negative binomial regression model with mass function

$$f(y_i;\theta_i,\mathbf{x}_i,\kappa) = \frac{\Gamma(y_i+1/\kappa)}{\Gamma(y_i+1)\Gamma(1/\kappa)}\frac{(\kappa\theta_i)^{y_i}}{(1+\kappa\theta_i)^{y_i+1/\kappa}} \tag{14.41}$$

where κ is a nonnegative parameter indicating the degree of overdispersion. The negative binomial model reduces to a Poisson model in the limit as $\kappa \to 0$. We emphasize this material throughout chapter 13.

We now assume that the response variable instead follows a generalized Poisson(GP) distribution with mass function given by

$$f(y_i;\theta_i,\delta) = \frac{\theta_i(\theta_i+\delta y_i)^{y_i-1}e^{-\theta_i-\delta y_i}}{y_i!} \tag{14.42}$$

for $y_i = 0,1,2,\ldots,\ \theta_i > 0,\ 0 \le \delta < 1$. Following Joe and Zhu (2005), we see that

$$\mathrm{E}[Y_i] = \mu_i = \frac{\theta_i}{1-\delta} \tag{14.43}$$

$$\mathrm{V}[Y_i] = \frac{\theta_i}{(1-\delta)^3} = \frac{1}{(1-\delta)^2}\mathrm{E}[Y_i] = \phi\mathrm{E}[Y_i] \tag{14.44}$$

More information on the generalized Poisson regression model is also presented in Consul and Famoye (1992).

The term $\phi = 1/(1-\delta)^2$ plays the role of a dispersion factor in the generalized Poisson mass function. Clearly, when $\delta = 0$, the GP distribution reduces to the usual Poisson distribution with parameter θ_i. Further, when $\delta < 0$ underdispersion is assumed in the model, and if $\delta > 0$, we get overdispersion. In this discussion, we are concerned with overdispersion, $\delta > 0$.

In chapter 13, we investigated the negative binomial model as an alternative to the Poisson for the length of stay data. Here we consider the generalized Poisson as an alternative. The software used to fit the generalized Poisson regression models parameterizes δ by using Fisher's Z transformation, thus forcing the dispersion parameter to be positive. Although the software does limit investigation of underdispersed models, we are interested only in overdispersed models. It is available at this book's web site.

```
. gpoisson los hmo white type2 type3, irr nolog
Generalized Poisson regression                    Number of obs   =       1495
                                                  LR chi2(4)      =      36.18
Dispersion     =  2.472845                        Prob > chi2     =     0.0000
Log likelihood = -4816.9845                       Pseudo R2       =     0.0037

------------------------------------------------------------------------------
         los |       IRR   Std. Err.      z    P>|z|     [95% Conf. Interval]
-------------+----------------------------------------------------------------
         hmo |  .9578067   .0470583    -0.88   0.380     .8698758    1.054626
       white |  .8838994   .0533205    -2.05   0.041     .7853348    .9948345
       type2 |   1.16412   .0528339     3.35   0.001     1.065039    1.272417
       type3 |  1.411021     .09608     5.06   0.000     1.234732    1.612478
-------------+----------------------------------------------------------------
  /tanhdelta |   .686312   .0151052                      .6567064    .7159177
-------------+----------------------------------------------------------------
       delta |  .5956076   .0097466                      .5761674    .6143742
------------------------------------------------------------------------------
Likelihood-ratio test of delta=0:  chibar2(1) = 4223.85 Prob>=chibar2 = 0.0000
```

Again we find that there is evidence of overdispersion such that the Poisson model is not flexible enough for these data. Now we are faced with choosing the generalized Poisson or the negative binomial. One can make such a choice between two nonnested models by following the arguments in Vuong (1989).

We investigated a Poisson model in chapter 12 for the number of deaths as a function of whether patients were smokers and their age category. The model is adjusted for person-years where previously we were able to specify the log of the offset. Here the command lacks the utility so that we generate a new variable and specify the offset.

```
. gen lpy = ln(pyears)
. gpoisson deaths smokes a2-a5, offset(lpy) irr nolog
Generalized Poisson regression                    Number of obs   =         10
                                                  LR chi2(5)      =      46.78
Dispersion     =  1.569606                        Prob > chi2     =     0.0000
Log likelihood = -36.714438                       Pseudo R2       =     0.3892

------------------------------------------------------------------------------
      deaths |       IRR   Std. Err.      z    P>|z|     [95% Conf. Interval]
-------------+----------------------------------------------------------------
      smokes |  1.130631   .1617845     0.86   0.391     .8541235    1.496653
          a2 |  4.664427   1.464581     4.90   0.000     2.520751    8.631111
          a3 |  14.11487    4.35194     8.59   0.000     7.713147    25.82985
          a4 |  30.61026   9.146816    11.45   0.000     17.04178    54.98182
          a5 |  43.20668   13.39753    12.15   0.000     23.52937    79.33988
         lpy |  (offset)
-------------+----------------------------------------------------------------
  /tanhdelta |   .380219   .1855922                      .0164649    .7439731
-------------+----------------------------------------------------------------
       delta |  .3628976   .1611507                      .0164634    .6315396
------------------------------------------------------------------------------
Likelihood-ratio test of delta=0:  chibar2(1) =    4.09 Prob>=chibar2 = 0.0216
```

Although the inference of the model is the same, in the Wald test the dispersion parameter is zero and is rejected in favor of the more general GP model. In a full analysis, we would now also fit a negative binomial model and then choose between these alternatives.

14.7 Censored count response models

One can amend both the Poisson and negative binomial models to allow censored observations. Censoring occurs when a case or observation is lost to the study. However, unlike truncation, censored observations make a known contribution to the model that is taken into account by the respective probability and log-likelihood functions.

Two parameterizations of censored count regression models are in the statistical literature. The traditional version of censoring, with respect to count response models, is found in the domains of econometrics and social science in general. Essentially, censoring is defined in terms of cutpoints. Left-censored observations range from 0 to a lower cutpoint, the value of which is taken by all left-censored observations. That is, left-censored observations are revalued to the value of the lower cutpoint. Right-censored observations, on the other hand, are revalued from the upper end of the range response counts to the value of the upper cutpoint. The probability function is redefined to account for the revalued observations. See Cameron and Trivedi (1998), Greene (2003), and Hilbe (2007) for the derivations and a full description of this econometric parameterization of censoring. Stata does not support this parameterization, but we illustrate these models here with user-written programs.

Traditional survival analyses are based on either nonparametric distributions such as the Cox proportional hazards model or parametric continuous responses such as exponential, Weibull, lognormal, log-logistic, gamma, or similar types of distributions. Moreover, parametric model probability functions are usually reparameterized as survivor functions or hazard functions, depending on whether survival or failure is being modeled. Cox models are likewise based on survivor or hazard functions; however, these functions are not based on a specific underlying probability function.

Censoring is an essential attribute of all survival models, with left-censored observations referring to the event of interest happening to a case before entry into the study. Admittedly, left censoring is relatively rare. Right-censored observations, on the other hand, are those that are lost to the study before it is concluded. Observations can be lost by simply leaving the study, by being disqualified from continuing in the study, or by some other means. Cleves, Gould, and Gutierrez (2004) provide an excellent introduction to survival models using Stata; see this source for more discussion of this type of survival modeling.

A second parameterization of censored count models derives from traditional parametric survival analysis. However, there are some major differences. Foremost, censored count models are not parameterized in terms of a survivor or hazard function. Rather, the model is similar to the traditional Poisson or negative binomial (NB2) model, except for allowing censored observations conceived of in the manner of survival models. There are no cutpoints and no revalued observations. An observation that is part of a count process can be lost to the study, in which case it is called right censored. For example, in a typical hospital length of stay study, the response is the count of days that a patient remains in the hospital. If we know that a patient has stayed in the hospital for 15 days but is transferred to a long-term facility before being discharged from the standard acute care hospital, that patient can be said to be right censored, particularly if later information about the patient is unavailable to the ongoing study at the acute site.

The log-likelihood function of the survival parameterization–censored Poisson is given by

$$\begin{aligned}(\boldsymbol{\beta};y) \quad = \quad & \sum_{i\in F} -\exp(\mathbf{x}_i\boldsymbol{\beta}) + y_i\mathbf{x}_i\boldsymbol{\beta} - \ln\Gamma(y_i+1) \quad + \\ & \sum_{i\in L} \ln\Gamma_I\left(y_i, \exp\ \mathbf{x}_i\boldsymbol{\beta}\ \right) \ + \\ & \sum_{i\in R} \ln\left[1-\ln\Gamma_I\left(y_i, \exp\ \mathbf{x}_i\boldsymbol{\beta}\ \right)\right] \end{aligned} \tag{14.45}$$

where F is the set of noncensored observations, L is the set of left-censored observations, R is the set of right-censored observations, and $\Gamma_I()$ is the two-parameter incomplete gamma function. Similarly, the survival parameterization–censored negative binomial is given by

$$\begin{aligned}(\boldsymbol{\beta},\alpha;y) \quad = \quad & \sum_{i\in F}\left\{y_i\ln\left(\frac{\mu_i}{1+\mu_i}\right) - \frac{1}{\alpha}\ln(1+\mu_i) + \ln\Gamma\left(y_i+\frac{1}{\alpha}\right) - \right. \\ & \left. \ln\Gamma(y_i+1) - \ln\Gamma(1/\alpha)\right\} + \\ & \sum_{i\in L} \ln B_I\left(y_i, n-y_i+1, \exp\ \mathbf{x}_i\boldsymbol{\beta}\ \right) \ + \\ & \sum_{i\in R} \ln B_I\left(y_i+1, n-y_i, \exp\ \mathbf{x}_i\boldsymbol{\beta}\ \right) \end{aligned} \tag{14.46}$$

where $\alpha = \exp(\tau)$, $\mu = \alpha\exp(\mathbf{x}\boldsymbol{\beta})$, and $B_I()$ is the three-parameter incomplete beta function.

The `cpoisson` command models censored Poisson data. It has been written so that specification of a censor variable is required to communicate which observations are censored and in what direction. The censor variable should be set such that 1 indicates no censoring, 0 indicates an observation that is left censored, and -1 indicates an observation that is right censored.

Here we illustrate how one can use these censored models with data. First we need to load a modification of the MedPar dataset used earlier.

```
. use http://www.stata-press.com/data/hh2/medparc, clear
. describe
Contains data from http://www.stata-press.com/data/hh2/medparc.dta
  obs:         1,495
 vars:            12                          11 Dec 2006 10:10
 size:        50,830 (99.5% of memory free)
-------------------------------------------------------------------------------
              storage  display     value
variable name   type   format      label      variable label
-------------------------------------------------------------------------------
provnum         str6   %9s                    Provider number
died            float  %9.0g                  1=Died; 0=Alive
white           float  %9.0g                  1=White
hmo             byte   %9.0g                  HMO/readmit'
los             int    %9.0g                  Length of Stay
age80           float  %9.0g                  1=age>=80
age             byte   %9.0g                  9 age groups
type1           byte   %8.0g                  Elective Admit
type2           byte   %8.0g                  Urgent Admit
type3           byte   %8.0g                  Emergency Admit
cen1            float  %9.0g                  censor=1:  no censoring
cenx            byte   %8.0g                  censor: -1 lost to study
-------------------------------------------------------------------------------
Sorted by:  provnum
```

We will use `cenx` as the censor variable. Here we illustrate those observations that are right censored.

```
. tab cenx
 censor: -1 |
    lost to |
      study |      Freq.     Percent        Cum.
------------+-----------------------------------
         -1 |          8        0.54        0.54
          1 |      1,487       99.46      100.00
------------+-----------------------------------
      Total |      1,495      100.00
```

There are 8 observations lost to study. To see which ones they are, we list the length of stay for those observations that are right censored.

```
. list los cenx if cenx == -1

      +------------+
      | los   cenx |
      |------------|
 943. |  24     -1 |
 970. |  24     -1 |
1066. |  24     -1 |
1127. |  24     -1 |
1450. |  24     -1 |
      |------------|
1452. | 116     -1 |
1466. |  91     -1 |
1481. |  74     -1 |
      +------------+
```

There are 8 censored observations with high lengths of stay, as well as observations for several patients who exited the study after being in the hospital for 24 days. We will next observe how many observations were not censored above 62 days. This will tell us how many left the study after 2 months of hospitalization.

```
. tab los if los > 62

  Length of
       Stay |      Freq.     Percent        Cum.
------------+-----------------------------------
         63 |          1       16.67       16.67
         65 |          1       16.67       33.33
         70 |          1       16.67       50.00
         74 |          1       16.67       66.67
         91 |          1       16.67       83.33
        116 |          1       16.67      100.00
------------+-----------------------------------
      Total |          6      100.00
```

All patients staying longer than 70 days were lost to the study. We know that the patients stayed for the days indicated but not what happened thereafter. Many outcome registries provide information on patient disposition—that is, whether patients died or were transferred to another facility or whether they developed symptoms that disqualified them from further participation in the study. We do not have this type of information here. Nevertheless, let us look at the observations from 20 days to 30.

```
. tab los if los > 20 & los < 30

  Length of
       Stay |      Freq.     Percent        Cum.
------------+-----------------------------------
         21 |         18       22.50       22.50
         22 |         15       18.75       41.25
         23 |         10       12.50       53.75
         24 |         11       13.75       67.50
         25 |          4        5.00       72.50
         26 |          7        8.75       81.25
         27 |          7        8.75       90.00
         28 |          5        6.25       96.25
         29 |          3        3.75      100.00
------------+-----------------------------------
      Total |         80      100.00
```

Five of 11 patients staying 24 days are lost to study. This result is common in real outcomes data. It probably indicates that patients having a given DRG (diagnostic-related group) are transferred to a long-term facility after 24 days. Patients having other DRGs remain.

We can now model the data. We model length of stay (`los`) by whether the patient is a member of a health maintenance organization (`hmo`) ethnic identity (white = 1; all others = 0), and type of admission (1 = elective, 2 = urgent, 3 = emergency). For this example, all patients are alive throughout the term of the study. Death would not change the model except that death is not a cause of censoring, even though death does mean that the patient is lost to the study. Censoring indicates nondeath loss to the study.

```
. cpoisson los hmo white type2 type3, cen(cenx) nolog
Censored Poisson Regression                    Number of obs   =       1495
                                               Wald chi2(4)    =     516.86
Log likelihood = -6625.8252                    Prob > chi2     =     0.0000

------------------------------------------------------------------------------
         los |      Coef.   Std. Err.      z    P>|z|     [95% Conf. Interval]
-------------+----------------------------------------------------------------
         hmo |  -.0635152   .0239736    -2.65   0.008    -.1105027   -.0165278
       white |   -.124207   .0282263    -4.40   0.000    -.1795296   -.0688844
       type2 |   .2245861   .0211375    10.62   0.000     .1831573    .2660148
       type3 |   .5682473   .0283422    20.05   0.000     .5126977     .623797
       _cons |   2.299732   .0279665    82.23   0.000     2.244919    2.354546
------------------------------------------------------------------------------
AIC Statistic =        8.871
LM Value      =    58313.181                    LM Chi2(1)     =       0.000
Score test OD =    30444.064                    Score Chi(1)   =       0.000
```

All predictors are significant; however, there appears to be substantial overdispersion in the data as indicated by the Lagrange multiplier test and score test for overdispersion. See Hilbe (2007) for a description of these tests. Modeling the data by using a censored negative binomial may be more appropriate.

```
. censornb los hmo white type2 type3, cen(cenx) nolog
Censored Negative Binomial Regression          Number of obs   =       1495
                                               Wald chi2(4)    =      76.72
Log likelihood = -4741.2083                    Prob > chi2     =     0.0000

------------------------------------------------------------------------------
         los |      Coef.   Std. Err.      z    P>|z|     [95% Conf. Interval]
-------------+----------------------------------------------------------------
xb           |
         hmo |  -.0616616   .0525967    -1.17   0.241    -.1647492    .0414261
       white |  -.1128446   .0680806    -1.66   0.097    -.2462801     .020591
       type2 |   .2232317   .0500492     4.46   0.000     .1251371    .3213262
       type3 |   .5673344   .0770629     7.36   0.000     .4162939     .718375
       _cons |    2.28947   .0673708    33.98   0.000     2.157426    2.421514
-------------+----------------------------------------------------------------
lnalpha      |
       _cons |  -.8401556   .0452084   -18.58   0.000    -.9287624   -.7515489
-------------+----------------------------------------------------------------
       alpha |   .4316433   .0195139                      .3950423    .4716355
------------------------------------------------------------------------------
AIC Statistic =        6.349
```

The `hmo` and `white` variables no longer appear to be significant predictors of length of stay. Given the substantial reduction in the value of the AIC statistic over that of the censored Poisson, we favor the censored negative binomial. We next see what difference the censored observations make to the model. Using a censor variable, `cen1`, consisting entirely of ones, we model the data by using a standard negative binomial (NB2).

```
. censornb los hmo white type2 type3, cen(cen1) nolog
Censored Negative Binomial Regression              Number of obs   =       1495
                                                   Wald chi2(4)    =     109.82
Log likelihood = -4797.4766                        Prob > chi2     =     0.0000

------------------------------------------------------------------------------
         los |      Coef.   Std. Err.      z    P>|z|     [95% Conf. Interval]
-------------+----------------------------------------------------------------
xb           |
         hmo |  -.0679552   .0532613    -1.28   0.202    -.1723455    .0364351
       white |  -.1290654   .0685418    -1.88   0.060    -.2634049     .005274
       type2 |    .221249   .0505925     4.37   0.000     .1220894    .3204085
       type3 |   .7061588   .0761311     9.28   0.000     .5569446     .855373
       _cons |   2.310279   .0679474    34.00   0.000     2.177105    2.443453
-------------+----------------------------------------------------------------
lnalpha      |
       _cons |   -.807982   .0444542   -18.18   0.000    -.8951107   -.7208533
-------------+----------------------------------------------------------------
       alpha |   .4457567   .0198158                      .4085624    .4863371
------------------------------------------------------------------------------
AIC Statistic =         6.425
```

As expected, there is little difference in the results. However, we do see the favorable effect of taking censoring into account. The censored model has a slightly reduced value for the AIC, and α is lower as well. Since only eight of the 1,495 observations are censored, we would not expect a significant change in the model results.

We use a robust variance estimator to the model to see whether there is a provider clustering effect. That is, perhaps length of stay depends partly on the provider; `los` is more highly correlated within providers than between them. We have also exponentiated the coefficients so that they may be interpreted as incidence-rate ratios.

```
. censornb los hmo white type2 type3, cen(cenx) cluster(provnum) irr nolog
Censored Negative Binomial Regression              Number of obs   =       1495
                                                   Wald chi2(4)    =      27.22
Log pseudolikelihood = -4741.2083                  Prob > chi2     =     0.0000
                               (Std. Err. adjusted for 54 clusters in provnum)
------------------------------------------------------------------------------
             |               Robust
         los |        IRR   Std. Err.      z    P>|z|     [95% Conf. Interval]
-------------+----------------------------------------------------------------
xb           |
         hmo |    .940201   .0480066    -1.21   0.227     .8506647    1.039161
       white |   .8932895   .0647576    -1.56   0.120     .7749715    1.029672
       type2 |    1.25011   .0755744     3.69   0.000     1.110426    1.407366
       type3 |    1.76356   .2863148     3.49   0.000     1.282913    2.424282
-------------+----------------------------------------------------------------
lnalpha      |
       _cons |  -.8401556   .0551285   -15.24   0.000    -.9482056   -.7321057
-------------+----------------------------------------------------------------
       alpha |   .4316433   .0237959                      .3874356    .4808953
------------------------------------------------------------------------------
AIC Statistic =         6.349
```

There does appear to be a clustering effect, although not large, so we leave the model in this form.

We next discuss polytomous response models, which include models having both ordered and unordered levels of response. Like the models we have addressed in this section, ordered and unordered response models are a variety of discrete response regression model. Ordered response models in particular have undergone some interesting enhancements in recent years. We will address many of these in the next chapter.

Part V

Multinomial Response Models

15 The ordered-response family

Contents

This chapter addresses data in which we record an integer response. Our responses are limited to a finite choice set, and there is some meaning to the order of the values assigned. The actual values are irrelevant but are interpreted to have meaning in that a larger value is considered higher in some sense. There is no interpretation in regard to the difference between the outcomes (the labels may be equally spaced but may represent different measures). Because the actual values are irrelevant, without loss of generality we may present a discussion of the ordered-outcome model, assuming that the outcomes are defined by the set $\{1, 2, \ldots, r\}$.

We begin by considering a model given by

$$y^* = \mathbf{X}\boldsymbol{\beta} + \epsilon \tag{15.1}$$

where y^* is unobserved. Instead, we observe an outcome y defined by

$$\begin{aligned} y &= 1 \text{ if } \kappa_0 < y^* \leq \kappa_1 & (15.2)\\ &= 2 \text{ if } \kappa_1 < y^* \leq \kappa_2 & (15.3)\\ &= 3 \text{ if } \kappa_2 < y^* \leq \kappa_3 & (15.4)\\ &\vdots \\ &= r \text{ if } \kappa_{r-1} < y^* < \kappa_r & (15.5)\end{aligned}$$

$\kappa_0, \kappa_1, \ldots, \kappa_r$ are called *cutpoints* that satisfy

$$-\infty = \kappa_0 < \kappa_1 < \kappa_2 < \cdots < \kappa_r = \infty \tag{15.6}$$

such that the cutpoints represent $r-1$ additional unknowns in the model. The probability of outcome i corresponds to the probability that the linear function plus the error is within the range of the associated cutpoints (κ_{i-1}, κ_i). For more reading, you can find an excellent review of models for ordered responses in Boes and Winkelmann (2006).

15.1 Ordered outcomes for general link

We assume that there are r total outcomes possible and that there are $r+1$ cutpoints, κ_i for $i = 0, \ldots, r$, where $\kappa_0 = -\infty$ and $\kappa_r = \infty$. We must therefore estimate $r-1$ cutpoints, in addition to obtaining an estimate of $\boldsymbol{\beta}$. This presentation does not include a constant in the model because the cutpoints take the place of the constant. Other presentations fit a constant and one less cutpoint. To emphasize: if we fit $r-1$ cutpoints, then the constant is not identifiable, but if we fit $r-2$ cutpoints, we may include a constant term. Regardless of this parameterization choice, the two approaches are identical and result in the same inference.

To present the ordered-outcome likelihood in the most general case, we let f denote the probability density function and F denote the cumulative distribution function. We may then derive the properties of the general model and substitute the logistic, probit, clog-log, or other densities where appropriate.

Below we list the quantities of interest (including analytic derivatives) to derive the log likelihood. Since the model of interest includes cutpoints in addition to the coefficient vector, we will examine this model only for maximum likelihood and include output from Stata's collection of model-specific commands.

$$I(a=b) = \begin{cases} 1 & \text{if } a=b \\ 0 & \text{otherwise} \end{cases} \tag{15.7}$$

$$\xi_i = \mathbf{x}_i\boldsymbol{\beta} + \text{offset}_i \tag{15.8}$$

$$\Pr(y_i = k) = \Pr(\kappa_{k-1} < \xi_i + \epsilon_i \le \kappa_k) \tag{15.9}$$

$$\Pr(\xi_i + \epsilon_i < \kappa) = F(\kappa - \xi_i) \tag{15.10}$$

$$\Pr(\xi_i + \epsilon_i > \kappa) = 1 - F(\kappa - \xi_i) \tag{15.11}$$

$$\Pr(\kappa_{k-1} < \xi_i + \epsilon_i < \kappa_k) = F(\kappa_k - \xi_i) - F(\kappa_{k-1} - \xi_i) \tag{15.12}$$

$$\mathcal{L} = \sum_{k=1}^{r}\sum_{i=1}^{n} \ln\left\{F(\kappa_k - \xi_i) - F(\kappa_{k-1} - \xi_i)\right\} I(y_i = k) \tag{15.13}$$

First derivatives are given by

$$\frac{\partial}{\partial \beta_t} = \sum_{i=1}^{n} x_{ti} \sum_{k=1}^{r} \left\{ \frac{-f(\kappa_k - \xi_i) + f(\kappa_{k-1} - \xi_i)}{F(\kappa_k - \xi_i) - F(\kappa_{k-1} - \xi_i)} \right\} I(y_i = k) \tag{15.14}$$

$$\begin{aligned} \frac{\partial}{\partial \kappa_t} = & \sum_{i=1}^{n} \left\{ \frac{f(\kappa_t - \xi_i)}{F(\kappa_t - \xi_i) - F(\kappa_{t-1} - \xi_i)} I(y_i = t) \right. \\ & \left. - \frac{f(\kappa_t - \xi_i)}{F(\kappa_{t+1} - \xi_i) - F(\kappa_t - \xi_i)} I(y_i = t+1) \right\} \end{aligned} \tag{15.15}$$

and second derivatives are given by

$$\begin{aligned} \frac{\partial^2}{\partial \beta_t \partial \beta_u} = & \sum_{i=1}^{n} x_{ti} x_{ui} \sum_{k=1}^{r} \left[\frac{f'(\kappa_k - \xi_i) - f'(\kappa_{k-1} - \xi_i)}{F(\kappa_k - \xi_i) - F(\kappa_{k-1} - \xi_i)} \right. \\ & \left. - \frac{-f(\kappa_k - \xi_i) + f(\kappa_{k-1} - \xi_i)\ ^2}{F(\kappa_k - \xi_i) - F(\kappa_{k-1} - \xi_i)\ ^2} \right] I(y_i = k) \end{aligned} \tag{15.16}$$

$$\begin{aligned} \frac{\partial^2}{\partial \kappa_t \partial \kappa_u} = & -\sum_{i=1}^{n} \frac{f(\kappa_t - \xi_i) f(\kappa_u - \xi_i)}{F(\kappa_t - \xi_i) - F(\kappa_u - \xi_i)\ ^2} \\ & I(y_i = \max\ t, u\ ) I(\ t - u\ = 1) \end{aligned} \tag{15.17}$$

$$\begin{aligned} \frac{\partial^2}{\partial \kappa_t \partial \kappa_t} = & \sum_{i=1}^{n} \left(\left[\frac{f'(\kappa_t - \xi_i)}{F(\kappa_t - \xi_i) - F(\kappa_{t-1} - \xi_i)} \right. \right. \\ & \left. - \frac{f(\kappa_t - \xi_i)\ f(\kappa_t - \xi_i) - f(\kappa_{t-1} - \xi_i)}{F(\kappa_t - \xi_i) - F(\kappa_{t-1} - \xi_i)\ ^2} \right] I(y_i = t) \\ & - \left[\frac{f'(\kappa_t - \xi_i)}{F(\kappa_{t+1} - \xi_i) - F(\kappa_t - \xi_i)} \right. \\ & \left. \left. - \frac{f(\kappa_t - \xi_i)\ f(\kappa_{t+1} - \xi_i) - f(\kappa_t - \xi_i)}{(F(\kappa_{t+1} - \xi_i) - F(\kappa_t - \xi_i))^2} \right] I(y_i = t+1) \right) \end{aligned} \tag{15.18}$$

$$\frac{\partial^2}{\partial \kappa_t \partial \beta_u} = -\sum_{i=1}^{n} x_{ui} \mathcal{A}_i I(y_i = t) - \mathcal{B}_i I(y_i = t+1) \tag{15.19}$$

where

$$\mathcal{A}_i = \frac{f(\kappa_t - \xi_i)\ F(\kappa_t - \xi_i) - F(\kappa_{t-1} - \xi_i) + f(\kappa_t - \xi_i) - f(\kappa_{t-1} - \xi_i)}{F(\kappa_t - \xi_i) - F(\kappa_{t-1} - \xi_i)\ ^2} \tag{15.20}$$

$$\mathcal{B}_i = \frac{f(\kappa_t - \xi_i)\ F(\kappa_{t+1} - \xi_i) - F(\kappa_t - \xi_i) + f(\kappa_{t+1} - \xi_i) - f(\kappa_t - \xi_i)}{F(\kappa_{t+1} - \xi_i) - F(\kappa_t - \xi_i)\ ^2} \tag{15.21}$$

In the derivation, we do not include a constant in the covariate list. Instead of absorbing the constant into the cutpoints, we may include the constant and then fit only $r-2$ cutpoints, as we mentioned previously. We have presented the material to match the output of the Stata supporting commands.

The ordered-outcome models include the proportional-odds assumption. This assumption dictates that the explanatory variables have the same effect on the odds of all levels of the response, thereby providing one coefficient for the table of parameter estimates with cutoff points defining ranges for the probability of being classified in a particular level of the response. The proportional-odds assumption is also called the parallel lines assumption.

A natural approach for the interpretation of ordered-response models is through marginal effects. This is a measure of how a marginal change in a predictor changes the predicted probabilities

$$\frac{\partial P(y=j|x)}{\partial x_\ell} = \{f(\kappa_{j-1}-\xi) - f(\kappa_j - \xi)\}\,\beta_\ell \tag{15.22}$$

However, the interpretation of these marginal effects clearly depends on the particular values of the observation's covariates. Thus, if we instead focus on the respective marginal effect, the result is then given by β_ℓ/β_m, which does not depend on the particular observations. The sign of the covariates can change only once as we proceed from the smallest to largest outcome categories. This limitation is the motivation for generalized ordered-outcome models.

15.2 Ordered outcomes for specific links

Herein, we outline some of the details for the commonly used links in ordered-outcome models. We point out specific functions available to Stata users for calculating necessary quantities in the log likelihood and derivatives. Interested users can use this information along with the Stata `ml` suite of programs to generate programs, though this is not necessary to fitting the models since Stata already provides commands.

15.2.1 Ordered logit

Ordered logit uses the previous derivations, substituting

$$F(y) = \frac{\exp(y)}{1+\exp(y)} \tag{15.23}$$

$$f(y) = \frac{\exp(y)}{\{1+\exp(y)\}^2} = F(y)\{1-F(y)\} \tag{15.24}$$

$$f'(y) = \frac{\exp(y)-\exp(2y)}{\{1+\exp(y)\}^3} = F(y)\{1-F(y)\}\left[F(y)-\{1-F(y)\}\right] \tag{15.25}$$

$F(y)$ is the inverse of the logit function evaluated at y. Stata has the built-in function `invlogit()`, which calculates $F(y)$, so that the calculations are straightforward for Stata users. The log likelihood is then

$$\mathcal{L} = \sum_{k=1}^{r}\sum_{i=1}^{n} \ln\{F(\kappa_k - \xi_i) - F(\kappa_{k-1} - \xi_i)\}\, I(y_i = k) \tag{15.26}$$

$$= \sum_{k=1}^{r}\sum_{i=1}^{n} \ln\{\texttt{invlogit}(\kappa_k - \xi_i) - \texttt{invlogit}(\kappa_{k-1} - \xi_i)\}\, I(y_i = k) \tag{15.27}$$

A foremost assumption of this model is that the coefficients do not vary while the thresholds, or cutoffs, differ across the response values. This is commonly known as the *parallel lines assumption.*

15.2.2 Ordered probit

Ordered probit uses the previous derivations, substituting

$$F(y) = \Phi(y) \tag{15.28}$$
$$f(y) = \phi(y) \tag{15.29}$$
$$f'(y) = -y\phi(y) \tag{15.30}$$

where $\phi(y)$ is the normal density function and $\Phi(y)$ is the cumulative normal distribution function.

$F(y)$ is the normal cumulative probability function evaluated at y and $f(y)$ is the normal density function evaluated at y. Stata has the built-in functions `normal()` and `normalden()`, which calculate $F()$ and $f(y)$, respectively, so that the calculations are straightforward for Stata users. The log likelihood is then

$$\mathcal{L} = \sum_{k=1}^{r}\sum_{i=1}^{n} \ln\{F(\kappa_k - \xi_i) - F(\kappa_{k-1} - \xi_i)\}\, I(y_i = k) \tag{15.31}$$

$$= \sum_{k=1}^{r}\sum_{i=1}^{n} \ln\{\texttt{normal}(\kappa_k - \xi_i) - \texttt{normal}(\kappa_{k-1} - \xi_i)\}\, I(y_i = k) \tag{15.32}$$

15.2.3 Ordered clog-log

Ordered clog-log uses the previous derivations, substituting

$$F(y) = 1 - \exp\{-\exp(y)\} \tag{15.33}$$
$$f(y) = \{F(y) - 1\}\ln\{1 - F(y)\} \tag{15.34}$$
$$f'(y) = f(y)\,[1 + \ln\{1 - F(y)\}] \tag{15.35}$$

$F(y)$ is the inverse clog-log function evaluated at y. Stata has the built-in function `invcloglog()` to calculate $F(y)$ so that the calculations are relatively straightforward.

The log likelihood is then

$$= \sum_{k=1}^{r}\sum_{i=1}^{n} \ln\left\{F(\kappa_k - \xi_i) - F(\kappa_{k-1} - \xi_i)\right\} I(y_i = k) \tag{15.36}$$

$$= \sum_{k=1}^{r}\sum_{i=1}^{n} \ln\left\{\texttt{invcloglog}(\kappa_k - \xi_i) - \texttt{invcloglog}(\kappa_{k-1} - \xi_i)\right\} I(y_i = k) \tag{15.37}$$

15.2.4 Ordered log-log

Ordered log-log uses the previous derivations, substituting

$$F(y) = \exp\{-\exp(-y)\} \tag{15.38}$$

$$f(y) = -F(y)\ln\{F(y)\} \tag{15.39}$$

$$f'(y) = f(y)\left[1 + \ln\{F(y)\}\right] \tag{15.40}$$

$F(y)$ is the one minus the inverse log-log function evaluated at $-y$; see (15.33). Stata has the built-in function `invcloglog()` to calculate $F(y)$ so that the calculations are relatively straightforward. The log likelihood is then

$$= \sum_{k=1}^{r}\sum_{i=1}^{n} \ln\left\{F(\kappa_k - \xi_i) - F(\kappa_{k-1} - \xi_i)\right\} I(y_i = k) \tag{15.41}$$

$$= \sum_{k=1}^{r}\sum_{i=1}^{n} \ln\left\{\texttt{invcloglog}(-\kappa_{k-1} + \xi_i) - \texttt{invcloglog}(-\kappa_k + \xi_i)\right\} I(y_i = k) \tag{15.42}$$

15.2.5 Ordered cauchit

Ordered cauchit, or ordered inverse Cauchy, uses the previous derivations, substituting

$$F(y) = .5 + \pi^{-1}\text{atan}(-y) \tag{15.43}$$

$$f(y) = -\frac{1}{\pi(1 + y^2)} \tag{15.44}$$

$$f'(y) = f(y)2\pi y \tag{15.45}$$

$F(y)$ is the cumulative distribution function for the Cauchy distribution. Although there are no built-in functions for this, the necessary geometric functions are available so that the calculations are straightforward.

Stata's official ordered-response models are limited to the ordered logit (`ologit`) and ordered probit (`oprobit`) procedures. However, Williams (2006) extended these to include the ordered clog-log (`ocloglog`) and a more general procedure called `oglm`, which allows the user to model all five previously listed ordered models, with the particular model selected as an option. Williams is also responsible for developing a multilinked generalized ordered program called `gologit2`, providing the user with a host of generalized ordered binomial modeling options. We give an overview of these types of models in the following section.

15.3 Generalized ordered outcome models

Here we investigate the model first introduced by Fu (1998). You can find a version of the software by Williams (2006) that fits additional models by searching for `gologit2`. Users will also benefit from the useful collection of programs commonly called SPost, which is described by the developers in Long and Freese (2006). Although the associated software (`gologit2`) uses the term *logit*, the command actually supports several link functions in the generalized ordered-outcome models.

The newest (2006 or later) version of `gologit2` provides the five links that we described above for standard ordered binomial models, plus many extremely useful extensions, including the ability to model partial proportional odds as well as constrained generalized ordered binomial models. One can also use the `gologit2` command to model standard ordered binomial models with the `pl` option. There are many other fitting enhancements that we will use to highlight examples for our discussion of generalized ordered binomial models.

The generalized ordered logit differs from the ordered logit model in that it relaxes the proportional-odds assumption. It allows that the explanatory variables may have a different effect on the odds that the outcome is above a cutpoint depending on how the outcomes are dichotomized. To accomplish this, the approach is similar to the unordered-outcome multinomial logit model in that there will be $r-1$ estimated coefficient vectors that correspond to the effect of changing from one set of outcomes to a higher outcome not in the set.

The sets of coefficient vectors are defined for the cutpoints between the r outcomes. These cutpoints partition the outcomes into two groups. Therefore, the first coefficient vector corresponds to partitioning the outcomes into the sets $\{1\}$ and $\{2,\ldots,r\}$. The second coefficient vector corresponds to partitioning the outcomes into the sets $\{1,2\}$ and $\{3,\ldots,r\}$. The $(r-1)$th coefficient vector corresponds to partitioning the outcomes into the sets $\{1,\ldots,r-1\}$ and $\{r\}$. To reflect this partitioning, we use the (nonstandard) subscript notation $\boldsymbol{\beta}_{\{1\},\{2,\ldots,r\}}$ to denote the coefficient vector that draws the partition between outcome $k=1$ and outcome $k=2$. In general, we will denote

$$\boldsymbol{\beta}_{\{1,\ldots,j\},\{j+1,\ldots,r\}} \tag{15.46}$$

for $j=1,\ldots,r-1$ as the coefficient vector that draws the partition between outcomes j and $j+1$.

The sets of coefficient vectors correspond to a set of cumulative distribution functions

$$\Pr(y \leq k) = F\left(-\mathbf{X}\boldsymbol{\beta}_{\{1,\ldots,k\},\{k+1,\ldots,r\}}\right) \tag{15.47}$$

for $k = 1, \ldots, r-1$. The distribution functions admit probabilities that are then defined as

$$\begin{aligned}
\Pr(y=1) &= F\left(-\mathbf{X}\boldsymbol{\beta}_{\{1\},\{2,\ldots,r\}}\right) && (15.48)\\
\Pr(y=2) &= F\left(-\mathbf{X}\boldsymbol{\beta}_{\{1,2\},\{3,\ldots,r\}}\right) - F\left(-\mathbf{X}\boldsymbol{\beta}_{\{1\},\{2,\ldots,r\}}\right) && (15.49)\\
&\vdots\\
\Pr(y=r) &= 1 - F\left(-\mathbf{X}\boldsymbol{\beta}_{\{1,\ldots,r-1\},\{r\}}\right) && (15.50)
\end{aligned}$$

In general, we may write

$$\Pr(y=j) = F\left(-\mathbf{X}\boldsymbol{\beta}_{\{1,\ldots,j\},\{j+1,\ldots,r\}}\right) - F\left(-\mathbf{X}\boldsymbol{\beta}_{\{1,\ldots,j-1\},\{j,\ldots,r\}}\right) \tag{15.51}$$

for $j = 1, \ldots, r$ if we also define

$$\begin{aligned}
F\left(-\mathbf{X}\boldsymbol{\beta}_{\{\},\{1,\ldots,r\}}\right) &= 0 && (15.52)\\
F\left(-\mathbf{X}\boldsymbol{\beta}_{\{1,\ldots,r\},\{\}}\right) &= 1 && (15.53)
\end{aligned}$$

The generalized ordered logit model assumes a logit function for F, but there is no reason that we could not use another link such as probit or clog-log. Fitting this model with other link functions such as the log link can be hard because the covariate values must restrict the above probabilities to the interval $[0,1]$. Out-of-sample predictions can produce out-of-range probabilities.

The generalized ordered logit fits $r-1$ simultaneous logistic regression models, where the dependent variables for these models are defined by collapsing the outcome variable into new binary dependent variables defined by the partitions described above.

15.4 Example: Synthetic data

In the following example, we first create a synthetic dataset for illustration. We follow the same approach for generating the synthetic dataset as that appearing in Greene (2002). Our synthetic dataset is created by setting the random-number seed so that the analyses may be recreated.

```
. set obs 100
obs was 0, now 100
. set seed 12345
. gen double x1 = 3*uniform()+1
. gen double x2 = 2*uniform()-1
```

```
. gen double y = 1 + .5*x1 + 1.2*x2 + invnormal(uniform())
. gen int ys = 1 if y <= 2.5
(40 missing values generated)
. replace ys = 2 if y <= 3 & y > 2.5
(15 real changes made)
. replace ys = 3 if y <= 4 & y > 3
(20 real changes made)
. replace ys = 4 if y > 4
(5 real changes made)
```

Our dataset defines an outcome variable in the set $\{1, 2, 3, 4\}$ that depends on two covariates and a constant. We first fit an ordered logit model and obtain predicted probabilities and predicted classifications according to the maximum probability. Since there are $r = 4$ outcomes, there are four predicted probabilities associated with the outcomes.

```
. ologit ys x1 x2, nolog
Ordered logistic regression                       Number of obs   =        100
                                                  LR chi2(2)      =      37.48
                                                  Prob > chi2     =     0.0000
Log likelihood = -87.533537                       Pseudo R2       =     0.1763
```

ys	Coef.	Std. Err.	z	P>|z|	[95% Conf.	Interval]
x1	.8340873	.2577068	3.24	0.001	.3289912	1.339183
x2	2.152467	.4408377	4.88	0.000	1.288441	3.016493
/cut1	2.702931	.7223316			1.287188	4.118675
/cut2	3.672531	.7675089			2.168241	5.176821
/cut3	5.913829	.9361559			4.078997	7.748661

```
. aic
AIC Statistic =    1.850671
. predict double (olpr1 olpr2 olpr3 olpr4), pr
```

Our parameterization of this model does not contain a constant. Other presentations of this model do include a constant that is equal to the negative of the first cutpoint, κ_1, labeled `/cut1` in the above output. The remaining cutpoints, κ_j^*, in the alternative (with constant) presentation are formed as

$$\kappa_j^* = \kappa_{j+1} - \kappa_1 \tag{15.54}$$

For comparison, we also fit ordered probit, clog-log, and log-log models together with estimating their respective predicted probabilities. These models are fitted using either specific commands such as `ologit` or the general `oglm` program.

The `oglm` command is a generalized ordered binomial procedure supporting six different link functions including logit, probit, clog-log, log-log, cauchit, and log. We will not model the cauchit or log links (we did not show the log link since we believe it to have undesirable properties when modeling ordered binomial models). Comparison of results with the ordered logit will provide an insight into how the models differ.

```
. oprobit ys x1 x2, nolog
Ordered probit regression                         Number of obs   =        100
                                                  LR chi2(2)      =      36.46
                                                  Prob > chi2     =     0.0000
Log likelihood = -88.044122                       Pseudo R2       =     0.1715

------------------------------------------------------------------------------
          ys |      Coef.   Std. Err.      z    P>|z|     [95% Conf. Interval]
-------------+----------------------------------------------------------------
          x1 |   .5028955   .1490696     3.37   0.001     .2107245    .7950666
          x2 |   1.189187   .2344599     5.07   0.000      .729654     1.64872
-------------+----------------------------------------------------------------
       /cut1 |   1.615977   .4130283                      .8064564    2.425497
       /cut2 |   2.177505   .4304469                      1.333844    3.021165
       /cut3 |   3.395485   .5020912                      2.411404    4.379566
------------------------------------------------------------------------------
. aic
AIC Statistic =     1.860882
. predict double (oppr1 oppr2 oppr3 oppr4), pr
```

We fit an ordered clog-log model via the `oglm` command, which fits ordered-outcome models through specifying a `link` option.

```
. oglm ys x1 x2, link(cloglog) nolog
Ordered Cloglog Regression                        Number of obs   =        100
                                                  LR chi2(2)      =      36.32
                                                  Prob > chi2     =     0.0000
Log likelihood = -88.114692                       Pseudo R2       =     0.1709

------------------------------------------------------------------------------
          ys |      Coef.   Std. Err.      z    P>|z|     [95% Conf. Interval]
-------------+----------------------------------------------------------------
          x1 |   .5921838   .1991983     2.97   0.003     .2017622    .9826054
          x2 |   1.635788   .3259243     5.02   0.000     .9969877    2.274588
-------------+----------------------------------------------------------------
       /cut1 |   2.408241   .5863804     4.11   0.000     1.258956    3.557525
       /cut2 |   3.113651   .6143953     5.07   0.000     1.909458    4.317844
       /cut3 |    4.99395   .7509111     6.65   0.000     3.522191    6.465708
------------------------------------------------------------------------------
. aic
AIC Statistic =     1.802294
. predict double (ocpr1 ocpr2 ocpr3 ocpr4), pr
```

We fit an ordered log-log model via the same `oglm` command this time with the option `link(loglog)`. The binary log-log model is rarely used in research; however, it is a viable model that should be checked against others when deciding on a binary-response probability model. We have found many examples that used a probit or clog-log model to model data when using a log-log model would have been preferable. The authors simply failed to try the model. The same motivation for model checking holds for the ordered log-log model.

The clog-log and log-log fitted values are asymmetric, unlike those from logit and probit. Remember this when interpreting predicted probabilities.

```
. oglm ys x1 x2, link(loglog) nolog
Ordered Loglog Regression                         Number of obs   =        100
                                                  LR chi2(2)      =      31.95
                                                  Prob > chi2     =     0.0000
Log likelihood = -90.300378                       Pseudo R2       =     0.1503

------------------------------------------------------------------------------
          ys |      Coef.   Std. Err.      z    P>|z|     [95% Conf. Interval]
-------------+----------------------------------------------------------------
          x1 |   .4839441   .1396078     3.47   0.001     .2103179    .7575703
          x2 |   .9360736   .1991001     4.70   0.000     .5458445    1.326303
-------------+----------------------------------------------------------------
       /cut1 |   1.166167   .3599873     3.24   0.001     .4606052     1.87173
       /cut2 |    1.70414   .3712522     4.59   0.000     .9764988    2.431781
       /cut3 |    2.68805   .4246582     6.33   0.000     1.855735    3.520365
------------------------------------------------------------------------------

. aic
AIC Statistic =    1.846008
. predict double (ozpr1 ozpr2 ozpr3 ozpr4), pr
```

Finally, we may also fit a generalized ordered logit model to the data. We do this for illustration only, since we generated the data ourselves without violating the assumption.

```
. gologit2 ys x1 x2, nolog
Generalized Ordered Logit Estimates               Number of obs   =        100
                                                  LR chi2(6)      =      38.61
                                                  Prob > chi2     =     0.0000
Log likelihood = -86.969883                       Pseudo R2       =     0.1816

------------------------------------------------------------------------------
          ys |      Coef.   Std. Err.      z    P>|z|     [95% Conf. Interval]
-------------+----------------------------------------------------------------
1            |
          x1 |   .7985641   .2749533     2.90   0.004     .2596656    1.337463
          x2 |   2.153229   .4754684     4.53   0.000     1.221328     3.08513
       _cons |  -2.617458    .758309    -3.45   0.001    -4.103716   -1.131199
-------------+----------------------------------------------------------------
2            |
          x1 |    .863976   .3177713     2.72   0.007     .2411556    1.486796
          x2 |   2.224649   .5346084     4.16   0.000     1.176836    3.272462
       _cons |  -3.776184   .9530848    -3.96   0.000    -5.644196   -1.908172
-------------+----------------------------------------------------------------
3            |
          x1 |   1.394936   .8413169     1.66   0.097    -.2540147    3.043887
          x2 |    1.72243   .9532961     1.81   0.071    -.1459959    3.590856
       _cons |  -7.370906   2.832295    -2.60   0.009     -12.9221   -1.819709
------------------------------------------------------------------------------

. aic
AIC Statistic =    1.859398
```

Here we test the proportional-odds assumption. If the proportional-odds assumption is valid, then the coefficient vectors should all be equal. We can use Stata's `test` command to investigate.

```
. test [1=2], notest
 ( 1)  [1]x1 - [2]x1 = 0
 ( 2)  [1]x2 - [2]x2 = 0
. test [1=3], accumulate
 ( 1)  [1]x1 - [2]x1 = 0
 ( 2)  [1]x2 - [2]x2 = 0
 ( 3)  [1]x1 - [3]x1 = 0
 ( 4)  [1]x2 - [3]x2 = 0
           chi2( 4) =    1.21
         Prob > chi2 =    0.8763
```

The χ^2 test fails to reject the proportional-odds assumption. Thus we prefer the ordered logit (and ordered probit) model over the generalized ordered logit model. Since we synthesized the data ourselves, we know that the proportional-odds assumption is valid, and the outcome of this test reflects that.

Since we rejected the generalized ordered probability model, we now compare the two remaining ordered probability models. The coefficients from the ordered logit and ordered probit models are different, but the predicted probabilities, and thus the predicted classifications, are similar. Choosing between the two models is then arbitrary. The pseudo-R^2 listed in the output for the models is the McFadden likelihood-ratio index. See section 4.6.4.3 for an alternative measure.

We also tested each of the above models by using the AIC statistic. Stata does not automatically calculate or display this statistic with its maximum likelihood models. We use a postestimation AIC command created specifically (Hilbe 2007) for comparing maximum likelihood models. Here the ordered logit model has the lowest AIC statistic, indicating a better-fitted model. However, none of the four statistics appears to significantly differ from one another.

We list the predicted probabilities created by the `predict` commands for each of the four ordered binomial models. We summarize the coding given to the models as

`olpr1`	logit	`oppr1`	probit
`ocpr1`	clog-log	`ozpr1`	log-log

Level 1 predicted probabilities are given by

```
. list ys olpr1 oppr1 ocpr1 ozpr1 in 55/62
```

	ys	olpr1	oppr1	ocpr1	ozpr1
55.	1	.72059288	.72576454	.72153677	.73021016
56.	4	.10715097	.11665252	.04734254	.22959935
57.	1	.90004242	.89496221	.88549951	.86811275
58.	3	.14792324	.16116606	.1002523	.26267433
59.	1	.79567335	.79308675	.7896648	.78137812
60.	1	.93437357	.93837838	.91497677	.9349882
61.	3	.33904474	.35871437	.35371489	.41196884
62.	2	.74352817	.73593559	.74918409	.7136501

Level 2 predicted probabilities are given by

```
. list ys olpr2 oppr2 ocpr2 ozpr2 in 55/62
```

	ys	olpr2	oppr2	ocpr2	ozpr2
55.	1	.15121107	.15153254	.12958763	.16371466
56.	4	.13323232	.1475768	.17432347	.13066765
57.	1	.05954241	.07026702	.05620795	.10074714
58.	3	.16609851	.17310521	.22084358	.14390514
59.	1	.11558267	.12292055	.10024583	.14462644
60.	1	.03968164	.0438813	.04208553	.05573897
61.	3	.23589771	.22040077	.2448032	.18522976
62.	2	.14079142	.14751074	.11789469	.16888194

Level 3 predicted probabilities are given by

```
. list ys olpr3 oppr3 ocpr3 ozpr3 in 55/62
```

	ys	olpr3	oppr3	ocpr3	ozpr3
55.	1	.11280263	.1140363	.12458575	.10359993
56.	4	.50813381	.45737699	.57300976	.33699491
57.	1	.0359572	.03355956	.04917247	.03104684
58.	3	.49749787	.45091564	.51979534	.34581024
59.	1	.07849593	.07928632	.09245479	.07305086
60.	1	.02312085	.01729154	.03626537	.00926918
61.	3	.35218261	.34273268	.32616807	.31496797
62.	2	.10196311	.10858556	.11139945	.1142161

Level 4 predicted probabilities are given by

```
. list ys olpr4 oppr4 ocpr4 ozpr4 in 55/62
```

	ys	olpr4	oppr4	ocpr4	ozpr4
55.	1	.01539342	.00866663	.02428985	.00247525
56.	4	.2514829	.2783937	.20532423	.30273809
57.	1	.00445797	.00121122	.00912008	.00009328
58.	3	.18848038	.21481309	.15910877	.2476103
59.	1	.01024805	.00470639	.01763457	.00094459
60.	1	.00282394	.00044878	.00667233	3.652e-06
61.	3	.07287494	.07815219	.07531384	.08783344
62.	2	.0137173	.0079681	.02152176	.00325186

15.5 Example: Automobile data

In the following analyses, we investigate the automobile dataset that is included with the Stata software. Like Fu (1998), we also collapse the `rep78` variable from five outcomes to three to ensure having enough of each outcome.

Our outcome variable is collapsed, and the data in memory are compressed using

```
. sysuse auto, clear
(1978 Automobile Data)
. replace rep78=3 if rep78<=3
(10 real changes made)
. drop if rep78==.
(5 observations deleted)
. label define replab 3 "poor-avg" 4 "good" 5 "best"
. label values rep replab
. tab rep78

     Repair |
Record 1978 |      Freq.     Percent        Cum.
------------+-----------------------------------
   poor-avg |         40       57.97       57.97
       good |         18       26.09       84.06
       best |         11       15.94      100.00
------------+-----------------------------------
      Total |         69      100.00
```

We also applied labels to the outcome variable to make the tabulation easier to read.

We wish to investigate the number of repairs made to a car (`rep78`) based on `foreign`, the foreign/domestic origin of the car; `length`, the length in inches of the car; `mpg`, the miles per gallon for the car; and `displacement`, the displacement of the engine in the car in cubic inches. Again we will fit all three models presented in this chapter for illustration. First, we fit an ordered logit model.

```
. ologit rep78 foreign length displacement, nolog
Ordered logistic regression                       Number of obs   =         69
                                                  LR chi2(3)      =      30.28
                                                  Prob > chi2     =     0.0000
Log likelihood = -51.054656                       Pseudo R2       =     0.2287

------------------------------------------------------------------------------
       rep78 |      Coef.   Std. Err.      z    P>|z|     [95% Conf. Interval]
-------------+----------------------------------------------------------------
     foreign |   2.831938   .8602209     3.29   0.001     1.145936     4.51794
      length |   .0333366   .0255179     1.31   0.191    -.0166776    .0833508
displacement |  -.0094944   .0069388    -1.37   0.171    -.0230941    .0041054
-------------+----------------------------------------------------------------
       /cut1 |   5.602613   4.022655                     -2.281645    13.48687
       /cut2 |   7.653871   4.087473                     -.3574295    15.66517
------------------------------------------------------------------------------
. aic
AIC Statistic =    1.624773
```

Next we fit an ordered probit model.

```
. oprobit rep78 foreign length displacement, nolog
Ordered probit regression                         Number of obs   =         69
                                                  LR chi2(3)      =      30.26
                                                  Prob > chi2     =     0.0000
Log likelihood = -51.064296                       Pseudo R2       =     0.2286

       rep78        Coef.   Std. Err.      z    P>|z|     [95% Conf. Interval]

     foreign     1.492901   .4309167     3.46   0.001     .6483198    2.337482
      length     .0171283   .0144129     1.19   0.235    -.0111206    .0453771
displacement     -.006107   .0040351    -1.51   0.130    -.0140156    .0018016

       /cut1     2.688937   2.167921                     -1.560111    6.937984
       /cut2     3.836788   2.182035                     -.4399217    8.113497

. aic
AIC Statistic =    1.625052
```

Both models indicate that foreign cars have a higher probability of having better repair records. They also indicate that longer cars and cars with smaller displacement engines might have a higher probability of better repair records, but the results are not significant.

To assess the proportional-odds assumption, we fit a generalized ordered logit model to the data where we removed the iteration log.

```
. gologit2 rep78 foreign length displacement
Generalized Ordered Logit Estimates               Number of obs   =         69
                                                  LR chi2(6)      =      40.65
                                                  Prob > chi2     =     0.0000
Log likelihood =    -45.8704                      Pseudo R2       =     0.3070

       rep78        Coef.   Std. Err.      z    P>|z|     [95% Conf. Interval]

poor-avg
     foreign     3.659307   1.029914     3.55   0.000     1.640713    5.677901
      length     .0372637   .0278299     1.34   0.181    -.0172819    .0918093
displacement    -.0068156   .0069198    -0.98   0.325     -.020378    .0067469
       _cons    -7.093281   4.459774    -1.59   0.112    -15.83428    1.647716

good
     foreign     .1758221   1.162968     0.15   0.880    -2.103553    2.455197
      length     .1879727   .0678996     2.77   0.006      .054892    .3210535
displacement    -.0946161   .0315209    -3.00   0.003    -.1563959   -.0328362
       _cons    -21.75103   7.993136    -2.72   0.007    -37.41729   -6.084771

WARNING! 6 in-sample cases have an outcome with a predicted probability that is
less than 0. See the gologit2 help section on Warning Messages for more
information.

. aic
AIC Statistic =    1.50349
```

Note the warning message. The output from the command reminds us of an oddity for generalized ordered models. The oddity is that a model may produce predicted probabilities outside $[0,1]$. If the number of such predictions is a large proportion of the total number of observations used in the estimation, then we would investigate further to see if there were a better model.

Next we test the proportional-odds assumption by obtaining a Wald test that the coefficient vectors are the same:

```
. test [poor-avg=good]
 ( 1)  [poor-avg]foreign - [good]foreign = 0
 ( 2)  [poor-avg]length - [good]length = 0
 ( 3)  [poor-avg]displacement - [good]displacement = 0
           chi2(  3) =    8.31
         Prob > chi2 =    0.0399
```

The results show that the data violate the proportional-odds assumption. Here we prefer the generalized ordered logit model, which takes into account the different effects of (1) moving from `poor-avg` to a higher classification and (2) moving from a lower classification to `best`.

Our results now indicate that the probability of having a better repair record if the car is foreign is important only when moving from `poor-avg` to a higher category. Also, the length of the car and engine displacement size have an important effect only when moving from `poor-avg` to `best` or from `good` to `best`.

Williams developed a generalized ordered binomial model with optional links of logit, probit, clog-log, log-log, and cauchit. The `gologit2` command is the generalized correlate of the `oglm` command. The default for `gologit2` is the generalized logit model, but it can model `oglm` models using the `pl` option.

Here we illustrate the odds ratio parameterization of the previous `gologit2` output.

```
. gologit2 rep78 foreign length displacement, link(logit) eform nolog
Generalized Ordered Logit Estimates               Number of obs   =         69
                                                  LR chi2(6)      =      40.65
                                                  Prob > chi2     =     0.0000
Log likelihood =   -45.8704                       Pseudo R2       =     0.3070

------------------------------------------------------------------------------
       rep78 |     exp(b)   Std. Err.      z    P>|z|     [95% Conf. Interval]
-------------+----------------------------------------------------------------
poor-avg     |
     foreign |   38.83443   39.99612     3.55   0.000     5.158847    292.3353
      length |   1.037967   .0288865     1.34   0.181     .9828665    1.096156
displacement |   .9932076   .0068728    -0.98   0.325     .9798282     1.00677
-------------+----------------------------------------------------------------
good         |
     foreign |   1.192226    1.38652     0.15   0.880     .1220222    11.64873
      length |   1.206801   .0819413     2.77   0.006     1.056426    1.378579
displacement |   .9097222   .0286753    -3.00   0.003     .8552206     .967697
------------------------------------------------------------------------------
WARNING! 6 in-sample cases have an outcome with a predicted probability that is
less than 0. See the gologit2 help section on Warning Messages for more
information.
```

Note the extremely high odds ratio for foreign cars in the poor-avg response level. Because of this, we find it advisable to try other links. We next model a generalized ordered probit model on the identical data used with gologit2 above but will return to showing nonexponentiated parameter estimates.

```
. gologit2 rep78 foreign length displacement, link(probit) nolog
Generalized Ordered Probit Estimates              Number of obs   =         69
                                                  LR chi2(6)      =      41.35
                                                  Prob > chi2     =     0.0000
Log likelihood = -45.520115                       Pseudo R2       =     0.3123

------------------------------------------------------------------------------
       rep78 |      Coef.   Std. Err.      z    P>|z|     [95% Conf. Interval]
-------------+----------------------------------------------------------------
poor-avg     |
     foreign |   2.166388   .5469985     3.96   0.000     1.094291    3.238486
      length |   .0212237   .0154616     1.37   0.170    -.0090805    .0515279
displacement |  -.0039511   .0040961    -0.96   0.335    -.0119793     .004077
       _cons |  -4.087608   2.399035    -1.70   0.088    -8.789631    .6144147
-------------+----------------------------------------------------------------
good         |
     foreign |   .1308595   .6611055     0.20   0.843    -1.164883    1.426602
      length |   .1096383   .0375006     2.92   0.003     .0361385    .1831382
displacement |  -.0551583    .017555    -3.14   0.002    -.0895655   -.0207511
       _cons |  -12.72151   4.369989    -2.91   0.004    -21.28653   -4.156485
------------------------------------------------------------------------------
WARNING! 6 in-sample cases have an outcome with a predicted probability that is
less than 0. See the gologit2 help section on Warning Messages for more
information.
. aic
AIC Statistic =    1.493337
```

The AIC statistic indicates that the generalized ordered probit may be a better-fitted model than the generalized ordered logit model. However, the difference is neither appreciable nor significant. The coefficient of foreign cars for the poor-avg response level (exponentiated coefficient equal to 8.73) does not indicate the problems implied by the large coefficient (and odds ratio) in the generalized ordered logit model.

We show the output for the clog-log and log-log links to demonstrate the difference between all the models. Also, we list the exponentiated coefficient for foreign cars in the poor-avg response level for comparison.

(*Continued on next page*)

```
. gologit2 rep78 foreign length displacement, link(cloglog) nolog
Generalized Ordered Cloglog Estimates             Number of obs   =         69
                                                  LR chi2(6)      =      39.13
                                                  Prob > chi2     =     0.0000
Log likelihood = -46.630058                       Pseudo R2       =     0.2956

------------------------------------------------------------------------------
       rep78 |      Coef.   Std. Err.      z    P>|z|     [95% Conf. Interval]
-------------+----------------------------------------------------------------
poor-avg     |
     foreign |   2.334334   .6826142     3.42   0.001     .9964343    3.672233
      length |   .0259075   .0191936     1.35   0.177    -.0117112    .0635263
displacement |  -.0047653   .0052544    -0.91   0.364    -.0150638    .0055332
       _cons |  -5.399207   3.135615    -1.72   0.085     -11.5449    .7464853
-------------+----------------------------------------------------------------
good         |
     foreign |   .2292758   1.110919     0.21   0.836    -1.948085    2.406637
      length |   .1276791   .0478079     2.67   0.008     .0339772    .2213809
displacement |  -.0656225   .0228494    -2.87   0.004    -.1104066   -.0208384
       _cons |  -15.28615   5.680261    -2.69   0.007    -26.41926   -4.153047
------------------------------------------------------------------------------

WARNING! 5 in-sample cases have an outcome with a predicted probability that is
less than 0. See the gologit2 help section on Warning Messages for more
information.

. aic
AIC Statistic =    1.525509

. display exp(2.334334) /* Exponentiated coefficient for Poor-Average foreign */
10.322583

. gologit2 rep78 foreign length displacement, link(loglog) nolog
Generalized Ordered Loglog Estimates              Number of obs   =         69
                                                  LR chi2(6)      =      42.80
                                                  Prob > chi2     =     0.0000
Log likelihood = -44.794841                       Pseudo R2       =     0.3233

------------------------------------------------------------------------------
       rep78 |      Coef.   Std. Err.      z    P>|z|     [95% Conf. Interval]
-------------+----------------------------------------------------------------
poor-avg     |
     foreign |   2.659394   .6925441     3.84   0.000     1.302033    4.016756
      length |   .0180377   .0143903     1.25   0.210    -.0101667    .0462421
displacement |  -.0035022   .0039727    -0.88   0.378    -.0112887    .0042842
       _cons |  -3.217335    2.11529    -1.52   0.128    -7.363228    .9285575
-------------+----------------------------------------------------------------
good         |
     foreign |   .0533791   .6130687     0.09   0.931    -1.148214    1.254972
      length |   .1187579   .0419275     2.83   0.005     .0365815    .2009343
displacement |  -.0598442   .0201469    -2.97   0.003    -.0993314    -.020357
       _cons |  -13.18729   4.665233    -2.83   0.005    -22.33098   -4.043603
------------------------------------------------------------------------------

WARNING! 6 in-sample cases have an outcome with a predicted probability that is
less than 0. See the gologit2 help section on Warning Messages for more
information.

. aic
AIC Statistic =    1.472314

. display exp(2.659394) /* Exponentiated coefficient for Poor-Average foreign */
14.287628
```

The log-log linked model had the lowest value for the AIC statistic. Although this indicates a better-fitted model, there is little difference between values for all four models. When evaluating models we would next look at the probabilities and consider which model is easiest to interpret. However, we must first assess whether the proportional-odds assumption has been met. We turn to this discussion in the following section.

15.6 Partial proportional-odds models

The generalized ordered logistic model is a viable alternative to the ordered logistic model when the assumption of proportional odds is untenable. In fact, the generalized model eliminates the restriction that the regression coefficients are the same for each outcome for all the variables. This may be a rather drastic solution when the proportional-odds assumption is violated by only one or a few of the regressors.

The `gologit2` command allows partial proportional-odds models. Such a model is somewhere between the ordered logistic model and the generalized ordered logistic model. Of particular use in determining the type of model we should fit to the data is the test described in Brant (1990) and supported in the command. The software provides both the global test of the proportional-odds assumption and a test of the proportional-odds assumption for each of the covariates.

Using Stata's automobile dataset, we now investigate an ordered-outcome model on categories of headroom size. We load the data by specifying `sysuse auto` and create the outcome variable of interest by typing

```
. sysuse auto, clear
(1978 Automobile Data)
. generate headsize = int(headroom+.5)
. replace headsize = 4 if headsize==5
(5 real changes made)
. tab headsize

   headsize |      Freq.     Percent        Cum.
------------+-----------------------------------
          2 |         17       22.97       22.97
          3 |         27       36.49       59.46
          4 |         30       40.54      100.00
------------+-----------------------------------
      Total |         74      100.00
```

A standard proportional-odds ordered logistic model may be fitted using the `ologit` command. The estimates and the results of Brant's test of parallel lines is given by

(*Continued on next page*)

```
. ologit headsize foreign mpg price displacement, nolog
Ordered logistic regression                       Number of obs   =         74
                                                  LR chi2(4)      =      23.03
                                                  Prob > chi2     =     0.0001
Log likelihood = -67.799401                       Pseudo R2       =     0.1452

------------------------------------------------------------------------------
    headsize |      Coef.   Std. Err.      z    P>|z|     [95% Conf. Interval]
-------------+----------------------------------------------------------------
     foreign |   .9710014   .7389785     1.31   0.189    -.4773698    2.419373
         mpg |  -.0782365   .0551093    -1.42   0.156    -.1862487    .0297757
       price |   -.000117   .0001062    -1.10   0.271    -.0003252    .0000912
displacement |   .0133369   .0051225     2.60   0.009      .003297    .0233768
-------------+----------------------------------------------------------------
       /cut1 |  -1.052366   1.896215                     -4.768879    2.664147
       /cut2 |   .9785758   1.887007                      -2.71989    4.677041
------------------------------------------------------------------------------

. brant
Brant Test of Parallel Regression Assumption
    Variable |      chi2   p>chi2    df
-------------+--------------------------
         All |     12.61    0.013     4
-------------+--------------------------
     foreign |      6.96    0.008     1
         mpg |      3.80    0.051     1
       price |      0.27    0.605     1
displacement |      2.05    0.152     1
----------------------------------------

A significant test statistic provides evidence that the parallel
regression assumption has been violated.
```

The test indicates that the parallel-lines assumption is violated. Use the `detail` option to request details of the test.

```
. brant, detail
Estimated coefficients from j-1 binary regressions
                    y>2          y>3
     foreign   2.129308   -1.2032583
         mpg -.01044161   -.24450781
       price -.00001008   -.00011063
displacement  .01663045    .00533099
       _cons -2.0935113     4.396156

Brant Test of Parallel Regression Assumption
    Variable |      chi2   p>chi2    df
-------------+--------------------------
         All |     12.61    0.013     4
-------------+--------------------------
     foreign |      6.96    0.008     1
         mpg |      3.80    0.051     1
       price |      0.27    0.605     1
displacement |      2.05    0.152     1
----------------------------------------

A significant test statistic provides evidence that the parallel
regression assumption has been violated.
```

The results of the test are shown again along with the fitted coefficients for the individual logistic regression models. Since we have an indication that the ordered logistic regression model assumptions are violated, we can fit a generalized ordered logistic regression model by using either the gologit or gologit2 command. This model is the least restrictive where all covariates are allowed to change for each outcome.

```
. gologit2 headsize foreign mpg price displacement
Generalized Ordered Logit Estimates               Number of obs   =         74
                                                  LR chi2(8)      =      39.26
                                                  Prob > chi2     =     0.0000
Log likelihood = -59.684516                       Pseudo R2       =     0.2475

------------------------------------------------------------------------------
    headsize |      Coef.   Std. Err.      z    P>|z|     [95% Conf. Interval]
-------------+----------------------------------------------------------------
2            |
     foreign |   1.978872   .9039067     2.19   0.029     .2072475    3.750497
         mpg |     -.0291   .0717361    -0.41   0.685    -.1697001    .1115002
       price |  -.0000388   .0001806    -0.22   0.830    -.0003927    .0003151
displacement |   .0140528   .0065791     2.14   0.033     .0011579    .0269476
       _cons |  -1.025763    2.46282    -0.42   0.677    -5.852801    3.801275
-------------+----------------------------------------------------------------
3            |
     foreign |  -.9988688   1.092489    -0.91   0.361    -3.140108     1.14237
         mpg |  -.1918918    .091883    -2.09   0.037    -.3719792   -.0118044
       price |  -.0001025   .0001248    -0.82   0.411    -.0003471    .0001421
displacement |   .0067022   .0056262     1.19   0.234    -.0043249    .0177294
       _cons |   2.948465   2.607146     1.13   0.258    -2.161447    8.058378
------------------------------------------------------------------------------
```

We can also fit a partial proportional-odds model where we allow the software to search for which of the covariates should be allowed to change across the outcomes. The gologit2 command will perform such a search:

```
. gologit2 headsize foreign mpg price displacement, autofit lrforce

------------------------------------------------------------------------------
Testing parallel lines assumption using the .05 level of significance...
Step  1:  Constraints for parallel lines imposed for price (P Value = 0.7293)
Step  2:  Constraints for parallel lines imposed for mpg (P Value = 0.1144)
Step  3:  Constraints for parallel lines imposed for displacement
          (P Value = 0.4775)
Step  4:  Constraints for parallel lines are not imposed for
          foreign (P Value = 0.00331)
Wald test of parallel lines assumption for the final model:
 ( 1)  [2]price - [3]price = 0
 ( 2)  [2]mpg - [3]mpg = 0
 ( 3)  [2]displacement - [3]displacement = 0
           chi2(  3) =    2.93
         Prob > chi2 =    0.4019
An insignificant test statistic indicates that the final model
does not violate the proportional odds/ parallel lines assumption
If you re-estimate this exact same model with gologit2, instead
of autofit you can save time by using the parameter
pl(price mpg displacement)
------------------------------------------------------------------------------
```

```
Generalized Ordered Logit Estimates               Number of obs   =         74
                                                  LR chi2(5)      =      35.23
                                                  Prob > chi2     =     0.0000
Log likelihood = -61.700114                       Pseudo R2       =     0.2221
 ( 1)  [2]price - [3]price = 0
 ( 2)  [2]mpg - [3]mpg = 0
 ( 3)  [2]displacement - [3]displacement = 0

    headsize        Coef.   Std. Err.      z    P>|z|     [95% Conf. Interval]

2
     foreign     1.942757    .8554556    2.27   0.023     .2660953     3.61942
         mpg    -.0936841    .0590846   -1.59   0.113    -.2094878    .0221196
       price    -.0000759    .0001113   -0.68   0.496     -.000294    .0001423
displacement     .0096062     .005053    1.90   0.057    -.0002975      .01951
       _cons     1.422395    1.946279    0.73   0.465    -2.392242    5.237032

3
     foreign    -.8960264    1.021318   -0.88   0.380    -2.897774    1.105721
         mpg    -.0936841    .0590846   -1.59   0.113    -.2094878    .0221196
       price    -.0000759    .0001113   -0.68   0.496     -.000294    .0001423
displacement     .0096062     .005053    1.90   0.057    -.0002975      .01951
       _cons     .2172205    1.942194    0.11   0.911     -3.58941    4.023851
```

Although this shortcut is useful, we also could have applied the knowledge gained from the Brant test. Applying such knowledge requires that we either stipulate the variables to hold constant (using the `pl` option) or stipulate the variables that are allowed to change (using the `npl` option). Here we show that the final model could have been obtained through such stipulation.

```
. gologit2 headsize foreign mpg price displacement, npl(foreign)
Generalized Ordered Logit Estimates               Number of obs   =         74
                                                  Wald chi2(5)    =      21.90
                                                  Prob > chi2     =     0.0005
Log likelihood = -61.700114                       Pseudo R2       =     0.2221
 ( 1)  [2]mpg - [3]mpg = 0
 ( 2)  [2]price - [3]price = 0
 ( 3)  [2]displacement - [3]displacement = 0

    headsize        Coef.   Std. Err.      z    P>|z|     [95% Conf. Interval]

2
     foreign     1.942757    .8554556    2.27   0.023     .2660953     3.61942
         mpg    -.0936841    .0590846   -1.59   0.113    -.2094878    .0221196
       price    -.0000759    .0001113   -0.68   0.496     -.000294    .0001423
displacement     .0096062     .005053    1.90   0.057    -.0002975      .01951
       _cons     1.422395    1.946279    0.73   0.465    -2.392242    5.237032

3
     foreign    -.8960264    1.021318   -0.88   0.380    -2.897774    1.105721
         mpg    -.0936841    .0590846   -1.59   0.113    -.2094878    .0221196
       price    -.0000759    .0001113   -0.68   0.496     -.000294    .0001423
displacement     .0096062     .005053    1.90   0.057    -.0002975      .01951
       _cons     .2172205    1.942194    0.11   0.911     -3.58941    4.023851
```

15.7 Continuation ratio models

The ordered models we have been discussing assume that the response categories are all proportional to one another, that none are assumed to be proportional, or that some are proportional. Regardless of these choices, categories are all assumed to have some type of order but with no specification of degree of difference. But what if a person must pass though a lower level before achieving, or being classified in, a current level? For instance, consider the ascending ranks within the classification of General in the United States Army. One starts as a Brigadier, or one-star, general, advances to a major general, lieutenant general, general, and then finally general of the Army. There are five ordered categories in all. However, except for the brigadier general, before achieving any given rank, one must have been previously classified as a lower-ranked general. The models we have thus far discussed do not take into account such a relationship between categories and therefore lose information intrinsic to an understanding of the response.

A continuation ratio ordered binomial model is designed to accommodate the above-described response. A Stata program called `ocratio` provides the ability to design ordered continuation ratio models with logit, probit, and clog-log links. Fitted probabilities for each category or level of response, crucial when assessing the model fit, are obtained using the postestimation `ocrpred` command.

We will demonstrate the model with a simple example. The data are from a study relating educational level to religiosity, controlling for age, having children, and gender. Three levels of education are tabulated as

```
. use http://www.stata-press.com/data/hh2/edreligion, clear
. tab educlevel
  educlevel |      Freq.     Percent        Cum.
------------+-----------------------------------
         AA |        205       34.11       34.11
         BA |        204       33.94       68.05
     MA/PhD |        192       31.95      100.00
------------+-----------------------------------
      Total |        601      100.00
```

The `age` predictor consists of nine age levels ranging from 17.5 to 57, in intervals of 5 years. The remaining predictors are binary, with the following counts, i.e., number of people in each level.

	Yes (= 1)	No (= 0)
religious	260	341
kids	430	171
male	286	315

We first model `educlevel` by using an ordered logistic regression model.

```
. ologit educlevel religious kids age male, nolog
Ordered logistic regression                       Number of obs   =        601
                                                  LR chi2(4)      =     102.85
                                                  Prob > chi2     =     0.0000
Log likelihood = -608.57623                       Pseudo R2       =     0.0779

------------------------------------------------------------------------------
   educlevel |      Coef.   Std. Err.      z    P>|z|     [95% Conf. Interval]
-------------+----------------------------------------------------------------
   religious |   -.377072   .1609609    -2.34   0.019    -.6925495   -.0615945
        kids |  -.3991817   .1897657    -2.10   0.035    -.7711157   -.0272477
         age |   .0288753   .0099135     2.91   0.004     .0094452    .0483054
        male |   1.452014   .1648725     8.81   0.000      1.12887    1.775158
-------------+----------------------------------------------------------------
       /cut1 |   .3890563   .2864184                     -.1723134    .9504259
       /cut2 |   2.022015   .2994218                      1.435159    2.608871
------------------------------------------------------------------------------
. aic
AIC Statistic =   2.045179
```

No other link produces a significantly better-fitting model. We assess the proportional-odds assumption of the model.

```
. brant, detail
Estimated coefficients from j-1 binary regressions
                  y>1          y>2
religious  -.47609015   -.25110501
     kids  -.65829292    .01479605
      age   .02295298    .03100181
     male    1.217344    1.6997914
    _cons   .09069369   -2.6307503
Brant Test of Parallel Regression Assumption
    Variable |      chi2   p>chi2    df
-------------+--------------------------
         All |     19.98    0.001     4
-------------+--------------------------
   religious |      1.24    0.266     1
        kids |      6.63    0.010     1
         age |      0.48    0.487     1
        male |      5.44    0.020     1
----------------------------------------
A significant test statistic provides evidence that the parallel
regression assumption has been violated.
```

Since the proportional-odds assumption of the model has been violated, we can run a partial proportional-odds model, holding those predictors constant across categories that are nonsignificant in the Brant table.

```
. gologit2 educlevel religious kids age male, autofit lrforce nolog
------------------------------------------------------------------------------
Testing parallel lines assumption using the .05 level of significance...
Step  1:  Constraints for parallel lines imposed for religious (P Value = 0.3627)
Step  2:  Constraints for parallel lines imposed for age (P Value = 0.2774)
Step  3:  Constraints for parallel lines are not imposed for
          kids (P Value = 0.00300)
          male (P Value = 0.01609)
Wald test of parallel lines assumption for the final model:
 ( 1)  [AA]religious - [BA]religious = 0
 ( 2)  [AA]age - [BA]age = 0
           chi2(  2) =    2.01
         Prob > chi2 =    0.3657
An insignificant test statistic indicates that the final model
does not violate the proportional odds/ parallel lines assumption
If you re-estimate this exact same model with gologit2, instead
of autofit you can save time by using the parameter
pl(religious age)
------------------------------------------------------------------------------
Generalized Ordered Logit Estimates               Number of obs   =        601
                                                  LR chi2(6)      =     121.01
                                                  Prob > chi2     =     0.0000
Log likelihood = -599.50008                       Pseudo R2       =     0.0917
 ( 1)  [AA]religious - [BA]religious = 0
 ( 2)  [AA]age - [BA]age = 0
------------------------------------------------------------------------------
   educlevel |      Coef.   Std. Err.      z    P>|z|     [95% Conf. Interval]
-------------+----------------------------------------------------------------
AA           |
   religious |  -.3898915   .1615269    -2.41   0.016    -.7064784   -.0733046
        kids |  -.7513673   .2251507    -3.34   0.001    -1.192655   -.3100801
         age |   .0296937   .0099014     3.00   0.003     .0102874       .0491
        male |   1.204864   .1893592     6.36   0.000     .8337264    1.576001
       _cons |  -.0874313   .3009205    -0.29   0.771    -.6772246    .5023621
-------------+----------------------------------------------------------------
BA           |
   religious |  -.3898915   .1615269    -2.41   0.016    -.7064784   -.0733046
        kids |  -.0362744   .2303568    -0.16   0.875    -.4877655    .4152166
         age |   .0296937   .0099014     3.00   0.003     .0102874       .0491
        male |   1.701747   .2004172     8.49   0.000     1.308937    2.094558
       _cons |  -2.489748   .3321644    -7.50   0.000    -3.140778   -1.838717
------------------------------------------------------------------------------
. aic
AIC Statistic =    2.014975
```

The model accounts for predictor odds that are both proportional and nonproportional across categories. The partial-proportional models result in a slightly lower AIC statistic, but we prefer it because of its accommodation of the proportionality violations found in the standard and generalized models.

However, the model did not take into account the continuation ratio nature of the response. We need to examine the results of modeling such a response and compare it with the partial-proportional model above.

```
. ocratio educlevel religious kids age male, link(logit)
Continuation-ratio logit Estimates                          Number of obs =       997
                                                            chi2(4)       =    109.20
                                                            Prob > chi2   =    0.0000
Log Likelihood = -605.4009                                  Pseudo R2     =    0.0827

------------------------------------------------------------------------------
   educlevel |      Coef.   Std. Err.      z    P>|z|     [95% Conf. Interval]
-------------+----------------------------------------------------------------
   religious |   -.320949    .143897    -2.23   0.026     -.602982   -.038916
        kids |  -.2768327   .1696096    -1.63   0.103    -.6092614    .0555959
         age |   .0276343   .0088566     3.12   0.002     .0102756    .0449929
        male |   1.311213   .1448946     9.05   0.000     1.027225    1.595201
-------------+----------------------------------------------------------------
  _cut1      |   .4384813   .2626932          (Ancillary parameters)
  _cut2      |   1.416228   .2820904
------------------------------------------------------------------------------

. aic
AIC Statistic =    1.222469
```

The number of physical observations in the dataset is 601 (as can be seen in the output from previous models). Here the number of observations listed is 997 and is a function of the number of physical observations as well as the number of levels in the response variable.

The model was run using alternative links, but none were significantly better than the logit model. Notice how much smaller this adjusted AIC is than previous models. It is smaller than the partial-proportional model and reflects what we know of the response—that a person with a higher level of education must have previously been classified with a lower level of education. The exception, of course, is the lowest level.

To facilitate interpretation, we next parameterize the model using the odds ratio form. We do this by applying the `eform` option.

```
. ocratio educlevel religious kids age male, link(logit) eform
Continuation-ratio logit Estimates                          Number of obs =       997
                                                            chi2(4)       =    109.20
                                                            Prob > chi2   =    0.0000
Log Likelihood = -605.4009                                  Pseudo R2     =    0.0827

------------------------------------------------------------------------------
   educlevel | Odds ratio   Std. Err.      z    P>|z|     [95% Conf. Interval]
-------------+----------------------------------------------------------------
   religious |   .7254602   .1043916    -2.23   0.026     .5471775    .9618315
        kids |   .7581813   .1285948    -1.63   0.103     .5437524     1.05717
         age |    1.02802   .0091048     3.12   0.002     1.010329     1.04602
        male |   3.710672   .5376563     9.05   0.000     2.793303     4.92932
-------------+----------------------------------------------------------------
  _cut1      |   .4384813   .2626932          (Ancillary parameters)
  _cut2      |   1.416228   .2820904
------------------------------------------------------------------------------

. aic
AIC Statistic =    1.222469
```

People in the study with higher education are some 27% less religious. Males had the higher education and those who were older tended to have more education. The relationship between having children and educational level is not significant. Each of these relationships is consistent with findings from other studies.

These data were better modeled using a continuation ratio model. Such a model reflects the nature of the response. Ordered categories are often modeled using standard ordered or even generalized ordered binomial models when a continuation ratio model would be better. Care must be taken to assess the relationship between ordered categories.

16 Unordered-response family

Contents

This chapter addresses data in which we record an integer response. Our responses are limited to a finite choice set, and there is no meaning to the values or to the relative magnitude of the outcomes that are assigned.

A useful tool for analyzing such data may be introduced using a utility argument. Assume that we collect data from respondents who provide one choice from a list of alternatives. We code these alternatives $1, 2, \ldots, r$ for convenience. Here we assume that the alternatives are unordered—that there is no indication of one alternative's being better or higher. For our $i = 1, \ldots, n$ respondents, we denote the selection

$$y^*_{ij} = \mathbf{x}_i \boldsymbol{\beta}_j + \epsilon_i \tag{16.1}$$

if respondent i makes choice j. Therefore, y_{ij} is the maximum among the r utilities $y^*_{i1}, y^*_{i2}, \ldots, y^*_{ir}$ such that

$$\Pr(y_{ij} > y^*_{ij}) \tag{16.2}$$

To use a parametric model, we must assume a distribution for ϵ. In the next two sections, we present the results of assuming logit and then probit.

16.1 The multinomial logit model

Sometimes referred to as *polytomous logistic regression* or the *nominal logit model*, the multinomial logit model assumes r equations for the r outcomes. One equation sets $\boldsymbol{\beta}$ to zero so that the problem is identifiable. The associated outcome is the base reference group. Below we assume that the outcomes are coded using the set $1, 2, \ldots, r$ and that the outcome of one is used as the base reference group. Our choice for which outcome to use as the base reference group will affect the estimated coefficients but not the predicted probabilities.

We may derive the log likelihood from

$$\xi_{ik} = \mathbf{x}_i\boldsymbol{\beta}_k + \text{offset}_{ki} \tag{16.3}$$

$$\Pr(y_i = k) = \frac{\exp(\xi_{ik})}{1 + \sum_{m=2}^{r} \exp(\xi_{im})} \qquad k = 2, 3, \ldots, r \tag{16.4}$$

$$\Pr(y_i = 1) = \frac{1}{1 + \sum_{m=2}^{r} \exp(\xi_{im})} \tag{16.5}$$

$$d_{ij} = \begin{cases} 1 & \text{if observation } i \text{ has outcome } j \text{ (if } y_i = j) \\ 0 & \text{otherwise} \end{cases} \tag{16.6}$$

$$= \sum_{i=1}^{n} \sum_{j=1}^{r} d_{ij} \ln \Pr(y_i = j) \tag{16.7}$$

$$\lambda_{ik} = d_{ik} - \Pr(y_i = k) \tag{16.8}$$

First and second derivatives of the log likelihood with respect to the coefficient vectors are given by

$$\frac{\partial}{\partial \beta_{kt}} = \sum_{i=1}^{n} w_i x_{ti} \lambda_{ik} \tag{16.9}$$

$$\frac{\partial^2}{\partial \beta_{kt} \partial \beta_{mu}} = -\sum_{i=1}^{n} \Pr(y_i = k) \;\; \mathrm{I}(k = m) - \Pr(y_i = m) \;\; w_i x_{ti} x_{ui} \tag{16.10}$$

16.1.1 Example: Relation to logistic regression

The multinomial logit model is a generalization of logistic regression to more than two outcomes. It is equal to logistic regression when there are only two outcomes.

Recall our example from chapter 9 in which we investigated the `heart` dataset. Here we show the equivalence of multinomial logit to logistic regression for two outcomes:

```
. use http://www.stata-press.com/data/hh2/heart
. mlogit death anterior hcabg kk2 kk3 age2-age4, nolog
Multinomial logistic regression                   Number of obs   =       4483
                                                  LR chi2(7)      =     211.35
                                                  Prob > chi2     =     0.0000
Log likelihood =  -636.6339                       Pseudo R2       =     0.1424

------------------------------------------------------------------------------
       death |      Coef.   Std. Err.      z    P>|z|     [95% Conf. Interval]
-------------+----------------------------------------------------------------
1            |
    anterior |   .6411581   .1672199     3.83   0.000     .3134131    .9689031
       hcabg |   .7450945   .3528945     2.11   0.035      .053434    1.436755
         kk2 |     .80282   .1671932     4.80   0.000     .4751274    1.130513
         kk3 |   2.659937   .3559227     7.47   0.000     1.962341    3.357532
        age2 |   .4923474   .3101561     1.59   0.112    -.1155473    1.100242
        age3 |   1.509629   .2659661     5.68   0.000     .9883453    2.030913
        age4 |   2.182796   .2711772     8.05   0.000     1.651298    2.714293
       _cons |  -5.049838   .2579719   -19.58   0.000    -5.555453   -4.544222
------------------------------------------------------------------------------
(death==0 is the base outcome)

. glm death anterior hcabg kk2 kk3 age2-age4, family(bin) nolog noheader
------------------------------------------------------------------------------
             |                 OIM
       death |      Coef.   Std. Err.      z    P>|z|     [95% Conf. Interval]
-------------+----------------------------------------------------------------
    anterior |   .6411581   .1672199     3.83   0.000     .3134131    .9689031
       hcabg |   .7450945   .3528945     2.11   0.035      .053434    1.436755
         kk2 |     .80282   .1671932     4.80   0.000     .4751274    1.130513
         kk3 |   2.659936   .3559226     7.47   0.000     1.962341    3.357532
        age2 |   .4923457    .310156     1.59   0.112    -.1155489     1.10024
        age3 |   1.509628    .265966     5.68   0.000     .9883438    2.030911
        age4 |   2.182794   .2711772     8.05   0.000     1.651296    2.714291
       _cons |  -5.049836   .2579718   -19.58   0.000    -5.555451    -4.54422
------------------------------------------------------------------------------
```

16.1.2 Example: Relation to conditional logistic regression

McFadden (1974) describes the analysis of multinomial data in which each respondent can choose only one from a common list of alternatives. The data may be organized such that there is 1 observation for each respondent for each possible choice. One choice is marked with 1 and $r-1$ choices are marked with 0. In conditional logistic regression, we assume that we have $i = 1, \ldots, n$ clusters (panels) of data where each cluster has n_i observations. In conditional logistic regression, none, some, or all of the observations in a cluster may be marked with a 1, so that the McFadden's choice model is a special case of conditional logistic regression.

Greene (2003) includes a 210-respondent dataset from Greene (2002) in which a choice model is applied to data recording travel mode between Sydney and Melbourne, Australia. The modes of travel are air, train, bus, and car. There are also two person-specific constants in the data: `hinc`, the household income, and `psize`, the party size in the chosen mode. The `hinc` and `psize` variables are constant for all choices for a given person (they vary between persons). The relation of the multinomial logit model to the conditional logistic model exists when we limit our covariates to person-specific constants.

For this example, we have 210 respondents and four choices per respondent, for a total of 840 observations. The variable `choice` is set to a member of the set $\{1, 2, 3, 4\}$, meaning {air, train, bus, car}. Indicator variables for the modes of travel are also created, as well as the interactions of the `hinc` and `psize` variables with the mode indicator variables `air`, `bus`, and `train`. Here we show the equivalence of the two commands:

First, we generate the necessary data

```
. use http://www.stata-press.com/data/hh2/tbl19-2
. gen cc      = mod(_n-1,4)+1
. gen g       = int((_n-1)/4)+1
. gen air     = cc==1
. gen bus     = cc==3
. gen train   = cc==2
. gen airinc  = air*hinc
. gen choice  = cc*mode
. gen hair    = hinc*air
. gen htrain  = hinc*train
. gen hbus    = hinc*bus
. gen psair   = psize*air
. gen pstrain = psize*train
. gen psbus   = psize*bus
```

and then fit the alternative model.

```
. clogit mode hair psair air htrain pstrain train hbus psbus bus, group(g) nolog
Conditional (fixed-effects) logistic regression   Number of obs   =        840
                                                  LR chi2(9)      =      75.56
                                                  Prob > chi2     =     0.0000
Log likelihood = -253.34085                       Pseudo R2       =     0.1298

------------------------------------------------------------------------------
        mode |      Coef.   Std. Err.      z    P>|z|     [95% Conf. Interval]
-------------+----------------------------------------------------------------
        hair |   .0035438   .0103047     0.34   0.731    -.0166531    .0237407
       psair |  -.6005541   .1992005    -3.01   0.003    -.9909798   -.2101284
         air |   .9434924    .549847     1.72   0.086    -.1341881    2.021173
      htrain |  -.0573078   .0118416    -4.84   0.000    -.0805169   -.0340987
     pstrain |  -.3098126   .1955598    -1.58   0.113    -.6931027    .0734775
       train |   2.493848   .5357211     4.66   0.000     1.443854    3.543842
        hbus |  -.0303253   .0132228    -2.29   0.022    -.0562415    -.004409
       psbus |  -.9404139   .3244532    -2.90   0.004     -1.57633   -.3044974
         bus |   1.977971    .671715     2.94   0.003     .6614336    3.294508
------------------------------------------------------------------------------
```

```
. mlogit choice hinc psize if choice>0, base(4) nolog
Multinomial logistic regression                   Number of obs   =        210
                                                  LR chi2(6)      =      60.84
                                                  Prob > chi2     =     0.0000
Log likelihood = -253.34085                       Pseudo R2       =     0.1072

------------------------------------------------------------------------------
      choice |      Coef.   Std. Err.      z    P>|z|     [95% Conf. Interval]
-------------+----------------------------------------------------------------
1            |
        hinc |   .0035438   .0103047     0.34   0.731    -.0166531    .0237407
       psize |  -.6005541   .1992005    -3.01   0.003    -.9909799   -.2101284
       _cons |   .9434923    .549847     1.72   0.086    -.1341881    2.021173
-------------+----------------------------------------------------------------
2            |
        hinc |  -.0573078   .0118416    -4.84   0.000    -.0805169   -.0340987
       psize |  -.3098126   .1955598    -1.58   0.113    -.6931028    .0734775
       _cons |   2.493848   .5357211     4.66   0.000     1.443854    3.543842
-------------+----------------------------------------------------------------
3            |
        hinc |  -.0303253   .0132228    -2.29   0.022    -.0562415    -.004409
       psize |   -.940414   .3244532    -2.90   0.004     -1.57633   -.3044974
       _cons |   1.977971    .671715     2.94   0.003     .6614334    3.294508
------------------------------------------------------------------------------
(choice==4 is the base outcome)
```

16.1.3 Example: Extensions with conditional logistic regression

In the previous section, we presented a pedagogical example illustrating the conditions under which the multinomial logit model and the conditional logistic regression model are the same. The conditions were restrictive and the example was contrived.

In the analysis presented in Greene (2003), the independent variables actually used in the data analysis are not respondent-specific constants. In fact, they are choice-specific covariates that cannot be used in the multinomial logit model—the dataset may not be collapsed by respondent since the values of the choice-specific covariates are not constant for the respondent. Here the conditional logit model is to be preferred. In the reference cited, the illustrated example is another model entirely (nested logit model). Although the tools to fit the nested logit model are now available in Stata, we will limit our discussion to a construction of the conditional logit model (called the unconditional model in the cited text).

We have choice-specific covariates: `gc`, the generalized cost constructed from measures on the in-vehicle cost and a product of a wage measure with the time spent traveling; `ttme`, reflecting the terminal time (there is zero waiting time for a car); `hinc`, the household income; and `airinc`, the interaction of the household income with the `fly` mode of travel. A conditional logistic model is preferred because of the presence of the choice-specific constants and is fitted below.

```
. clogit mode air bus train gc ttme airinc, group(g) nolog
Conditional (fixed-effects) logistic regression   Number of obs   =        840
                                                  LR chi2(6)      =     183.99
                                                  Prob > chi2     =     0.0000
Log likelihood = -199.12837                       Pseudo R2       =     0.3160

------------------------------------------------------------------------------
        mode |      Coef.   Std. Err.      z    P>|z|     [95% Conf. Interval]
-------------+----------------------------------------------------------------
         air |   5.207443   .7790551     6.68   0.000     3.680523    6.734363
         bus |   3.163194   .4502659     7.03   0.000     2.280689    4.045699
       train |   3.869043   .4431269     8.73   0.000      3.00053    4.737555
          gc |  -.0155015    .004408    -3.52   0.000     -.024141   -.006862
        ttme |  -.0961248   .0104398    -9.21   0.000    -.1165865   -.0756631
      airinc |    .013287   .0102624     1.29   0.195    -.0068269     .033401
------------------------------------------------------------------------------
```

16.1.4 The independence of irrelevant alternatives

The multinomial logit model carries with it the assumption of the independence of irrelevant alternatives (IIA). A humorous example of violating this assumption is illustrated in a story about Groucho Marx, a famous American comedian. Marx was dining in a posh restaurant when the waiter informed him that the specials for the evening were steak, fish, and chicken. Groucho ordered the steak. The waiter returned later and apologized that there was no fish that evening. Groucho replied, "In that case, I'll have the chicken."[1]

A common application of the multinomial logit model is to model choices in available modes of transportation for commuters. Let us assume, for the sake of emphasizing the role of the IIA assumption, that commuters may choose from the following:

- Driver of a car
- Passenger in a car
- Passenger on a train
- Bicyclist
- Pedestrian

In fitting a model, let us further assume that we use car drivers as our reference group. We obtain a risk ratio of 5 in comparing the propensity with ride the train compared with the propensity to drive a car. The IIA assumption says that this preference is unaffected by the presence of the other choices (not involved in the ratio). If all other modes of commuting vanished and the associated people had to choose between driving a car and riding a train, they would do so in such a way that the risk ratio of 5 would hold constant.

1. Unfortunately, we have no reference for this story and fear that it may be urban legend. In any case, we hope that it is a memorable way to introduce the topic at hand.

Because this is a rather strong assumption, researchers using this model are interested in available tests to evaluate the assumption. Hausman (1978) introduced such a test. As we might guess, the test is built from fitting two multinomial logit models. The first estimation is the full model, and the second estimation deletes all observations for one of the choices in the outcome variable. Equivalently, the second model eliminates one choice from the outcome set. If the IIA assumption holds, then the full model is more efficient than the second model, but both models are consistent and we expect to see no systematic change in the coefficients that are common to both models. The test statistic is given by

$$\chi^2_\nu = (\boldsymbol{\beta}_s - \boldsymbol{\beta}_f)^{\mathrm{T}} (\mathrm{V}_s - \mathrm{V}_f)^{-1} (\boldsymbol{\beta}_s - \boldsymbol{\beta}_f) \tag{16.11}$$

where the s subscript is the subset (second) model, the f subscript is the full (first) model, and the ν subscript denotes the degrees of freedom of the test statistic.

There may be numeric problems in calculating the inverse of the difference in the variance matrices, and so the inverse term in the middle of the equation is a generalized inverse. The degrees of freedom of the test statistic are taken to be the number of common coefficients in the two estimation models. When a generalized inverse is needed, the degrees of freedom are taken to be the rank of the difference matrix.

16.1.5 Example: Assessing the IIA

Let us return to our first analysis of the travel mode data and assess the IIA assumption. Stata provides the `hausman` test. To use this command, we must fit two models. The first model is the full model and the second model is a subset where we remove one of the choices.

```
. mlogit choice hinc psize if choice>0, base(4) nolog
Multinomial logistic regression                   Number of obs   =        210
                                                  LR chi2(6)      =      60.84
                                                  Prob > chi2     =     0.0000
Log likelihood = -253.34085                       Pseudo R2       =     0.1072

------------------------------------------------------------------------------
      choice |      Coef.   Std. Err.      z    P>|z|     [95% Conf. Interval]
-------------+----------------------------------------------------------------
1            |
        hinc |   .0035438   .0103047     0.34   0.731    -.0166531    .0237407
       psize |  -.6005541   .1992005    -3.01   0.003    -.9909799   -.2101284
       _cons |   .9434923    .549847     1.72   0.086    -.1341881    2.021173
-------------+----------------------------------------------------------------
2            |
        hinc |  -.0573078   .0118416    -4.84   0.000    -.0805169   -.0340987
       psize |  -.3098126   .1955598    -1.58   0.113    -.6931028    .0734775
       _cons |   2.493848   .5357211     4.66   0.000     1.443854    3.543842
-------------+----------------------------------------------------------------
3            |
        hinc |  -.0303253   .0132228    -2.29   0.022    -.0562415    -.004409
       psize |   -.940414   .3244532    -2.90   0.004     -1.57633   -.3044974
       _cons |   1.977971    .671715     2.94   0.003     .6614334    3.294508
------------------------------------------------------------------------------
(choice==4 is the base outcome)
```

```
. estimates store all
. mlogit choice hinc psize if choice!=0 & choice!=2, base(4) nolog
Multinomial logistic regression                   Number of obs   =        147
                                                  LR chi2(4)      =      27.02
                                                  Prob > chi2     =     0.0000
Log likelihood =  -141.9678                       Pseudo R2       =     0.0869

------------------------------------------------------------------------------
      choice |      Coef.   Std. Err.      z    P>|z|     [95% Conf. Interval]
-------------+----------------------------------------------------------------
1            |
        hinc |   .0037738   .0106795     0.35   0.724    -.0171577    .0247053
       psize |  -.5861604     .19851    -2.95   0.003    -.9752328    -.197088
       _cons |   .9074443   .5592261     1.62   0.105    -.1886188    2.003507
-------------+----------------------------------------------------------------
3            |
        hinc |    -.03247   .0138722    -2.34   0.019    -.0596589    -.005281
       psize |  -.9604403   .3423713    -2.81   0.005    -1.631476   -.2894049
       _cons |   2.074974   .7159842     2.90   0.004     .6716704    3.478277
------------------------------------------------------------------------------
(choice==4 is the base outcome)
. estimates store partial
. hausman partial all, alleqs constant
                 ---- Coefficients ----
             |      (b)          (B)            (b-B)     sqrt(diag(V_b-V_B))
             |    partial        all         Difference          S.E.
-------------+----------------------------------------------------------------
1            |
        hinc |    .0037738     .0035438          .00023         .0028044
       psize |   -.5861604    -.6005541        .0143938                .
       _cons |    .9074443     .9434923       -.0360481         .1019906
-------------+----------------------------------------------------------------
3            |
        hinc |     -.03247    -.0303253       -.0021447         .0041945
       psize |   -.9604403     -.940414       -.0200263          .109308
       _cons |    2.074974     1.977971         .097003         .2478555
------------------------------------------------------------------------------
                         b = consistent under Ho and Ha; obtained from mlogit
          B = inconsistent under Ha, efficient under Ho; obtained from mlogit
    Test:  Ho:  difference in coefficients not systematic
                  chi2(6) = (b-B)'[(V_b-V_B)^(-1)](b-B)
                          =        0.26
                Prob>chi2 =      0.9997
                (V_b-V_B is not positive definite)
```

The results show that the IIA assumption is not violated for our data.

One cannot apply the Hausman test with blind faith. The test does not require us to choose any particular outcome to elide from the nested model and, annoyingly, one can see conflicting results from the test depending on which outcome we remove. For the variance matrix used in the Hausman test to be singular is also common. These problems are usually associated with models that are either ill fitting or based on small samples (or both). In these cases, a more comprehensive review of the model is in order, for which a reliable application of this test may not be possible.

16.1.6 Interpreting coefficients

Interpreting the coefficients is hard because of the nonlinearity of the link function and the incorporation of a base reference group. As in section 10.6, we can motivate an alternative metric that admits a transformation of the coefficients for easier interpretation.

Since the model is fitted using one of the outcomes as a base reference group, the probabilities that we calculate are relative to that base group. We can define a relative-risk measure for an observation i as the probability of outcome k over the probability of the reference outcome; see (16.4) and (16.5).

$$\text{Relative risk for outcome } k \quad = \quad \exp(\xi_{ik}) \tag{16.12}$$

This measure can be calculated for each outcome and each covariate.

For illustration of the interpretation of a given coefficient, we will assume a model with 2 covariates $\mathbf{x}_1$ and $\mathbf{x}_2$ along with a constant. The relative-risk ratio for $\mathbf{x}_1$ and outcome k is then calculated as the ratio of relative-risk measures. The numerator relative risk is calculated such that the specific covariate is incremented by one relative to the value used in the denominator. The relative-risk ratio for $\mathbf{x}_1$ and outcome k is calculated as

$$\begin{aligned}
\text{Relative} - \text{risk ratio for } \mathbf{x}_1 &= \frac{\exp(\beta_0 + (x_{1ki}+1)\beta_1 + x_{2ki}\beta_2)}{\exp(\beta_0 + x_{1ki}\beta_1 + x_{2ki}\beta_2)} && (16.13)\\
&= \exp(\beta_1) && (16.14)
\end{aligned}$$

The calculation of the relative-risk ratio simplifies such that there is no dependence on a particular observation. The relative-risk ratio is therefore constant—it is independent of the particular values of the covariates.

Interpreting exponentiation coefficients is straightforward for multinomial logit models. These exponentiated coefficients represent relative-risk ratios. A relative-risk ratio of 2 indicates that an outcome is twice as likely relative to the base reference group if the associated covariate is increased by one.

16.1.7 Example: Medical admissions—introduction

We now wish to demonstrate the development of a multinomial model from the medical data used earlier in the text. We must first convert the three `type` variables to one response variable called `admit`. We then tabulate `admit`, showing the number of cases in each type of admission.

```
. use http://www.stata-press.com/data/hh2/medpar, clear
. gen byte admit = type1 + 2*type2 + 3*type3
. label define admitlab 1 "Elective" 2 "Urgent" 3 "Emergency"
. label values admit admitlab
. label define whitelab 1 "White" 0 "Other"
```

```
. label values white whitelab
. tab admit
```

admit	Freq.	Percent	Cum.
Elective	1,134	75.85	75.85
Urgent	265	17.73	93.58
Emergency	96	6.42	100.00
Total	1,495	100.00	

As expected, emergency admissions are less common than elective or urgent admissions. We next summarize variables `white` (1 = white patient; 0 = otherwise), and `los` (hospital length of stay in days) within each level or category of `admit`. The mean value of `white` indicates the percentage of white patients in each level of `admit`. Ninety-two percent of the patients being admitted as elective are white; 8% are nonwhite.

We next model `admit` on `white`.

```
. tab admit white
```

	white		
admit	Other	White	Total
Elective	80	1,054	1,134
Urgent	37	228	265
Emergency	10	86	96
Total	127	1,368	1,495

The odds ratios for the two levels, with one (elective) as the reference, can be determined directly:

$$\frac{(80)(228)}{(37)(1054)} = 0.46771629 \quad (16.15)$$

$$\frac{(80)(86)}{(10)(1054)} = 0.65275142 \quad (16.16)$$

Directly modeling the relationship between the measures by using multinomial logistic regression confirms the direct calculations.

```
. mlogit admit white, rrr nolog

Multinomial logistic regression                   Number of obs   =       1495
                                                  LR chi2(2)      =      12.30
                                                  Prob > chi2     =     0.0021
Log likelihood = -1029.3204                       Pseudo R2       =     0.0059

------------------------------------------------------------------------------
       admit |        RRR   Std. Err.      z    P>|z|     [95% Conf. Interval]
-------------+----------------------------------------------------------------
Urgent       |
       white |   .4677163   .0990651    -3.59   0.000     .3088112    .7083893
-------------+----------------------------------------------------------------
Emergency    |
       white |   .6527514   .2308533    -1.21   0.228       .32637    1.305526
------------------------------------------------------------------------------
(admit==Elective is the base outcome)
```

The printed relative-risk ratios agree with the direct calculations. Moreover, for admit=3 versus admit=2, we can calculate

$$\frac{(37)(86)}{(10)(228)} = 1.395614 \tag{16.17}$$

The coefficient of the model is $\log(1.395614) = 0.33333446$. You can also obtain the coefficient by subtracting coefficients. Here we run mlogit again but this time using admit=2 as the reference category.

```
. mlogit admit white, base(2) nolog

Multinomial logistic regression                   Number of obs   =       1495
                                                  LR chi2(2)      =      12.30
                                                  Prob > chi2     =     0.0021
Log likelihood = -1029.3204                       Pseudo R2       =     0.0059

------------------------------------------------------------------------------
       admit |      Coef.   Std. Err.      z    P>|z|     [95% Conf. Interval]
-------------+----------------------------------------------------------------
Elective     |
       white |   .7598934   .2118059     3.59   0.000     .3447615    1.175025
       _cons |   .7711087    .198814     3.88   0.000     .3814405    1.160777
-------------+----------------------------------------------------------------
Emergency    |
       white |   .3333345   .3782075     0.88   0.378    -.4079386    1.074608
       _cons |  -1.308333   .3564085    -3.67   0.000    -2.006881    -.609785
------------------------------------------------------------------------------
(admit==Urgent is the base outcome)
```

In the next section, we illustrate interpreting the relative-risk ratios and coefficients.

16.1.8 Example: Medical admissions—summary

We will next develop a more complex model by adding an additional predictor with the aim of interpreting and testing the resulting model. We use the data introduced in the previous section with the same admit response variable. We begin by summarizing the mean, standard deviation, and minimum and maximum values for each predictor, white and los, for all three levels of the outcome measure admit.

```
. by admit, sort: summarize white los

-> admit = Elective
    Variable |       Obs        Mean    Std. Dev.       Min        Max
-------------+--------------------------------------------------------
       white |      1134    .9294533    .2561792          0          1
         los |      1134    8.830688    6.456009          1         60

-> admit = Urgent
    Variable |       Obs        Mean    Std. Dev.       Min        Max
-------------+--------------------------------------------------------
       white |       265    .8603774    .3472509          0          1
         los |       265    11.19623    8.824852          1         63

-> admit = Emergency
    Variable |       Obs        Mean    Std. Dev.       Min        Max
-------------+--------------------------------------------------------
       white |        96    .8958333    .3070802          0          1
         los |        96    18.23958    20.61259          1        116
```

We next proceed to model `admit` on `white` and `los` by using a multinomial model. There is an order to the levels, but a rather loose one (qualitative rather than quantitative). We will evaluate the ordinality of the model afterward.

```
. mlogit admit white los, nolog
Multinomial logistic regression                   Number of obs   =       1495
                                                  LR chi2(4)      =      87.02
                                                  Prob > chi2     =     0.0000
Log likelihood = -991.96238                       Pseudo R2       =     0.0420

------------------------------------------------------------------------------
       admit |      Coef.   Std. Err.      z    P>|z|     [95% Conf. Interval]
-------------+----------------------------------------------------------------
Urgent       |
       white |  -.7086604   .2140645    -3.31   0.001    -1.128219   -.2891017
         los |   .0375103   .0084408     4.44   0.000     .0209666     .054054
       _cons |  -1.187453   .2227772    -5.33   0.000    -1.624088   -.7508174
-------------+----------------------------------------------------------------
Emergency    |
       white |  -.2267416    .377415    -0.60   0.548    -.9664613    .5129782
         los |   .0780083   .0097884     7.97   0.000     .0588233    .0971932
       _cons |  -3.188991   .3917186    -8.14   0.000    -3.956746   -2.421237
------------------------------------------------------------------------------
(admit==Elective is the base outcome)

. aic
AIC Statistic =    1.335067
```

Interpretation is facilitated by expressing the coefficients in exponential form, using relative-risk ratios.

```
. mlogit admit white los, rrr nolog
Multinomial logistic regression                   Number of obs   =       1495
                                                  LR chi2(4)      =      87.02
                                                  Prob > chi2     =     0.0000
Log likelihood = -991.96238                       Pseudo R2       =     0.0420

------------------------------------------------------------------------------
       admit |        RRR   Std. Err.      z    P>|z|     [95% Conf. Interval]
-------------+----------------------------------------------------------------
Urgent       |
       white |   .4923032   .1053847    -3.31   0.001      .323609    .7489361
         los |   1.038223   .0087634     4.44   0.000     1.021188    1.055542
-------------+----------------------------------------------------------------
Emergency    |
       white |   .7971268   .3008476    -0.60   0.548     .3804269    1.670258
         los |   1.081132   .0105826     7.97   0.000     1.060588    1.102073
------------------------------------------------------------------------------
(admit==Elective is the base outcome)
. predict p1 p2 p3
(option pr assumed; predicted probabilities)
```

Elective admission is the reference level; both the other levels are interpreted relative to the reference level. Therefore, we may state, on the basis of the model, that whites are half as likely to be admitted as an urgent admission than as an elective admission. Whites are approximately 20% less likely to be admitted as an emergency admission than as elective. We can also say that for a one-unit increase in the variable `los`, the relative risk of having an urgent admission rather than an elective admission increases by some 4%. Likewise, for a one-unit change in the variable `los`, the relative risk of having an emergency admission rather than an elective admission increases by about 8%.

We can express the results of the above multinomial regression as probabilities. If `los` is set at its mean value for each level, we can calculate probabilities for each level for both the white and nonwhite categories. We may do this by using the `prtab` command:

```
. prtab white
mlogit: Predicted probabilities for admit
Predicted probability of outcome 2 (Urgent)

----------------------
    white | Prediction
----------+-----------
    Other |     0.2884
    White |     0.1687
----------------------

Predicted probability of outcome 3 (Emergency)

----------------------
    white | Prediction
----------+-----------
    Other |     0.0581
    White |     0.0550
----------------------
```

```
Predicted probability of outcome 1 (Elective)

--------------------------
     white |  Prediction
-----------+--------------
     Other |      0.6535
     White |      0.7763
--------------------------

         white         los
x=   .91505017   9.8541806
```

We can perform a variety of tests on both the predictors as well as omnibus tests on the model by using the all option.

```
. mlogtest, all
**** Likelihood-ratio tests for independent variables (N=1495)
 Ho: All coefficients associated with given variable(s) are 0.
             |       chi2   df   P>chi2
-------------+-------------------------
       white |     10.161    2    0.006
         los |     74.716    2    0.000
---------------------------------------

**** Wald tests for independent variables (N=1495)
 Ho: All coefficients associated with given variable(s) are 0.
             |       chi2   df   P>chi2
-------------+-------------------------
       white |     10.963    2    0.004
         los |     64.757    2    0.000
---------------------------------------

**** Hausman tests of IIA assumption (N=1495)
 Ho: Odds(Outcome-J vs Outcome-K) are independent of other alternatives.
 Omitted |      chi2   df   P>chi2   evidence
---------+------------------------------------
  Urgent |     1.400    3    0.706   for Ho
Emergenc |     2.546    3    0.467   for Ho
----------------------------------------------

**** Small-Hsiao tests of IIA assumption (N=1495)
 Ho: Odds(Outcome-J vs Outcome-K) are independent of other alternatives.
 Omitted |  lnL(full)  lnL(omit)    chi2   df   P>chi2   evidence
---------+--------------------------------------------------------
  Urgent |   -129.878   -127.490   4.775    3    0.189   for Ho
Emergenc |   -349.799   -348.335   2.928    3    0.403   for Ho
------------------------------------------------------------------

**** Wald tests for combining alternatives (N=1495)
 Ho: All coefficients except intercepts associated with a given pair
     of alternatives are 0 (i.e., alternatives can be combined).
Alternatives tested|      chi2   df   P>chi2
-------------------+------------------------
   Urgent-Emergenc |    17.799    2    0.000
   Urgent-Elective |    32.105    2    0.000
 Emergenc-Elective |    64.854    2    0.000
--------------------------------------------
```

```
**** LR tests for combining alternatives (N=1495)
 Ho: All coefficients except intercepts associated with a given pair
     of alternatives are 0 (i.e., alternatives can be collapsed).
Alternatives tested|       chi2    df    P>chi2
-------------------+---------------------------
   Urgent-Emergenc |     20.660     2     0.000
   Urgent-Elective |     30.752     2     0.000
 Emergenc-Elective |     70.508     2     0.000
-------------------+---------------------------
```

Finally, we can graph the relationship of the length of stay and race variables for each level by using the specifications below. The three respective graphs are given in figures 16.1, 16.2, and 16.3.

```
. line p1 los if white==0 || line p1 los if white==1,
> legend(order(1 "nonwhite" 2 "white")) title(Medicare LOS vs Admission Type)
> l1title("Admit=1: Elective Admission") clpattern("-###")
```

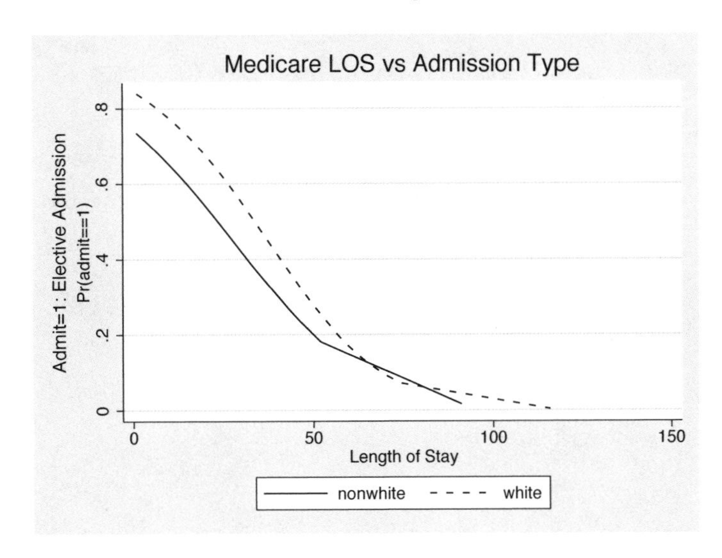

Figure 16.1: Length of stay versus admission type for elective admissions

```
. line p2 los if white==0 || line p2 los if white==1,
> legend(order(1 "nonwhite" 2 "white")) title(Medicare LOS vs Admission Type)
> l1title("Admit=2: Urgent Admission") clpattern("-###")
```

(Continued on next page)

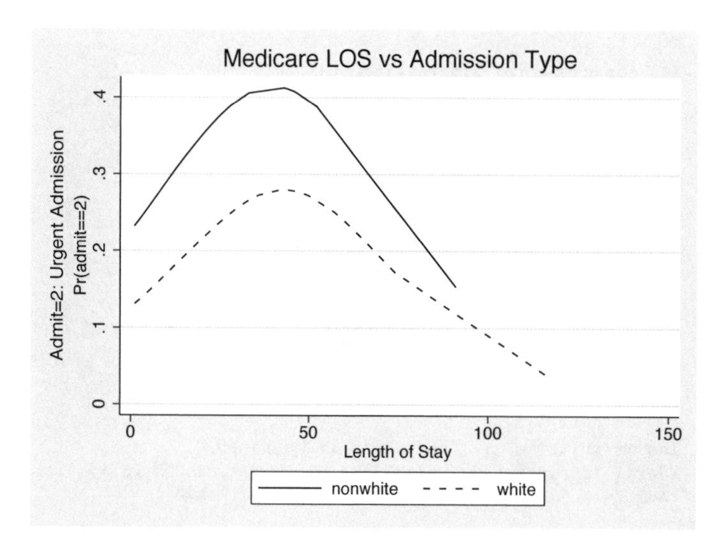

Figure 16.2: Length of stay versus admission type for urgent admissions

```
. line p3 los if white==0 || line p3 los if white==1,
> legend(order(1 "nonwhite" 2 "white")) title(Medicare LOS vs Admission Type)
> l1title("Admit=3: Emergency Admission") clpattern("-###")
```

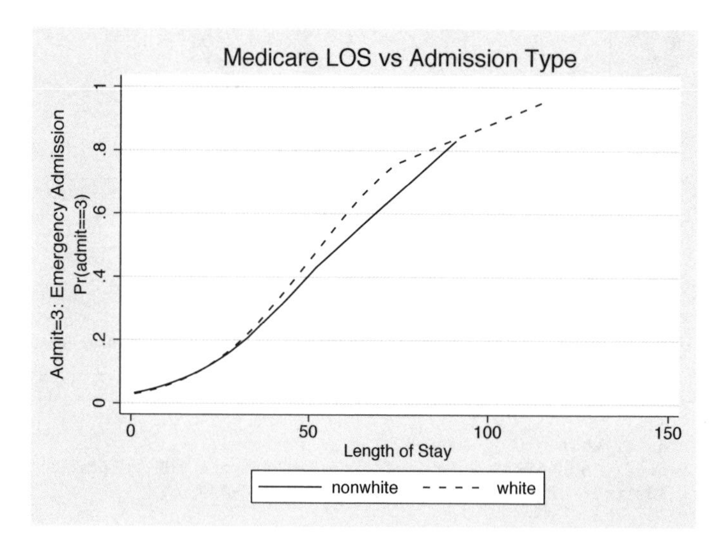

Figure 16.3: Length of stay versus admission type for emergency admissions

In particular, the figures illustrate the quadratic relationship for each race related to hospital stay.

Earlier, we foreshadowed our intent to test the ordinality of the response. Here we investigate this topic through comparison of an ordered and an unordered model.

```
. quietly mlogit admit white los
. estimates store mlogit
. quietly ologit admit white los
. lrtest mlogit, force
Likelihood-ratio test                                 LR chi2(2)  =      3.99
(Assumption: . nested in mlogit)                      Prob > chi2 =    0.1361
```

Since the models are not recognized as being nested because of the naming conventions of equations, we used the `force` option. Because the hypothesis is not rejected, we conclude that the outcome is not significantly ordered; that is, the response is best modeled as unordered.

16.2 The multinomial probit model

The multinomial probit model was first described by Albright, Lerman, and Manski (1977) and examined in full by Daganzo (1979). It is a computationally difficult model to fit and has been found in commercial software applications only within the past few years. It entered into official Stata with version 9.

The multinomial probit model is to the binary probit as the multinomial logit is to the binary logit model. The multinomial probit model is often referred to as the random utility model in that an individual's choice among r alternatives is that choice having the maximum utility. The utility U_{ji} is the utility of alternative j for individual i and is assumed to be structured as a linear combination of observed alternative choices.

Unlike the multinomial logit model, the multinomial probit model does not carry with it the assumption of irrelevant alternatives, because it does not assume r different equations. However, it is a much more difficult model to fit, requiring computation of multivariate normal probabilities.

To use a parametric multinomial probit model, we assume that $y_i = j$ if $U_{ji} = \max\{U_{1i}, \ldots, U_{ri}\}$; that is, the outcome of a particular observation is determined by the maximum utility. Thus,

$$\Pr(y_i = j) = \Pr(U_{ji} > U_{j'i}) \tag{16.18}$$

for all $j' \neq j$. In the multinomial probit model, the utility for a choice j given by U_{ji} is parameterized as a linear predictor from a matrix of covariates and coefficients introduced as $\mathbf{x}_i\boldsymbol{\beta}$.

A generalization of this approach allows the covariates to differ for each choice $j = 1, \ldots, r$ such that the linear predictor used to parameterize the utility is given by $\mathbf{x}_{ji}\boldsymbol{\beta}$. This model may be estimated in Stata by using the `asmprobit` command.

Under the usual multinomial probit specification, we assume that the outcome may take one or r values with r unobserved errors $\epsilon_{1i}, \ldots, \epsilon_{ri}$. For the multinomial probit model, we assume that these errors are distributed as multivariate normal

$$[\epsilon_{1i}, \epsilon_{2i}, \ldots, \epsilon_{ri}] \sim \mathrm{N}(\mathbf{0}_{r\times 1}, \mathbf{\Sigma}_{r\times r}) \tag{16.19}$$

Instead of normalizing one of the coefficient vectors as we did for the multinomial logit model, we instead must normalize the covariance matrix entries. Since we are making comparisons of the variances, we normalize the variances (diagonal elements of $\mathbf{\Sigma}$) to one. And since we are making comparisons of the outcomes, we must normalize $r-1$ of the covariances to zero. The size of $\mathbf{\Sigma}$ is $r\times r$, leaving $(r-1)(r-2)/2$ unrestricted covariances. Thus, for a model with three choices ($r = 3$), all three probabilities are based on the cumulative bivariate normal distribution where the variance matrix has ones on the diagonal and one $(1 = (3-1)(3-2)/2)$ covariance parameter.

As mentioned previously, the difficulty in implementing this model is the need to calculate the multivariate normal probabilities. Lerman and Manski (1981) demonstrate how one can estimate these probabilities via simulation. When the first edition of this text was written, there were no programs to fit these models in Stata. This is no longer a limitation since users now have the `mprobit` command at their disposal. Furthermore, an extension allowing choice-specific covariates may also be fitted using the `asmprobit` command.

```
. mprobit choice hinc psize if choice>0, base(4) nolog
Multinomial probit regression                     Number of obs   =        210
                                                  Wald chi2(6)    =      50.53
Log likelihood = -254.33587                       Prob > chi2     =     0.0000

------------------------------------------------------------------------------
      choice |      Coef.   Std. Err.      z    P>|z|     [95% Conf. Interval]
-------------+----------------------------------------------------------------
_outcome_1   |
        hinc |   .0022036   .0076576     0.29   0.774     -.012805    .0172122
       psize |  -.4375164    .148394    -2.95   0.003    -.7283632   -.1466695
       _cons |   .6958218   .3920542     1.77   0.076    -.0725904    1.464234
-------------+----------------------------------------------------------------
_outcome_2   |
        hinc |  -.0404515     .00823    -4.92   0.000    -.0565821   -.0243209
       psize |  -.2391164   .1500077    -1.59   0.111    -.5331261    .0548933
       _cons |   1.815538   .3813136     4.76   0.000     1.068177    2.562899
-------------+----------------------------------------------------------------
_outcome_3   |
        hinc |  -.0211674   .0090003    -2.35   0.019    -.0388077   -.0035272
       psize |    -.63597    .203865    -3.12   0.002    -1.035538    -.236402
       _cons |   1.347706   .4441984     3.03   0.002     .4770932    2.218319
------------------------------------------------------------------------------
(choice=4 is the base outcome)
```

The last example illustrates that we can compare the multinomial probit fit with the earlier model fit by using multinomial logit. There is a similarity in the coefficients from these models that we will now explore.

16.2.1 Example: A comparison of the models

We repeat a model of `admit` on `white` and `los` for comparing results with `mprobit`. Note the comparison of the log-likelihood functions, AIC and BIC GOF statistics, predicted probabilities for each level of `admit` from observations 1–10, and correlations of the probabilities for each level. Also compare the respective parameter estimates and their *p*-values. All the values are similar, indicating that neither model is preferred over the other. Because the multinomial logit has many accepted fit statistics associated with the model, we would recommend it. However, if the theory underlying the data prefers normality, then multinomial probit should be selected.

The estimates for the multinomial logit model are given by

```
. mlogit admit white los, nolog
Multinomial logistic regression                   Number of obs   =       1495
                                                  LR chi2(4)      =      87.02
                                                  Prob > chi2     =     0.0000
Log likelihood = -991.96238                       Pseudo R2       =     0.0420

------------------------------------------------------------------------------
       admit |      Coef.   Std. Err.      z    P>|z|     [95% Conf. Interval]
-------------+----------------------------------------------------------------
Urgent       |
       white |  -.7086604   .2140645    -3.31   0.001    -1.128219   -.2891017
         los |   .0375103   .0084408     4.44   0.000     .0209666     .054054
       _cons |  -1.187453   .2227772    -5.33   0.000    -1.624088   -.7508174
-------------+----------------------------------------------------------------
Emergency    |
       white |  -.2267416    .377415    -0.60   0.548    -.9664613    .5129782
         los |   .0780083   .0097884     7.97   0.000     .0588233    .0971932
       _cons |  -3.188991   .3917186    -8.14   0.000    -3.956746   -2.421237
------------------------------------------------------------------------------
(admit==Elective is the base outcome)
. aic
AIC Statistic =    1.335067
. predict pl1 pl2 pl3
(option pr assumed; predicted probabilities)
```

For comparison, the estimates for the multinomial probit are given by

(Continued on next page)

```
. mprobit admit white los, nolog
Multinomial probit regression                     Number of obs   =       1495
                                                  Wald chi2(4)    =      73.36
Log likelihood = -993.34447                       Prob > chi2     =     0.0000

------------------------------------------------------------------------------
       admit |      Coef.   Std. Err.      z    P>|z|     [95% Conf. Interval]
-------------+----------------------------------------------------------------
Urgent       |
       white |  -.5706817    .1771321    -3.22   0.001    -.9178542   -.2235091
         los |   .0302881    .0064955     4.66   0.000     .0175572    .0430189
       _cons |  -.9956175     .183421    -5.43   0.000    -1.355116    -.636119
-------------+----------------------------------------------------------------
Emergency    |
       white |  -.2238527    .2454722    -0.91   0.362    -.7049694     .257264
         los |   .0549894    .0070543     7.80   0.000     .0411632    .0688157
       _cons |  -2.283496    .2523319    -9.05   0.000    -2.778057   -1.788935
------------------------------------------------------------------------------
(admit=Elective is the base outcome)
. aic
AIC Statistic =    1.336916
. predict pp1 pp2 pp3
(option pr assumed; predicted probabilities)
```

The predicted probabilities are similar, as we would expect, since the two approaches are modeling the same outcomes.

```
. list pp1 pl1 pp2 pl2 pp3 pl3 in 1/10

     +-----------------------------------------------------------------+
     |      pp1        pl1        pp2        pl2        pp3        pl3 |
     |-----------------------------------------------------------------|
  1. | .7318475   .7346393   .2366323   .2326272   .0315202   .0327334 |
  2. | .7318475   .7346393   .2366323   .2326272   .0315202   .0327334 |
  3. | .7318475   .7346393   .2366323   .2326272   .0315202   .0327334 |
  4. | .7318475   .7346393   .2366323   .2326272   .0315202   .0327334 |
  5. | .7318475   .7346393   .2366323   .2326272   .0315202   .0327334 |
     |-----------------------------------------------------------------|
  6. | .7234864    .726253   .2424414   .2387618   .0340723   .0349852 |
  7. | .7234864    .726253   .2424414   .2387618   .0340723   .0349852 |
  8. | .7234864    .726253   .2424414   .2387618   .0340723   .0349852 |
  9. | .7234864    .726253   .2424414   .2387618   .0340723   .0349852 |
 10. | .7234864    .726253   .2424414   .2387618   .0340723   .0349852 |
     +-----------------------------------------------------------------+
```

The pairwise correlations of the predicted probabilities (for level 1, 2, and 3 admission types) for the multinomial logit and multinomial probit models are given below. Not surprisingly, the correlations are very high.

```
. corr pp1 pl1
(obs=1495)

             |      pp1      pl1
-------------+------------------
         pp1 |   1.0000
         pl1 |   0.9997   1.0000
```

```
. corr pp2 pl2
(obs=1495)

             |      pp2      pl2
-------------+------------------
         pp2 |   1.0000
         pl2 |   0.9955   1.0000

. corr pp3 pl3
(obs=1495)

             |      pp3      pl3
-------------+------------------
         pp3 |   1.0000
         pl3 |   0.9976   1.0000
```

16.2.2 Example: Comparing probit and multinomial probit

We may suspect that if we were to combine two categories of a three-category response variable, and then model the two responses on predictors using both probit and multinomial probit, that the results would be identical. However, as described by Long (1997) and Long and Freese (2006), this is not the case. The multinomial probit parameter estimates are greater—by a scale factor of $\sqrt{2} \approx 1.414$.

To illustrate this relationship, we combine categories 2 and 3, defining them as outcome zero, but we keep the original category one the same.

```
. tab admit, nolabel

      admit |      Freq.     Percent        Cum.
------------+-----------------------------------
          1 |      1,134       75.85       75.85
          2 |        265       17.73       93.58
          3 |         96        6.42      100.00
------------+-----------------------------------
      Total |      1,495      100.00

. gen admitbin = (admit==1)  /* Create a binary variable */

. tab admitbin

   admitbin |      Freq.     Percent        Cum.
------------+-----------------------------------
          0 |        361       24.15       24.15
          1 |      1,134       75.85      100.00
------------+-----------------------------------
      Total |      1,495      100.00
```

We next model the binary outcome admitbin by using a probit model.

(Continued on next page)

```
. probit admitbin white los, nolog
Probit regression                               Number of obs   =       1495
                                                LR chi2(2)      =      65.65
                                                Prob > chi2     =     0.0000
Log likelihood = -793.56443                     Pseudo R2       =     0.0397

------------------------------------------------------------------------------
    admitbin |      Coef.   Std. Err.      z    P>|z|     [95% Conf. Interval]
-------------+----------------------------------------------------------------
       white |   .3588006   .1223274     2.93   0.003     .1190433    .5985579
         los |  -.0300373   .0043505    -6.90   0.000    -.0385641   -.0215105
       _cons |   .6863979   .1261822     5.44   0.000     .4390854    .9337105
------------------------------------------------------------------------------
```

Then we model the same data by using multinomial probit, making sure to constrain the reference category or level to zero, just as probit does automatically.

```
. mprobit admitbin white los, baseoutcome(0) nolog
Multinomial probit regression                   Number of obs   =       1495
                                                Wald chi2(2)    =      58.02
Log likelihood = -793.56443                     Prob > chi2     =     0.0000

------------------------------------------------------------------------------
    admitbin |      Coef.   Std. Err.      z    P>|z|     [95% Conf. Interval]
-------------+----------------------------------------------------------------
_outcome_2   |
       white |   .5074207   .1729971     2.93   0.003     .1683526    .8464887
         los |  -.0424792   .0061525    -6.90   0.000    -.0545379   -.0304204
       _cons |   .9707133   .1784486     5.44   0.000     .6209605    1.320466
------------------------------------------------------------------------------
(admitbin=0 is the base outcome)
```

Now we illustrate that dividing the multinomial probit coefficient by the square root of 2 produces the original probit coefficients.

```
. display "white: " _b[white]/sqrt(2) _n "los  : " _b[los]/sqrt(2) _n
> "_cons: " _b[_cons]/sqrt(2)
white: .3588006
los  : -.0300373
_cons: .68639794
```

So, the relationship holds as stated. The label above the list of coefficient names in the output table says `_outcome_2`. This is a simple shift in the frame of reference. `_outcome_1` is the reference level, which is zero for the binary and multinomial probit models, but with zero meaning a combined `admit=2` or `admit=3`.

We now look at the relationship of predicted probabilities. We again model `admitbin` on `white` and `los` by using probit regression, calculating probabilities for level 1, that is, the probability of success, which here is the probability of an elective admission. How the previous model has been reparameterized should be clear.

```
. probit admitbin white los, nolog

Probit regression                                 Number of obs   =       1495
                                                  LR chi2(2)      =      65.65
                                                  Prob > chi2     =     0.0000
Log likelihood = -793.56443                       Pseudo R2       =     0.0397

------------------------------------------------------------------------------
    admitbin |      Coef.   Std. Err.      z    P>|z|     [95% Conf. Interval]
-------------+----------------------------------------------------------------
       white |   .3588006   .1223274     2.93   0.003     .1190433    .5985579
         los |  -.0300373   .0043505    -6.90   0.000    -.0385641   -.0215105
       _cons |   .6863979   .1261822     5.44   0.000     .4390854    .9337105
------------------------------------------------------------------------------

. predict prob1
(option p assumed; Pr(admitbin))
```

We then model the same data by using `mprobit` and forcing the reference level to be zero. We also have Stata adjust the model by the square root of 2 through specifying the `probitparam` option. The coefficients are now the same as those obtained from the `probit` command. After fitting the model, we obtain the predicted probabilities.

```
. mprobit admitbin white los, probitparam baseoutcome(0) nolog

Multinomial probit regression                     Number of obs   =       1495
                                                  Wald chi2(2)    =      58.02
Log likelihood = -793.56443                       Prob > chi2     =     0.0000

------------------------------------------------------------------------------
    admitbin |      Coef.   Std. Err.      z    P>|z|     [95% Conf. Interval]
-------------+----------------------------------------------------------------
_outcome_2   |
       white |   .3588006   .1223274     2.93   0.003     .1190433    .5985579
         los |  -.0300373   .0043505    -6.90   0.000    -.0385641   -.0215105
       _cons |   .6863979   .1261822     5.44   0.000     .4390854    .9337105
------------------------------------------------------------------------------
(admitbin=0 is the base outcome)

. predict mprob1 mprob2
(option pr assumed; predicted probabilities)
```

A quick check of the probabilities confirms that the probabilities from the multinomial model match those from the binary probit model.

```
. corr prob1 mprob1 mprob2
(obs=1495)

             |    prob1   mprob1   mprob2
-------------+---------------------------
       prob1 |   1.0000
      mprob1 |  -1.0000   1.0000
      mprob2 |   1.0000  -1.0000   1.0000
```

A listing illustrates how the multinomial predicted probabilities relate to the predicted probability from the binary model.

```
. list prob1 mprob1 mprob2 in 1/10

     +--------------------------------+
     |    prob1     mprob1     mprob2 |
     |--------------------------------|
  1. | .7442039    .255796   .7442039 |
  2. | .7442039    .255796   .7442039 |
  3. | .7442039    .255796   .7442039 |
  4. | .7442039    .255796   .7442039 |
  5. | .7442039    .255796   .7442039 |
     |--------------------------------|
  6. | .7344486   .2655514   .7344486 |
  7. | .7344486   .2655514   .7344486 |
  8. | .7344486   .2655514   .7344486 |
  9. | .7344486   .2655514   .7344486 |
 10. | .7344486   .2655514   .7344486 |
     +--------------------------------+
```

The `mprob1` probabilities are calculated as `1.0 - mprob2`, which makes sense when we recall that, for a given variable, the probability of 1 and the probability of 0 must sum to 1.0. The `mprob1` probabilities correlate perfectly, but inversely, with `mprob2` and with `prob1`.

Of ancillary interest, the probabilities appear the same regardless of whether we use the `probitparam` option. The coefficients, however, differ.

```
. quietly mprobit admitbin white los, baseoutcome(0) nolog

. predict pm1 pm2
(option pr assumed; predicted probabilities)

. corr prob1 mprob1 mprob2 pm1 pm2
(obs=1495)

             |    prob1   mprob1   mprob2      pm1      pm2
-------------+---------------------------------------------
       prob1 |   1.0000
      mprob1 |  -1.0000   1.0000
      mprob2 |   1.0000  -1.0000   1.0000
         pm1 |  -1.0000   1.0000  -1.0000   1.0000
         pm2 |   1.0000  -1.0000   1.0000  -1.0000   1.0000

. list prob1 mprob1 mprob2 pm1 pm2 in 1/5

     +------------------------------------------------------+
     |    prob1    mprob1     mprob2       pm1          pm2 |
     |------------------------------------------------------|
  1. | .7442039   .255796   .7442039   .255796     .7442039 |
  2. | .7442039   .255796   .7442039   .255796     .7442039 |
  3. | .7442039   .255796   .7442039   .255796     .7442039 |
  4. | .7442039   .255796   .7442039   .255796     .7442039 |
  5. | .7442039   .255796   .7442039   .255796     .7442039 |
     +------------------------------------------------------+
```

16.2.3 Example: Concluding remarks

Many think that the advantage of multinomial probit over multinomial logit is that the former is immune from violation of the IIA. However, this is not the case. As Long and Freese (2006) point out, the parameter estimates and probabilities from the two models are nearly always very close, particularly when we adjust for changes of scale—$\pi^2/6$ for multinomial logit models and $\sqrt{2}$ for multinomial probit.

We have not discussed fit statistics here but many of the same tests that have been given to multinomial logit models can likewise be given to evaluate multinomial probit models. Exponentiated probit coefficients are not odds ratios, even though prima facie they appear be. For probit models, exponentiated coefficients have no statistically interesting interpretation. Therefore, the multinomial tests related to odds ratios are not to be used for evaluating multinomial probit models. However, commands such as `prvalue`, which is an extremely useful utility, and several others that are found in Long and Freese (2006), can well be used to evaluate these models.

You can find many other categorical response models in the statistical literature. Stata has several more, including stereotype logistic regression, first discussed by Anderson (1984); alternative-specific multinomial probit; conditional logistic regression, which we briefly looked at when comparing it with multinomial models; and rank-ordered logistic regression. Long and Freese (2006) also discuss each of the above-mentioned models.

Part VI

Extensions to the GLM

17 Extending the likelihood

Contents

This chapter includes examples of fitting quasilikelihood models by using the accompanying software. Because quasilikelihood variance functions are not included in the software, the user must program appropriate functions to fit the model. See chapter 19 for information on the programs, and review the Stata documentation for programming assistance.

17.1 The quasilikelihood

We have previously presented the algorithms for fitting GLMs. We also mentioned that although our illustrations generally begin from a specified distribution, we need only specify the first two moments. Our original exponential-family density was specified as

$$f(y) = \exp\left\{\frac{y\theta - b(\theta)}{a(\phi)} + c(y,\phi)\right\} \tag{17.1}$$

with the log likelihood given by

$$\mathcal{L} = \sum_{i=1}^{n}\left\{\frac{y_i\theta_i - b(\theta_i)}{a(\phi)} + c(y_i,\phi)\right\} \tag{17.2}$$

We further showed that estimates could be obtained by solving the estimating equation given by

$$\frac{\partial\mathcal{L}}{\partial\mu} = \frac{\partial\mathcal{L}}{\partial\theta}\frac{\partial\theta}{\partial\mu} = \frac{y - b'(\theta)}{a(\phi)}\frac{1}{b''(\theta)} = \frac{y-\mu}{\mathrm{V}(y)} \tag{17.3}$$

We assume that there is a quasilikelihood defined by

$$Q(y;\mu) = \int_y^\mu \frac{y-\mu}{\phi\mathrm{V}(\mu)}d\mu \tag{17.4}$$

If there exists a distribution (from the exponential family of distributions) that has $V(\mu)$ as its variance function, then the quasilikelihood reduces to a standard log likelihood. If not, then the resulting estimates are more properly called maximum quasilikelihood (MQL) estimates. A corresponding deviance called the quasideviance may also be defined as

$$D = 2\int_{\widehat{\mu}}^{y} \frac{y-\mu}{V(\mu)} d\mu \tag{17.5}$$

More extensions (not covered here) to the GLM approach include methods based on the quasilikelihood that allow for estimation of the variance function. Nelder and Pregibon (1987) describe a method for estimating the mean-variance relationship by using what they call the extended quasilikelihood. The extension adds a term to the quasilikelihood that allows comparison of quasideviances calculated from distributions with different variance functions. It does this by normalizing the quasilikelihood so that it has the properties of a log likelihood as a function of the scalar variance. This normalization is a saddle point approximation of the exponential family members derived, assuming that the distribution exists.

17.2 Example: Wedderburn's leaf blotch data

Here we present a quasilikelihood analysis of data from Wedderburn (1974). This analysis parallels the analysis presented in McCullagh and Nelder (1989). The data represent the proportion of leaf blotch on 10 varieties of barley grown at nine sites in 1965. The amount of leaf blotch is recorded as a percentage, and thus we can analyze the data by using binomial family variance with a logistic link.

```
. use http://www.stata-press.com/data/hh2/accident
. xi: glm y I.site I.variety, family(binom) link(logit) nolog
I.site            _Isite_1-9          (naturally coded; _Isite_1 omitted)
I.variety         _Ivariety_1-10      (naturally coded; _Ivariety_1 omitted)
note: y has non-integer values

Generalized linear models                          No. of obs      =         90
Optimization     : ML                              Residual df     =         72
                                                   Scale parameter =          1
Deviance         =  6.125989621                    (1/df) Deviance =   .0850832
Pearson          =  6.391999102                    (1/df) Pearson  =   .0887778

Variance function: V(u) = u*(1-u/1)                [Binomial]
Link function    : g(u) = ln(u/(1-u))              [Logit]

                                                   AIC             =   .8620201
Log likelihood   = -20.79090427                    BIC             =  -317.8603

------------------------------------------------------------------------------
             |                 OIM
           y |      Coef.   Std. Err.      z    P>|z|     [95% Conf. Interval]
-------------+----------------------------------------------------------------
    _Isite_2 |   1.639066   4.843975     0.34   0.735     -7.85495    11.13308
    _Isite_3 |   3.326515   4.528246     0.73   0.463    -5.548685    12.20171
    _Isite_4 |    3.58223   4.512192     0.79   0.427    -5.261504    12.42596
    _Isite_5 |   3.583056   4.512147     0.79   0.427    -5.260591     12.4267
    _Isite_6 |   3.893253   4.498006     0.87   0.387    -4.922677    12.70918
```

```
     _Isite_7 |   4.730018   4.479785     1.06   0.291    -4.050199    13.51023
     _Isite_8 |    5.52269   4.479259     1.23   0.218    -3.256497    14.30188
     _Isite_9 |   6.794584    4.49963     1.51   0.131    -2.024528     15.6137
  _Ivariety_2 |   .1500848   2.428801     0.06   0.951    -4.610278    4.910448
  _Ivariety_3 |   .6894659   2.256572     0.31   0.760    -3.733334    5.112265
  _Ivariety_4 |   1.048159   2.179627     0.48   0.631    -3.223831     5.32015
  _Ivariety_5 |   1.614706   2.099848     0.77   0.442    -2.500922    5.730333
  _Ivariety_6 |   2.371158   2.043985     1.16   0.246    -1.634978    6.377294
  _Ivariety_7 |   2.570528    2.03547     1.26   0.207    -1.418919    6.559975
  _Ivariety_8 |   3.342041   2.018879     1.66   0.098    -.6148901    7.298972
  _Ivariety_9 |   3.500049   2.018248     1.73   0.083     -.455645    7.455743
 _Ivariety_10 |   4.253008   2.027916     2.10   0.036      .278365    8.227651
        _cons |  -8.054648    4.77236    -1.69   0.091     -17.4083    1.299006
------------------------------------------------------------------------------
```

Since the data are proportions instead of counts, there is no a priori theoretical reason that ϕ should be near one. Because of this, we use a quasilikelihood approach where we allow the Pearson deviance to scale the standard errors.

```
. xi: glm y I.site I.variety, family(binom) link(logit) scale(x2) nolog
I.site            _Isite_1-9          (naturally coded; _Isite_1 omitted)
I.variety         _Ivariety_1-10      (naturally coded; _Ivariety_1 omitted)
note: y has non-integer values

Generalized linear models                          No. of obs      =        90
Optimization     : ML                              Residual df     =        72
                                                   Scale parameter =         1
Deviance         =  6.125989621                    (1/df) Deviance =  .0850832
Pearson          =  6.391999102                    (1/df) Pearson  =  .0887778

Variance function: V(u) = u*(1-u/1)                [Binomial]
Link function    : g(u) = ln(u/(1-u))              [Logit]

                                                   AIC             =  .8620201
Log likelihood   = -20.79090427                    BIC             = -317.8603

------------------------------------------------------------------------------
             |                 OIM
           y |      Coef.   Std. Err.      z    P>|z|     [95% Conf. Interval]
-------------+----------------------------------------------------------------
     _Isite_2 |   1.639066   1.443291     1.14   0.256    -1.189733    4.467865
     _Isite_3 |   3.326515   1.349218     2.47   0.014      .682096    5.970934
     _Isite_4 |    3.58223   1.344435     2.66   0.008     .9471863    6.217273
     _Isite_5 |   3.583056   1.344421     2.67   0.008     .9480384    6.218073
     _Isite_6 |   3.893253   1.340208     2.90   0.004     1.266494    6.520012
     _Isite_7 |   4.730018   1.334779     3.54   0.000       2.1139    7.346136
     _Isite_8 |    5.52269   1.334622     4.14   0.000     2.906879    8.138501
     _Isite_9 |   6.794584   1.340692     5.07   0.000     4.166877    9.422291
  _Ivariety_2 |   .1500848   .7236759     0.21   0.836    -1.268294    1.568464
  _Ivariety_3 |   .6894659   .6723591     1.03   0.305    -.6283337    2.007265
  _Ivariety_4 |   1.048159   .6494329     1.61   0.107    -.2247057    2.321024
  _Ivariety_5 |   1.614706   .6256624     2.58   0.010       .38843    2.840982
  _Ivariety_6 |   2.371158   .6090174     3.89   0.000     1.177506     3.56481
  _Ivariety_7 |   2.570528   .6064804     4.24   0.000     1.381849    3.759208
  _Ivariety_8 |   3.342041   .6015372     5.56   0.000      2.16305    4.521032
  _Ivariety_9 |   3.500049   .6013491     5.82   0.000     2.321426    4.678672
 _Ivariety_10 |   4.253008   .6042298     7.04   0.000     3.068739    5.437276
        _cons |  -8.054648   1.421953    -5.66   0.000    -10.84163   -5.267671
------------------------------------------------------------------------------
(Standard errors scaled using square root of Pearson X2-based dispersion)
```

To investigate the fit, we graph the Pearson residuals versus the fitted values and the Pearson residuals versus the log of the variance function.

```
. predict double eta, eta
. predict double pearson, pearson
. predict double mu
(option mu assumed; predicted mean y)
. gen double logvar = log(mu*mu*(1-mu)*(1-mu))
. label var pearson "Pearson residuals"
. label var eta     "Linear predictor"
. label var logvar  "Log(variance)"
. twoway scatter pearson eta, yline(0)
. twoway scatter pearson logvar, yline(0)
```

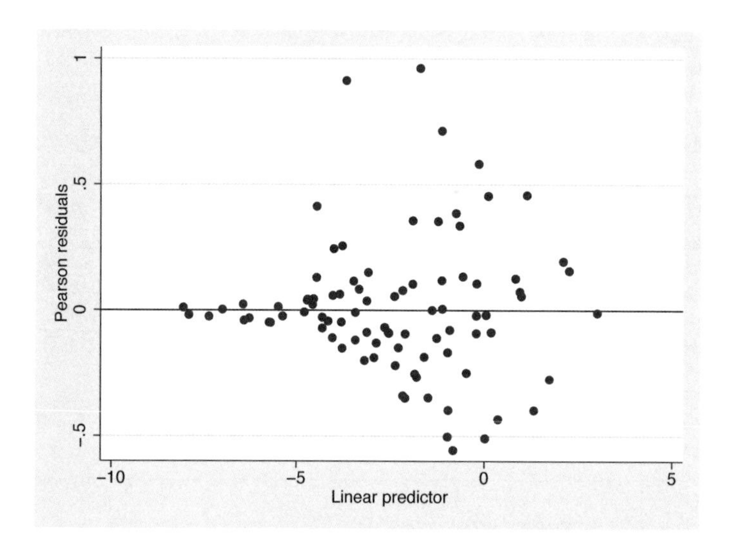

Figure 17.1: Pearson residuals versus linear predictor

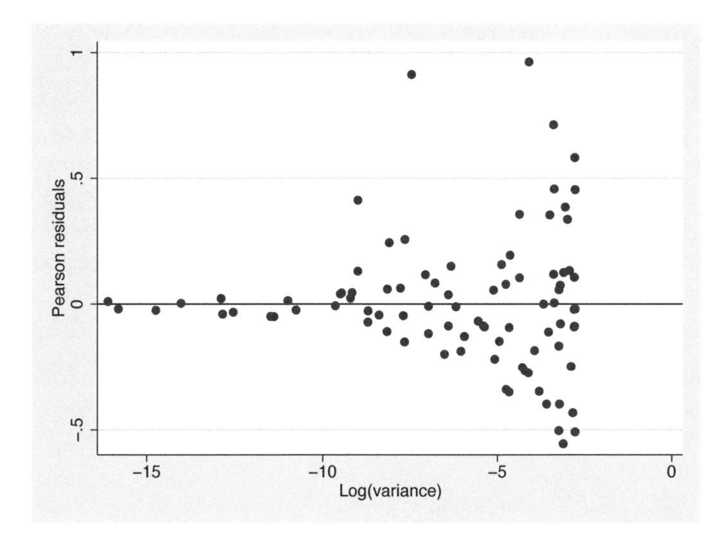

Figure 17.2: Pearson residuals versus log(variance)

Our diagnostic graphs in figures 17.1 and 17.2 indicate that there is a lack of fit for very small and very large values. Wedderburn suggests using a variance function with form

$$\mathrm{V}(\mu) = \mu^2(1-\mu)^2 \tag{17.6}$$

This is the square of the usual binomial variance function. Implementing this solution with the accompanying software uses the most sophisticated support mechanisms of the `glm` command. To successfully implement a solution, we must program the respective support. For this particular case, the quasideviance function cannot be defined because of the zeros in the data. This will ultimately mean that our solution will not support the `irls` option of the command. The quasilikelihood is then given as

$$Q(\mu; y) = (2y-1)\ln\left(\frac{\mu}{1-\mu}\right) - \frac{y}{\mu} - \frac{1-y}{1-\mu} \tag{17.7}$$

The easiest path to the solution is to build on the fact that the variance function of interest is the square of the usual binomial variance. We copy the `glim_v2.ado` file (which defines the support for the binomial variance function) to the file `binsq.ado` and edit the eight sections of the command. The eight sections correspond to various support functions for the IRLS and N–R methods. Our responsibility will be to edit these sections to calculate appropriate quantities for the target variance function. In actuality, we need only edit the sections for the variance function calculation, the derivative of the variance, and the likelihood. The remaining sections either are removed (and thus not supported) or are left the same.

The support file binsq.ado is listed below.

Listing 17.1: Skeleton code for a user-written variance program

```
1  program define binsq
2      args todo eta mu return touse
3
4      if `todo' == -1 {
5          if "$SGLM_m" != "1" {                  /* Title */
6              global SGLM_vt "Quasi"
7              global SGLM_vf "u^2*(1-u/$SGLM_m)^2"
8          }
9          else {
10             global SGLM_vt "Quasi"
11             global SGLM_vf "u^2*(1-u)^2"
12         }
13
14         local y      "$SGLM_y"
15         local m      "$SGLM_m"
16         local touse "`eta'"
17
18         capture assert `m'>0 if `touse'        /* sic, > */
19         if _rc {
20             di as error `"`m' has nonpositive values"'
21             exit 499
22         }
23         capture assert `m'==int(`m') if `touse'
24         if _rc {
25             di as error`"`m' has noninteger values"'
26             exit 499
27         }
28         capture assert `y'>=0 if `touse'
29         if _rc {
30             di as error `"dependent variable `y'
31                 has negative values"'
32             exit 499
33         }
34         capture assert `y'<=`m' if `touse'
35         if _rc {
36             noi di as error `"`y'>`m' in cases"'
37             exit 499
38         }
39
40         global SGLM_mu "glim_mu 0 $SGLM_m"
41         exit
42     }
43     if `todo' == 0 {               /* Initialize eta */
44         if "$SGLM_L" == "glim_l01" {          /* Identity */
45             gen double `eta' = `mu'/$SGLM_m
46         }
```

```
47          else if "$SGLM_L" == "glim_l02" {    /* Logit */
48              gen double `eta' = ln(`mu'/($SGLM_m-`mu'))
49          }
50          else if "$SGLM_L" == "glim_l03" {    /* Log */
51              gen double `eta' = ln(`mu'/$SGLM_m)
52          }
53          else if "$SGLM_L" == "glim_l05" {    /* Log compl */
54              gen double `eta' = ln(1-`mu'/$SGLM_m)
55          }
56          else if "$SGLM_L" == "glim_l06" {    /* Log-log */
57              gen double `eta' = -ln(-ln(`mu'/$SGLM_m))
58          }
59          else if "$SGLM_L" == "glim_l07" {    /* Clog-log */
60              gen double `eta' = ln(-ln(1-`mu'/$SGLM_m))
61          }
62          else if "$SGLM_L" == "glim_l08" {    /* Probit */
63              gen double `eta' = invnormal(`mu'/$SGLM_m)
64          }
65          else if "$SGLM_L" == "glim_l09" {    /* Reciprocal */
66              gen double `eta' = $SGLM_m/`mu'
67          }
68          else if "$SGLM_L" == "glim_l10" {    /* Power(-2) */
69              gen double `eta' = ($SGLM_m/`mu')^2
70          }
71          else if "$SGLM_L" == "glim_l11" {    /* Power(a) */
72              gen double `eta' = (`mu'/$SGLM_m)^$SGLM_a
73          }
74          else if "$SGLM_L" == "glim_l12" {    /* OPower(a) */
75              gen double `eta' = /*
76                  */ ((`mu'/($SGLM_m-`mu'))^$SGLM_a-1) / $SGLM_a
77          }
78          else {
79              gen double `eta' = $SGLM_m*($SGLM_y+.5)/($SGLM_m+1)
80          }
81          exit
82      }
83      if `todo' == 1 {                /* V(mu) */
84          gen double `return' =  `mu'*`mu'*(1-`mu'/$SGLM_m) /*
85              */ *(1-`mu'/$SGLM_m)
86          exit
87      }
88      if `todo' == 2 {                /* (d V)/(d mu) */
89          gen double `return' =  2*`mu' - /*
90              */ 6*`mu'*`mu'/$SGLM_m + /*
91              */ 4*`mu'*`mu'*`mu'/($SGLM_m*$SGLM_m)
92          exit
93      }
94      if `todo' == 3 {                /* deviance */
95          noi di as error "deviance calculation not supported"
```

```
 96          gen double `return' = 0
 97          exit
 98      }
 99      if `todo' == 4 {              /* Anscombe */
100          noi di as error "Anscombe residuals not supported"
101          exit 198
102      }
103      if `todo' == 5 {              /* ln-likelihood */
104          local y "$SGLM_y"
105          local m "$SGLM_m"
106          if "`y'" == "" {
107              local y "`e(depvar)'"
108          }
109          if "`m'" == "" {
110              local m "`e(m)'"
111          }
112          gen double `return' = (2*`y'-1)* /*
113              */ ln(`mu'/(`m'-`mu')) - `y'/`mu' - /*
114              */ (`m'-`y')/(`m'-`mu')
115          exit
116      }
117      if `todo' == 6 {
118          noi di as error "adj. deviance residuals not supported"
119          exit 198
120      }
121      noi di as error "Unknown call to glim variance function"
122      error 198
123  end
```

We may then proceed to specify this variance function with the `glm` command and then obtain the same graphs as before to analyze the fit of the resulting model.

```
. xi: glm y I.site I.variety, family(binsq) link(logit) nolog
I.site           _Isite_1-9          (naturally coded; _Isite_1 omitted)
I.variety        _Ivariety_1-10      (naturally coded; _Ivariety_1 omitted)
deviance calculation not supported

Generalized linear models                          No. of obs      =        90
Optimization     : ML                              Residual df     =        72
                                                   Scale parameter =  .9885464
Deviance         =                0                (1/df) Deviance =         0
Pearson          =  71.17534179                    (1/df) Pearson  =  .9885464

Variance function: V(u) = u^2*(1-u)^2              [Quasi]
Link function    : g(u) = ln(u/(1-u))              [Logit]

                                                   AIC             = -.5527161
Log likelihood   =  42.87222528                    BIC             = -323.9863

------------------------------------------------------------------------------
             |                 OIM
           y |      Coef.   Std. Err.      z    P>|z|     [95% Conf. Interval]
-------------+----------------------------------------------------------------
    _Isite_2 |   1.383119     .455533     3.04   0.002     .4902907    2.275947
    _Isite_3 |   3.860061    .4516254     8.55   0.000     2.974891     4.74523
    _Isite_4 |      3.557    .4577697     7.77   0.000     2.659788    4.454212
    _Isite_5 |    4.10786    .4495166     9.14   0.000     3.226824    4.988897
    _Isite_6 |   4.305356    .4611785     9.34   0.000     3.401463    5.209249
    _Isite_7 |   4.918099    .4451046    11.05   0.000      4.04571    5.790488
    _Isite_8 |   5.694892    .4538692    12.55   0.000     4.805325    6.584459
    _Isite_9 |   7.067632    .4512644    15.66   0.000      6.18317    7.952094
 _Ivariety_2 |  -.4673533    .4877992    -0.96   0.338    -1.423422    .4887157
 _Ivariety_3 |   .0788063    .4795614     0.16   0.869    -.8611168    1.018729
 _Ivariety_4 |   .9540754    .4765712     2.00   0.045     .0200131    1.888138
 _Ivariety_5 |    1.35263    .4661998     2.90   0.004      .438895    2.266365
 _Ivariety_6 |   1.328541    .5237571     2.54   0.011     .3019957    2.355086
 _Ivariety_7 |   2.340071    .4502318     5.20   0.000     1.457633    3.222509
 _Ivariety_8 |   3.262581    .4553686     7.16   0.000     2.370075    4.155087
 _Ivariety_9 |   3.135486    .4829722     6.49   0.000     2.188878    4.082094
_Ivariety_10 |   3.887267    .4612816     8.43   0.000     2.983171    4.791362
       _cons |  -7.922378    .4418132   -17.93   0.000    -8.788316    -7.05644
------------------------------------------------------------------------------
```

The resulting graphs are obtained using the following commands:

```
. predict double eta, eta
. predict double pearson, pearson
. predict double mu
(option mu assumed; predicted mean y)
. gen double logvar = log(mu*mu*(1-mu)*(1-mu))
. label var eta     "Linear predictor"
. label var pearson "Pearson residuals"
. label var logvar  "Log(variance)"
. twoway scatter pearson eta, yline(0)
. twoway scatter pearson logvar, yline(0)
```

We can see in figures 17.3 and 17.4 that the effect of the extreme fitted values has been substantially reduced using this model.

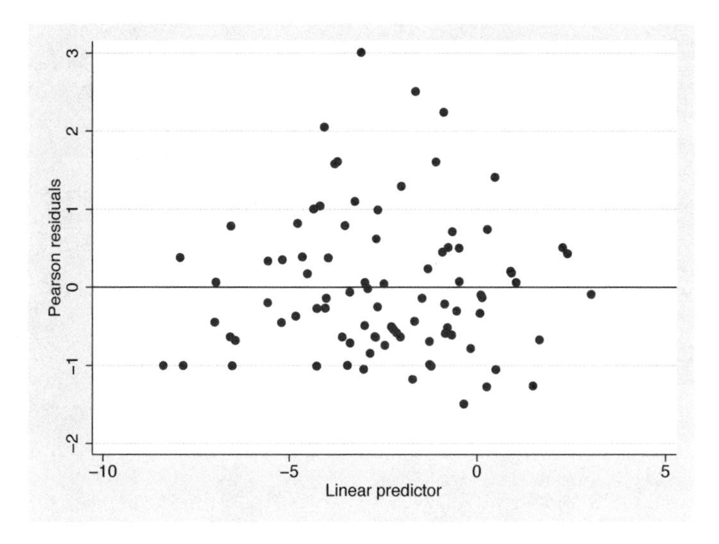

Figure 17.3: Pearson residuals versus linear predictor

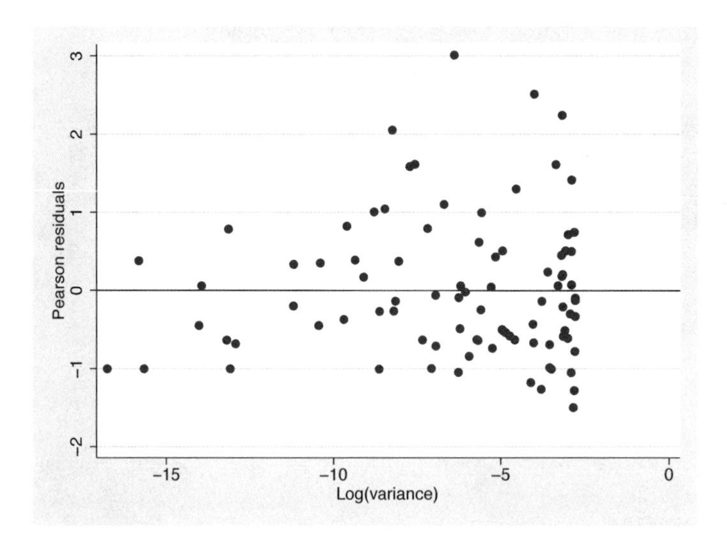

Figure 17.4: Pearson residuals versus log(variance)

17.3 Generalized additive models

Hastie and Tibshirani (1986) discuss a technique for generalizing the link function of the GLM called generalized additive models (GAMs). In a GAM approach, we attempt to overcome the inability of our subset of covariates to adequately model the response surface of our model. With this shortcoming, the fitted values deviate from our response surface in some systematic manner. The GAM modification is to replace the linear predictor with the additive model given by

$$\eta = \eta_0 + \sum_j f_j(x_j) \tag{17.8}$$

where the $f_j(x_j)$ are smooth functions that are estimated from the data. Our model is still additive in the covariates but is no longer linear, hence the change in the name of the analysis to GAM. The smooth functions are identifiable only up to a constant, which we denote η_0.

The examples included in the first edition of this book were generated with user-written software that is no longer available for modern operating systems. There is, as yet, no official program for fitting GAMs, but this could change quickly when the Stata community sets its mind on development. Periodically use the `net search` capabilities, along with keeping your copy of Stata up to date, to learn of associated enhancements to the base software.

18 Clustered data

Contents

This chapter introduces several extensions to GLM that address clustered (panel) data. The topic of the chapter actually deserves its own book, so our goal here is to relay introductory information to give an overview of these modern methods. We hope that by introducing the topics here under a notation consistent with our treatment of GLM, the conceptual relationships will be clear even if the implementation details are not.

18.1 Generalization from individual to clustered data

There are several methods to address clustered data. Clustered data occur when there is a natural classification to observations such that data may be organized according to generation or sampling from units. For example, we may collect data on loans where we have multiple observations from different banks. Addressing the dependence in the data on the bank itself would be natural, and there are several methods that we might use to take this dependence into account.

Each method for addressing the panel structure of the data has advantages and limitations, and recognizing the assumptions and inferences that are available benefits the researcher.

In a clustered dataset, we assume that we have $i = 1, \ldots, n$ panels (clusters) where each panel has $t = 1, \ldots, n_i$ correlated observations. Our exponential-family notation is amended to read

$$\exp\left\{\frac{y_{it}\theta_{it} - b(\theta_{it})}{a(\phi)} - c(y_{it}, \phi)\right\} \tag{18.1}$$

We will focus on the Poisson model as an example. We will use a ship accident dataset from McCullagh and Nelder (1989), listed on page 205 of that text. This dataset contains a count of the number of reported damage incidents, `incident`; the aggregate months of service by ship type, `months`; the period of operation, `op`; the year of construction, `co`; and the ship type, `ship`. We use `months` as the exposure in our model. We generate indicator variables `op_##_##` to reflect definition by the starting and ending years of operation, and we generate indicator variables `co_##_##` to reflect definition by the starting and ending years of construction. For example, the indicator variable `op_75_79` records whether the observation is for a ship in operation from 1975 to 1979, and the indicator variable `co_65_69` records whether the ship was in construction from 1965 to 1969.

Our goal is to focus on the methods and techniques for deriving the associated models with the methods, so we include output from the examples without the associated inference or usual diagnostic plots.

18.2 Pooled estimators

A simple approach to modeling panel data is simply to ignore the panel dependence that might be present. The result of this approach is called a *pooled* estimator since the data are simply pooled together without regard to which cluster the data naturally belong. The resulting estimated coefficient vector is consistent, but it is not efficient. As a result, the estimated (naive) standard errors will not be a reliable measure for testing purposes. To address the standard errors, we may use a modified sandwich (see sec. 3.6.4) estimate of variance or another variance estimate that adjusts for the clusters that naturally occur in the data.

When using a pooled estimator with a modified variance estimate, remember that using the modified variance estimate is a declaration that the underlying likelihood is not correct. This is obvious since there is no alteration of the optimized estimating equation and since the modification of the results is limited to modifying the estimated variance of the coefficient vector.

The modified sandwich estimate of variance is robust to correlation within panels and is robust to *any* type of correlation. This variance estimate may result in smaller or larger standard errors than the naive (Hessian) standard errors of the pooled estimator, depending on the type of correlation present. The calculation of the modified sandwich estimate of variance uses the sums of the residuals from each cluster. If the residuals are negatively correlated, the sums will be small and the modified sandwich estimate of variance will produce smaller standard errors than the naive estimator.

With this approach, our pooled estimator for a Poisson model results in a log likelihood given by

$$\mathcal{L} = \sum_{i=1}^{n} \sum_{t=1}^{n_i} \left[y_{it} \ln\{\exp(\mathbf{x}_{it}\boldsymbol{\beta})\} - \exp(\mathbf{x}_{it}\boldsymbol{\beta}) - \ln\Gamma(y_{it}+1) \right] \tag{18.2}$$

which we may solve using the usual GLM methods presented in this text. Suspecting that there may be correlation within panels, we fit our model of interest by specifying the modified sandwich estimate of variance. Our estimator is consistent but is inefficient, and our standard errors are calculated to be robust to correlation within panels.

Applying the pooled estimator model to the ship accident data provides estimates given by

```
. glm incident op_75_79 co_65_69 co_70_74 co_75_79, family(poiss) link(log)
> cluster(ship) offset(exposure) eform nolog

Generalized linear models                          No. of obs      =        34
Optimization     : ML                              Residual df     =        29
                                                   Scale parameter =         1
Deviance         =  62.36534078                    (1/df) Deviance =  2.150529
Pearson          =  82.73714004                    (1/df) Pearson  =  2.853005

Variance function: V(u) = u                        [Poisson]
Link function    : g(u) = ln(u)                    [Log]

                                                   AIC             =  5.006819
Log pseudolikelihood = -80.11591605                BIC             = -39.89911
                                  (Std. Err. adjusted for 5 clusters in ship)

------------------------------------------------------------------------------
             |               Robust
    incident |        IRR   Std. Err.      z    P>|z|     [95% Conf. Interval]
-------------+----------------------------------------------------------------
    op_75_79 |    1.47324   .1287036     4.44   0.000       1.2414    1.748377
    co_65_69 |   2.125914   .2850531     5.62   0.000     1.634603    2.764897
    co_70_74 |   2.860138   .6213563     4.84   0.000     1.868384    4.378325
    co_75_79 |   2.021926   .4265285     3.34   0.001     1.337221    3.057227
    exposure |   (offset)
------------------------------------------------------------------------------
```

In addition to applying the modified sandwich estimate of variance, we might apply the variable jackknife or grouped bootstrap estimate of variance. The weighted sandwich estimates of variance are not available for grouped data in the `glm` command. The application of these variance estimates for our example is for illustration only. There are only five ships in this dataset! The asymptotic properties of these variance estimates depend on the number of groups being large.

18.3 Fixed effects

Ignoring the panel structure of our data, we may assume that there is some fixed panel-specific effect. To address the panel structure in our data, we may include an effect for each panel in our estimating equation. We may assume that these effects are fixed effects or random effects. With regard to assuming fixed effects, we may use conditional or unconditional fixed-effects estimators. Unconditional fixed-effects estimators simply include an indicator variable for the panel in our estimation. Conditional fixed-effects estimators alter the likelihood to a conditional likelihood, which removes the fixed effects from the estimation.

18.3.1 Unconditional fixed-effects estimators

If there are a finite number of panels in a population and each panel is represented in our sample, we could use an unconditional fixed-effects estimator. If there are an infinite number of panels (or effectively uncountable), then we would use a conditional fixed-effects estimator, because using an unconditional fixed-effects estimator would result in biased estimates. Applying the unconditional estimator to our Poisson model results in a log likelihood given by

$$\mathcal{L} = \sum_{i=1}^{n}\sum_{t=1}^{n_i}\left[y_{it}\ln\{\exp(\mathbf{x}_{it}\boldsymbol{\beta}+\nu_i)\} - \exp(\mathbf{x}_{it}\boldsymbol{\beta}) - \ln\Gamma(y_{it}+1)\right] \tag{18.3}$$

Fitting this likelihood to our ship accident dataset provides the following results:

```
. xi: glm incident op_75_79 co_65_69 co_70_74 co_75_79 i.ship, family(poiss)
> link(log) offset(exposure) eform nolog
i.ship            _Iship_1-5          (naturally coded; _Iship_1 omitted)

Generalized linear models                          No. of obs      =        34
Optimization     : ML                              Residual df     =        25
                                                   Scale parameter =         1
Deviance         =  38.69505154                    (1/df) Deviance =  1.547802
Pearson          =  42.27525312                    (1/df) Pearson  =   1.69101

Variance function: V(u) = u                        [Poisson]
Link function    : g(u) = ln(u)                    [Log]

                                                   AIC             =  4.545928
Log likelihood   = -68.28077143                    BIC             = -49.46396

------------------------------------------------------------------------------
             |                 OIM
    incident |        IRR   Std. Err.      z    P>|z|     [95% Conf. Interval]
-------------+----------------------------------------------------------------
    op_75_79 |   1.468831   .1737218     3.25   0.001     1.164926    1.852019
    co_65_69 |   2.008002   .3004803     4.66   0.000     1.497577    2.692398
    co_70_74 |    2.26693    .384865     4.82   0.000     1.625274    3.161912
    co_75_79 |   1.573695   .3669393     1.94   0.052     .9964273    2.485397
    _Iship_2 |   .5808026   .1031447    -3.06   0.002     .4100754    .8226088
    _Iship_3 |    .502881   .1654716    -2.09   0.037     .2638638    .9584087
    _Iship_4 |    .926852   .2693234    -0.26   0.794     .5244081    1.638141
    _Iship_5 |   1.384833   .3266535     1.38   0.168     .8722007    2.198762
    exposure |   (offset)
------------------------------------------------------------------------------
```

18.3.2 Conditional fixed-effects estimators

A conditional fixed-effects estimator is formed by conditioning out the fixed effects from the estimation. This allows a much more efficient estimator at the cost of placing constraints on inference in the form of the conditioning imposed on the likelihood.

To derive the conditional fixed-effects Poisson model, we begin by specifying the associated probability of a specific outcome defined as

$$\begin{aligned} \Pr(Y_{it} = y_{it}) &= \exp\{-\exp(\nu_i + \mathbf{x}_{it}\boldsymbol{\beta})\}\exp(\nu_i + \mathbf{x}_{it}\boldsymbol{\beta})^{y_{it}}/y_{it}! && (18.4) \\ &= \frac{1}{y_{it}!}\exp\{-\exp(\nu_i)\exp(\mathbf{x}_{it}\boldsymbol{\beta}) + \nu_i y_{it}\}\exp(\mathbf{x}_{it}\boldsymbol{\beta})^{y_{it}} && (18.5) \end{aligned}$$

Since we know that the observations are independent, we may write the joint probability for the observations within a panel as

$$\begin{aligned} &\Pr(Y_{i1} = y_{i1}, \ldots, Y_{in_i} = y_{in_i}) = \\ &\prod_{t=1}^{n_i} \frac{1}{y_{it}!}\exp\{-\exp(\nu_i)\exp(\mathbf{x}_{it}\boldsymbol{\beta}) + \nu_i y_{it}\}\exp(\mathbf{x}_{it}\boldsymbol{\beta})^{y_{it}} && (18.6) \\ &= \prod_{t=1}^{n_i} \frac{\exp(\mathbf{x}_{it}\boldsymbol{\beta})^{y_{it}}}{y_{it}!}\exp\left\{-\exp(\nu_i)\sum_t \exp(\mathbf{x}_{it}\boldsymbol{\beta}) + \nu_i\sum_t y_{it}\right\} && (18.7) \end{aligned}$$

We also know that the sum of n_i Poisson independent random variables, each with parameter λ, is distributed as Poisson with parameter $n_i\lambda$ so that we have

$$\Pr\left(\sum_t y_{it}\right) = \frac{1}{(\sum_t y_{it})!}\exp\left\{-\exp(\nu_i)\sum_t \exp(\mathbf{x}_{it}\boldsymbol{\beta}) + \nu_i\sum_t y_{it}\right\}\left\{\sum_t \exp(\mathbf{x}_{it}\boldsymbol{\beta})\right\}^{\sum_t y_{it}} \quad (18.8)$$

So, the conditional log likelihood is conditioned on the sum of the outcomes in the panel (since that is the sufficient statistic for ν_i). The appropriate conditional probability is given by

$$\Pr\left(Y_{i1} = y_{i1}, \ldots, Y_{in_i} = y_{in_i}\middle|\nu_i, \boldsymbol{\beta}, \sum_t y_{it}\right) = \left(\sum_t y_{it}\right)!\prod_{t=1}^{n_i}\frac{\exp(\mathbf{x}_{it}\boldsymbol{\beta})^{y_{it}}}{y_{it}!\{\sum_k \exp(\mathbf{x}_{ik}\boldsymbol{\beta})\}^{y_{it}}} \quad (18.9)$$

which is free of the fixed effect ν_i. The conditional log likelihood (with offsets) is then derived as

$$\mathcal{L}_c = \ln\left[\prod_{i=1}^{n}\left(\sum_{t=1}^{n_i} y_{it}\right)! \prod_{t=1}^{n_i} \frac{\exp(\mathbf{x}_{it}\boldsymbol{\beta} + \text{offset}_{it})^{y_{it}}}{y_{it}!\,\{\sum_{\ell=1}^{n_\ell}\exp(\mathbf{x}_{i\ell}\boldsymbol{\beta} + \text{offset}_{i\ell})\}^{y_{it}}}\right] \tag{18.10}$$

$$= \ln\left\{\prod_{i=1}^{n} \frac{(\sum_t y_{it})!}{\prod_{t=1}^{n_i} y_{it}!} \prod_{t=1}^{n_i} p_{it}^{y_{it}}\right\} \tag{18.11}$$

$$= \sum_{i=1}^{n}\left[\ln\Gamma\left(\sum_{t=1}^{n_i} y_{it} + 1\right) - \sum_{t=1}^{n_i}\ln\Gamma(y_{it}+1) + \sum_{t=1}^{n_i}\left\{y_{it}(\mathbf{x}_{it}\boldsymbol{\beta} + \text{offset}_{it}) - y_{it}\ln\sum_{\ell=1}^{n_i}\exp(\mathbf{x}_{i\ell}\boldsymbol{\beta} + \text{offset}_{i\ell})\right\}\right] \tag{18.12}$$

$$p_{it} = e^{\mathbf{x}_{it}\boldsymbol{\beta} + \text{offset}_{it}} \Big/ \sum_{\ell} e^{\mathbf{x}_{i\ell}\boldsymbol{\beta} + \text{offset}_{i\ell}} \tag{18.13}$$

Applying this estimator to our ship accident dataset provides the following results:

```
. xtpoisson incident op_75_79 co_65_69 co_70_74 co_75_79, fe i(ship)
> offset(exposure) eform nolog

Conditional fixed-effects Poisson regression     Number of obs      =        34
Group variable (i): ship                         Number of groups   =         5

                                                 Obs per group: min =         6
                                                                avg =       6.8
                                                                max =         7

                                                 Wald chi2(4)       =     48.44
Log likelihood  = -54.641859                     Prob > chi2        =    0.0000

------------------------------------------------------------------------------
    incident |        IRR   Std. Err.      z    P>|z|     [95% Conf. Interval]
-------------+----------------------------------------------------------------
    op_75_79 |   1.468831   .1737218     3.25   0.001     1.164926    1.852019
    co_65_69 |   2.008003   .3004803     4.66   0.000     1.497577    2.692398
    co_70_74 |    2.26693    .384865     4.82   0.000     1.625274    3.161912
    co_75_79 |   1.573695   .3669393     1.94   0.052     .9964273    2.485397
    exposure |   (offset)
------------------------------------------------------------------------------
```

Earlier, we saw a special case of conditional fixed-effects logistic regression wherein only one member of each group had a positive choice. With this type of data, one can investigate using either the `mlogit` or the `clogit` command. Whereas the `clogit` model assumes that only the best choice is marked for a group, another model rank orders all the choices. Such a model is commonly referred to as an exploding logit model, a Plackett–Luce model, or a rank-ordered logit model. Stata can fit such models through the `rologit` command.

18.4 Random effects

A random-effects estimator parameterizes the random effects according to an assumed distribution for which the parameters of the distribution are estimated. These models are called subject-specific models, as the likelihood models the individual observations instead of the panels.

18.4.1 Maximum likelihood estimation

A maximum likelihood solution to GLMs with random effects is possible depending on the assumed distribution of the random effect. Stata has several random-effects estimators for individual models that fall within the GLM framework.

A standard approach assumes a normally distributed random effect. We can then build the likelihood on the basis of the joint distribution of the outcome and the random effect. With the normal distribution, we write out the likelihood, complete the square, and obtain a likelihood involving the product of functions of the form

$$\int_{-\infty}^{\infty} e^{-z^2} f(z)dz \tag{18.14}$$

This is numerically approximated using Gauss–Hermite quadrature. The accuracy of the approximation is affected by the number of points used in the quadrature calculation and the smoothness of the function $f(z)$. Stata provides the `quadchk` command for users to assess the effect of the number of points used in the quadrature. Increasing the number of points used in the quadrature attempts to overcome a poorly behaved function $f(z)$, but it is not a panacea. The number of observations in a panel affects the smoothness of $f(z)$, and the *Stata Reference Manual* recommends (from cited experience) that we should be suspicious of results where panel sizes are greater than 50.

The derivation of the approximate log likelihood $\mathcal{L}_a$ for a Gaussian random-effects model is given by

$$\mathcal{L} = \ln \prod_{i=1}^{n} \int_{-\infty}^{\infty} \frac{e^{-\nu^2/2\sigma_\nu^2}}{\sqrt{2\pi}\sigma_\nu} \left\{ \prod_{t=1}^{n_i} \mathcal{F}(\mathbf{x}_{it}\boldsymbol{\beta} + \nu) \right\} d\nu \tag{18.15}$$

$$\int_{-\infty}^{\infty} e^{-x^2} f(x)dx \approx \sum_{m=1}^{M} w_m^* f(x_m^*) \tag{18.16}$$

$$z = \frac{\nu}{\sqrt{2}\sigma_\nu} \tag{18.17}$$

$$\rho = \frac{\sigma_\nu^2}{\sigma_\nu^2 + \sigma_\epsilon^2} \tag{18.18}$$

$$\nu = z\sqrt{2\frac{\rho}{1-\rho}} \tag{18.19}$$

$$dz = dv/\sqrt{2\frac{\rho}{1-\rho}} \tag{18.20}$$

$$\mathcal{L} = \ln\prod_{i=1}^{n}\left\{\int_{-\infty}^{\infty} e^{-z^2}\frac{1}{\sqrt{\pi}}\prod_{t=1}^{n_i}\mathcal{F}\left(\mathbf{x}_{it}\boldsymbol{\beta}+\sqrt{2\frac{\rho}{1-\rho}}\,z\right)dz\right\}^{w_i} \tag{18.21}$$

$$\approx \ln\prod_{i=1}^{n}\left\{\sum_{m=1}^{M} w_m^*\frac{1}{\sqrt{\pi}}\prod_{t=1}^{n_i}\mathcal{F}\left(\mathbf{x}_{it}\boldsymbol{\beta}+\sqrt{2\frac{\rho}{1-\rho}}\,x_m^*\right)\right\}^{w_i} \tag{18.22}$$

$$\mathcal{L}_a = \sum_{i=1}^{n} w_i\ln\frac{1}{\sqrt{\pi}}\sum_{m=1}^{M} w_m^*\prod_{t=1}^{n_i}\mathcal{F}\left(\mathbf{x}_{it}\boldsymbol{\beta}+\sqrt{2\frac{\rho}{1-\rho}}\,x_m^*\right) \tag{18.23}$$

where (w_m^*, x_m^*) are the weights and abscissas of the Gauss–Hermite quadrature. We have not explicitly given the form of the function $\mathcal{F}(\cdot)$. This generalization facilitates deriving specific models. Stata has random-effects models for logit, probit, clog-log, Poisson, regression, tobit, and interval regression.

Specifically, for the Gaussian random-effects Poisson model, we may substitute

$$\mathcal{F}(z) = \exp\{-\exp(z)\}\exp(z)^{y_{it}}/y_{it}! \tag{18.24}$$

Applying this model specification to our ship accident dataset results in

```
. xtpoisson incident op_75_79 co_65_69 co_70_74 co_75_79, re normal i(ship)
> offset(exposure) eform nolog
(6 missing values generated)

Random-effects Poisson regression               Number of obs      =        34
Group variable (i): ship                        Number of groups   =         5

Random effects u_i ~ Gaussian                   Obs per group: min =         6
                                                               avg =       6.8
                                                               max =         7

                                                LR chi2(4)         =     55.25
Log likelihood  = -74.781731                    Prob > chi2        =    0.0000

------------------------------------------------------------------------------
    incident |        IRR   Std. Err.      z    P>|z|     [95% Conf. Interval]
-------------+----------------------------------------------------------------
    op_75_79 |   1.466695    .173441     3.24   0.001     1.163276    1.849255
    co_65_69 |   2.032729   .3040807     4.74   0.000     1.516164     2.72529
    co_70_74 |   2.357714   .3992821     5.06   0.000     1.691761    3.285816
    co_75_79 |   1.647449   .3817246     2.15   0.031     1.046124    2.594423
    exposure |   (offset)
-------------+----------------------------------------------------------------
    /lnsig2u |  -2.353033   .8583269    -2.74   0.006    -4.035323   -.6707431
-------------+----------------------------------------------------------------
     sigma_u |    .308351    .132333                      .1329661    .7150724
------------------------------------------------------------------------------
Likelihood-ratio test of sigma_u=0: chibar2(01) =    10.67 Pr>=chibar2 = 0.001
```

Other random-effects models may also be derived assuming a distribution other than Gaussian for the random effect. Alternative distributions are chosen so that the joint likelihood may be analytically solved. For example, assuming gamma random effects in a Poisson model allows us to derive a model that may be calculated without the need for numeric approximations. Stata's `xtpoisson` command will therefore fit either Gaussian random effects or gamma random effects.

In the usual Poisson model, we hypothesize that the mean of the outcome variable y is given by $\lambda_{it} = \exp(\mathbf{x}_{it}\boldsymbol{\beta})$. In the panel setting, we assume that each panel has a different mean that is given by $\exp(\mathbf{x}_{it}\boldsymbol{\beta} + \eta_i) = \lambda_{it}\nu_i$. So we refer to the random effect as entering multiplicatively rather than additively as in random-effects linear regression.

Since the random effect $\nu_i = \exp(\eta_i)$ is positive, we select a gamma distribution and add the restriction that the mean of the random effects equals one so that we have only one more parameter, θ, to estimate.

$$f(\nu_i) = \frac{\theta^\theta}{\Gamma(\theta)}\nu_i^{\theta-1}\exp(-\theta\nu_i) \tag{18.25}$$

Since the conditional mean of the outcome given the random effect is Poisson and the random effect is gamma(θ, θ), we have that the joint density function for a panel is given by

$$f(\nu_i, \epsilon_{i1}, \epsilon_{i2}, \ldots, \epsilon_{in_i}) = \frac{\theta^\theta}{\Gamma(\theta)}\nu_i^{\theta-1}\exp(-\theta\nu_i)\prod_{t=1}^{n_i}\exp(-\nu_i\lambda_{it})(\nu_i\lambda_{it})^{y_{it}}/y_{it}! \tag{18.26}$$

and the joint density for all the panels is the product since the panels are independent.

The log likelihood for gamma-distributed random effects may then be derived by integrating over ν_i. The log likelihood is derived as

$$\begin{aligned}
\mathcal{L} &= \ln\prod_{i=1}^{n}\left\{\int_0^\infty \frac{\theta^\theta}{\Gamma(\theta)}\nu_i^{\theta-1}\exp(-\theta\nu_i)\prod_{t=1}^{n_i}\frac{(\nu_i\lambda_{it})^{y_{it}}}{y_{it}!}\exp(-\nu_i\lambda_{it})d\nu_i\right\} && (18.27)\\
&= \sum_{i=1}^{n}\left\{\ln\Gamma\left(\theta + \sum_{t=1}^{n_i} y_{it}\right) - \ln\Gamma(\theta) - \sum_{t=1}^{n_i}\ln\Gamma(y_{it}+1) + \theta\ln u_i\right.\\
&\quad \left. + \left(\sum_{t=1}^{n_i} y_{it}\right)\ln(1-u_i) - \left(\sum_{t=1}^{n_i} y_{it}\right)\ln\left(\sum_{t=1}^{n_i}\lambda_{it}\right) + \sum_{t=1}^{n_i} y_{it}\xi_{it}\right\} && (18.28)\\
u_i &= \frac{\theta}{\theta + \sum_{t=1}^{n_i}\lambda_{it}} && (18.29)\\
\lambda_{it} &= \exp(\mathbf{x}_{it}\boldsymbol{\beta}) && (18.30)
\end{aligned}$$

Applying this model specification to our ship accident dataset results in

```
. xtpoisson incident op_75_79 co_65_69 co_70_74 co_75_79, re i(ship)
> offset(exposure) eform nolog

Random-effects Poisson regression               Number of obs      =        34
Group variable (i): ship                        Number of groups   =         5

Random effects u_i ~ Gamma                      Obs per group: min =         6
                                                               avg =       6.8
                                                               max =         7

                                                Wald chi2(4)       =     50.90
Log likelihood  = -74.811217                    Prob > chi2        =    0.0000

------------------------------------------------------------------------------
    incident |        IRR   Std. Err.      z    P>|z|     [95% Conf. Interval]
-------------+----------------------------------------------------------------
    op_75_79 |   1.466305   .1734005     3.24   0.001     1.162957    1.848777
    co_65_69 |   2.032543    .304083     4.74   0.000     1.515982     2.72512
    co_70_74 |   2.356853   .3999259     5.05   0.000     1.690033    3.286774
    co_75_79 |   1.641913   .3811398     2.14   0.033      1.04174     2.58786
    exposure |   (offset)
-------------+----------------------------------------------------------------
    /lnalpha |  -2.368406   .8474597                     -4.029397   -.7074155
-------------+----------------------------------------------------------------
       alpha |   .0936298   .0793475                      .0177851    .4929165
------------------------------------------------------------------------------
Likelihood-ratio test of alpha=0: chibar2(01) =    10.61 Prob>=chibar2 = 0.001
```

The above command is implemented in Stata in internal (fast) code. It has fast execution times for small to medium datasets and is specific to the Poisson model. Stata has other commands that are specific to other family–link combinations.

Stata users also have available a command for fitting general linear latent and mixed models: Rabe-Hesketh, Pickles, and Taylor (2000). With this approach, we may specify an arbitrarily sophisticated model including GLM with random-effects models. The command will also allow for fitting the ordered and unordered outcomes not available as part of the glm command. The method of estimation includes Gaussian quadrature. Although not available as part of the Stata package, it is a powerful user-written (free) addition that one can obtain using Stata's Internet capabilities:

```
. net search gllamm
```

The only downside to the command is that it is implemented using a derivative-free method. The resulting need for calculating accurate numeric derivatives results in slow execution times.

We can use this command to fit a Gaussian random-effects Poisson model for our ship accident dataset by using Gaussian quadrature for the random-effects estimation. The results from this are

```
. gllamm incident op_75_79 co_65_69 co_70_74 co_75_79, link(log) family(poisson)
> i(ship) offset(exposure) eform
Iteration 0:   log likelihood = -78.987823
Iteration 1:   log likelihood = -73.980982
Iteration 2:   log likelihood = -73.889377
Iteration 3:   log likelihood = -73.889346
Iteration 4:   log likelihood = -73.889346

number of level 1 units = 34
number of level 2 units = 5

Condition Number = 6.7997693

gllamm model

log likelihood = -73.889346

------------------------------------------------------------------------------
    incident |     exp(b)   Std. Err.      z    P>|z|     [95% Conf. Interval]
-------------+----------------------------------------------------------------
    op_75_79 |   1.469583   .1737318     3.26   0.001     1.165645    1.852772
    co_65_69 |   2.025639    .302931     4.72   0.000     1.511007    2.715551
    co_70_74 |   2.333301    .394834     5.01   0.000     1.674685    3.250936
    co_75_79 |   1.638866   .3771262     2.15   0.032     1.043927    2.572864
    exposure |   (offset)
------------------------------------------------------------------------------


Variances and covariances of random effects
------------------------------------------------------------------------------


***level 2 (ship)

    var(1): .17662891 (.09378639)
------------------------------------------------------------------------------
```

Results from this derivative-free implementation will not match the Gaussian random effects `xtpois` command. Calculating numeric derivatives will result in a reliable comparison of approximately four digits between the two methods.

18.4.2 Gibbs sampling

Instead of analytically deriving or numerically approximating (via quadrature) the joint distribution of the model and random effect, we can also fit the model using Gibbs sampling. This approach takes a Bayesian view of the problem and focuses on the conditional distributions. It then uses Monte Carlo calculation techniques to estimate the conditional probabilities.

Zeger and Karim (1991) describe a general implementation for fitting GLMs with random effects. An advantage of their approach is that we are not limited to one random effect. This generalization costs in computational efficiency: implementation will require much greater execution times. However, the Gibbs sampling approach will fit models that occasionally cause numeric problems for quadrature-implemented methods.

In the following, we assume that there is one random effect to facilitate comparison to the other models in this chapter. As usual, we assume that y_{it} follows an exponential-family form

$$f(y_{it};\nu_i) = \exp\left\{\frac{y_{it}\theta_{it} - b(\theta_{it})}{a(\phi)} - c(y_{it},\phi)\right\} \tag{18.31}$$

and that our link function is given by

$$g(\mu_{it}) = \mathbf{x}_{it}\boldsymbol{\beta} + \nu_i \tag{18.32}$$

We further assume that the random effects ν_i are independent observations from some parametric distribution $\mathcal{F}$ with zero mean and variance σ^2_ν. Our objective is to derive the posterior distribution $f(\boldsymbol{\beta},\sigma^2_\nu)$ given by

$$f(\boldsymbol{\beta},\sigma^2_\nu) = \frac{\prod_{i=1}^n \int f(y_i;\nu_i,\boldsymbol{\beta})g(\nu_i;\sigma^2_\nu)p(\boldsymbol{\beta},\sigma^2_\nu)\, d\nu_i}{\int \prod_{i=1}^n \int f(y_i;\nu_i,\boldsymbol{\beta})g(\beta_b;\sigma^2_\nu)p(\boldsymbol{\beta},\sigma^2_\nu)\, d\nu_i\, d\boldsymbol{\beta}\, d\sigma^2_\nu} \tag{18.33}$$

See Zeger and Karim (1991) for the implementation details of the Monte Carlo methods based on the above-defined conditional distributions. A summary of these models is presented in Hilbe and Hardin (2007) or Hardin and Hilbe (2002) present an entire text for interested readers. There are no facilities in Stata at this point for using this approach.

18.5 GEEs

Recall our derivation of the GLM estimating algorithm, in which we wrote

$$\boldsymbol{\beta}^{(r)} = \boldsymbol{\beta}^{(r-1)} - \left\{\mathcal{L}''\left(\boldsymbol{\beta}^{(r-1)}\right)\right\}^{-1} \mathcal{L}'\left(\boldsymbol{\beta}^{(r-1)}\right) \tag{18.34}$$

for $r = 1, 2, \ldots$ and a reasonable starting value $\boldsymbol{\beta}^{(0)}$. We further have that

$$\begin{aligned}
\mathcal{L}'(\boldsymbol{\beta}^{(r-1)}) &= \mathbf{X}^{\mathrm{T}}\mathbf{W}^{(r-1)}\mathbf{z}^{(r-1)} && (18.35)\\
\mathcal{L}''(\boldsymbol{\beta}^{(r-1)}) &= \mathbf{X}^{\mathrm{T}}\mathbf{W}^{(r-1)}\mathbf{X} && (18.36)\\
\mathbf{W}^{(r-1)} &= D\left\{\frac{1}{\mathrm{V}(\mu)a(\phi)}\left(\frac{\partial\mu}{\partial\eta}\right)^2\right\} && (18.37)\\
\mathbf{z}^{(r-1)} &= (y-\mu)\left(\frac{\partial\eta}{\partial\mu}\right) + \left(\eta^{(r-1)} - \text{offset}\right) && (18.38)
\end{aligned}$$

where $D()$ denotes a diagonal matrix.

In that original derivation, we assumed that we had n independent observations. Here we investigate how we might estimate $\boldsymbol{\beta}$ if we have n panels (clusters), where each panel $i = 1, 2, \ldots, n$ has n_i *correlated* observations. We have n independent panels, but the observations within panels are correlated.

We may rewrite the (updating) estimating equation given in (18.34) as

$$\begin{aligned}\boldsymbol{\beta}^{(r)} = \boldsymbol{\beta}^{(r-1)} - &\left[\mathbf{X}^{\mathrm{T}} D\left\{(\mathrm{V}(\mu_i)a(\phi))^{-1}\left(\frac{\partial\mu}{\partial\eta}\right)_i^2\right\}\mathbf{X}\right]^{-1}\\ &\times\ \mathbf{X}^{\mathrm{T}} D\left\{\{\mathrm{V}(\mu_i)a(\phi)\}^{-1}\left(\frac{\partial\mu}{\partial\eta}\right)_i^2\right\}\\ &\times\ \left\{(y_i-\mu_i)\left(\frac{\partial\mu}{\partial\eta}\right)_i + \left(\eta_i^{(r-1)} - \text{offset}_i\right)\right\}\end{aligned} \tag{18.39}$$

We now assume that we have a panel dataset. However, for a pooled dataset, we assume that the observations within panel are independent. First, we can rewrite the variance as

$$\begin{aligned}\mathrm{V}(\mathbf{y}_i) &= D\{\mathrm{V}(\mu_{it})a(\phi)\} &(18.40)\\ &= D\{\mathrm{V}(\mu_{it})a(\phi)\}^{1/2}\ \mathbf{I}\ D\{\mathrm{V}(\mu_{it})a(\phi)\}^{1/2} &(18.41)\end{aligned}$$

without loss of generality. Making this substitution, our pooled GLM estimator may be found from the estimating equation given by

$$\begin{aligned}&\boldsymbol{\beta}^{(r)} = \boldsymbol{\beta}^{(r-1)}\\ &\quad-\left[\sum_{i=1}^{n}\mathbf{X}_i^{\mathrm{T}} D\left(\frac{\partial\mu}{\partial\eta}\right)_{it} D\{\mathrm{V}(\mu_{it})a(\phi)\}^{-1/2}\,\mathbf{I}\,D\{\mathrm{V}(\mu_{it})a(\phi)\}^{-1/2} D\left(\frac{\partial\mu}{\partial\eta}\right)_{it}\mathbf{X}_i\right]^{-1}\\ &\quad\times\sum_{i=1}^{n}\left\{\mathbf{X}_i^{\mathrm{T}} D\left(\frac{\partial\mu}{\partial\eta}\right)_{it} D\{\mathrm{V}(\mu_{it})a(\phi)\}^{-1/2}\,\mathbf{I}\,D\{\mathrm{V}(\mu_{it})a(\phi)\}^{-1/2} D\left(\frac{\partial\mu}{\partial\eta}\right)_{it}\right.\\ &\quad\times\left.\left[(y_{it}-\mu_{it})\left(\frac{\partial\mu}{\partial\eta}\right)_{it} + \left(\eta_{it}^{(r-1)} - \text{offset}_{it}\right)\right]\right\}\end{aligned} \tag{18.42}$$

Liang and Zeger (1986) present an extension to the usual GLM method. What they show is a construction of the covariance matrix for a panel of observations in terms of the usual variance function of the mean and an introduced "working" correlation matrix. The correlation matrix is then treated as a matrix of ancillary parameters, so standard errors are not calculated for the estimated correlations. The covariance matrix for the entire set of observations is then block diagonal with the individual blocks for $i = 1, \ldots, n$ defined by

$$\mathrm{V}(\mathbf{y}_i) = D\{\mathrm{V}(\mu_i)a(\phi)\}^{1/2}\ \mathbf{R}\ D\{\mathrm{V}(\mu_i)a(\phi)\}^{1/2} \tag{18.43}$$

where $\mathbf{R}$ is the within-panel correlation matrix (generalized from our identity matrix for a pooled GLM).

We can then write our GEE estimation algorithm, assuming that there are $i = 1, \ldots, n$ panels where each panel has $t = 1, \ldots, n_i$ observations

$$\begin{aligned}\boldsymbol{\beta}^{(r)} &= \boldsymbol{\beta}^{(r-1)} \\ &- \left[\sum_{i=1}^{n} \mathbf{X}_i^{\mathrm{T}} D\left(\frac{\partial\mu}{\partial\eta}\right)_{it} D\left\{\mathrm{V}(\mu_{it})a(\phi)\right\}^{-1/2} \mathbf{R}\, D\left\{\mathrm{V}(\mu_{it})a(\phi)\right\}^{-1/2} D\left(\frac{\partial\mu}{\partial\eta}\right)_{it} \mathbf{X}_i\right]^{-1} \\ &\times \sum_{i=1}^{n} \left[\mathbf{X}_i^{\mathrm{T}} D\left(\frac{\partial\mu}{\partial\eta}\right)_{it} D\left\{\mathrm{V}(\mu_{it})a(\phi)\right\}^{-1/2} \mathbf{R}\, D\left\{\mathrm{V}(\mu_{it})a(\phi)\right\}^{-1/2} D\left(\frac{\partial\mu}{\partial\eta}\right)_{it} \right.\\ &\left. \times \left\{(y_{it} - \mu_{it})\left(\frac{\partial\mu}{\partial\eta}\right)_{it} + \left(\eta_{it}^{(r-1)} - \mathrm{offset}_{it}\right)\right\}\right] \end{aligned} \tag{18.44}$$

The estimation proceeds alternating between updating $\boldsymbol{\beta}$ and updating an estimate of the correlation matrix $\mathbf{R}$ (which is usually parameterized to have some user-specified structure). The form of the GEE shows that the focus is on the marginal distribution and that the estimator sums the panels after account for within-panel correlation. As such, this is called a *population-averaged* estimator. It is also easy to see that when the correlation is assumed to be independent, we have an ordinary pooled GLM estimator.

Zeger, Liang, and Albert (1988), as well as other sources, provide further theoretical insight into this model. One can find good comparisons of panel-data estimators in Neuhaus, Kalbfleisch, and Hauck (1991) or Neuhaus (1992). Users can find complete coverage of GEE and its extensions in Hardin and Hilbe (2002).

Recall from chapter 3 the general form of the modified sandwich estimate of variance. There we presented arguments for modifying the usual sandwich estimate of variance that would allow for independent panels (clusters) that had some unspecified form of intrapanel correlation. That is exactly the case here. Therefore, we may specify a modified sandwich estimate of variance by using

$$\widehat{B}_{\mathrm{MS}} = \left(\sum_{i=1}^{n} a(\phi)^2 \mathbf{X}_i \mathcal{S}\mathcal{S}^{\mathrm{T}} \mathbf{X}_i^{\mathrm{T}}\right) \tag{18.45}$$

$$\mathcal{S} = D\left(\frac{\partial\mu}{\partial\eta}\right) D\left\{\mathrm{V}(\mu_{it})a(\phi)\right\}^{-1/2} \mathbf{R}^{-1} D\left\{\mathrm{V}(\mu_{it})a(\phi)\right\}^{-1/2} D\left(y_{it} - \mu_{it}\right) \tag{18.46}$$

as the middle of the sandwich and allowing that $\mathrm{V}_{\mathrm{H}}^{-1}$ is the first term of the product in (18.44). This is $\mathrm{V}_{\mathrm{EH}}^{-1}$ and is the usual choice in software (including Stata) since it is the variance matrix implemented in the modified IRLS optimization algorithm.

$$\widehat{\mathrm{V}}_{\mathrm{MS}} = \widehat{\mathrm{V}}_{\mathrm{H}}^{-1} \widehat{B}_{\mathrm{MS}} \widehat{\mathrm{V}}_{\mathrm{H}}^{-1} \tag{18.47}$$

We assume a specific form of correlation in estimating the coefficient vector where we also estimate that specified correlation structure. This allows us to obtain a more efficient estimator of $\boldsymbol{\beta}$ if the true correlation follows the structure of our specified model. The modified sandwich estimate of variance provides standard errors that are robust to possible misspecification of this correlation structure. We therefore recommend that the modified sandwich estimate of variance always be used with this model. In Stata, this means that you should always specify the `robust` option when using the `xtgee` command or any command that calls `xtgee`, as we illustrate below.

Applying this model specification to our ship accident dataset results in

```
. xtpoisson incident op_75_79 co_65_69 co_70_74 co_75_79, pa i(ship)
> offset(exposure) vce(robust) eform nolog
GEE population-averaged model                   Number of obs      =        34
Group variable:                       ship      Number of groups   =         5
Link:                                  log      Obs per group: min =         6
Family:                            Poisson                     avg =       6.8
Correlation:                  exchangeable                     max =         7
                                                Wald chi2(3)       =    181.55
Scale parameter:                         1      Prob > chi2        =    0.0000
                                   (Std. Err. adjusted for clustering on ship)
```

incident	IRR	Semi-robust Std. Err.	z	P>|z|	[95% Conf.	Interval]
op_75_79	1.483299	.1197901	4.88	0.000	1.266153	1.737685
co_65_69	2.038477	.1809524	8.02	0.000	1.712955	2.425859
co_70_74	2.643467	.4093947	6.28	0.000	1.951407	3.580962
co_75_79	1.876656	.33075	3.57	0.000	1.328511	2.650966
exposure	(offset)					

18.6 Other models

Since the first edition of this book, Stata has continued development of clustered-data commands. It has also developed a powerful matrix programming language that can be used in direct matrix manipulations, as well as through indirect definition of matrices from data; see the documentation on `mata`. If we limit discussion to linear models, another approach to models is through the specification of various levels of correlation. Such specification amounts to a mixed model in which various fixed and random effects are posited for a specific model. Stata's `xtmixed` command will fit such models.

Through a National Institutes of Health–funded effort with academia, Stata also developed extensions to GLMs that allow specification of measurement error. Though not strictly for clustered data, this software includes various approaches to model estimation including regression calibration and simulation extrapolation. The software developed from this effort includes commands `qvf`, `rcal`, and `simex`. Full description of the software and theory behind the estimators were published in the *Stata Journal*; see Hardin and Carroll (2003) for details. The associated collection of software can be found using `net search st0049`. This collection of software basically replicates the `glm` command, introducing various support mechanisms for consideration of measurement error in one or more predictor variables. Stata users may also use the `gllamm` command for such models.

To motivate interest in the approaches afforded by these extensions to GLMs, we present a slightly expanded presentation of an example in the previously cited paper. Here we will synthesize data to illustrate the issues faced when covariates are measured with error.

We will generate data for a linear model. That is, the data will be for an identity link from the Gaussian family. We begin with

```
. set seed 1
. set more off
. set obs 500
obs was 40, now 500
. gen x1 = uniform()*10
. gen x2 = uniform()*5
. gen x3 = uniform()
. gen x4 = uniform()
. gen x5 = uniform()
. gen err = invnormal(uniform())
. gen y =  1*x1 + 2*x2 + 3*x3 + 4*x4 + 5 + err
. gen a1 = x3 + 0.25*invnormal(uniform())
. gen a2 = x3 + 0.25*invnormal(uniform())
. gen b1 = x4 + 0.25*invnormal(uniform())
. gen b2 = x4 + 0.25*invnormal(uniform())
```

Thus we have generated data for a linear regression model. We now assume that instead of actually obtaining the x3 and x4 variables, we could obtain only two error-prone proxy measures for each of these covariates. Instead of observing x3, we instead obtained 2 observations a1 and a2. Likewise, instead of observing x4, we have the two error-prone proxy measures b1 and b2. To be clear, had we actually obtained the x3 and x4 covariates, we would simply fit

```
. glm y x1 x2 x3 x4, noheader
Iteration 0:   log likelihood = -656.80701
```

y	Coef.	OIM Std. Err.	z	P>|z|	[95% Conf.	Interval]
x1	1.011518	.0138241	73.17	0.000	.984423	1.038613
x2	2.058875	.0275402	74.76	0.000	2.004897	2.112853
x3	3.017885	.1446245	20.87	0.000	2.734426	3.301344
x4	4.033718	.1424256	28.32	0.000	3.754569	4.312867
_cons	4.796872	.144516	33.19	0.000	4.513626	5.080118

Note how the estimated coefficients are near the values used in synthesizing the data, just as we would expect. However, faced with the data at hand, we may not know exactly how to proceed. One might consider simply using one of the two proxy measures for each of the "unobserved" covariates.

```
. glm y x1 x2 a1 b1, noheader
Iteration 0:   log likelihood = -842.19414
```

y	Coef.	OIM Std. Err.	z	P>|z|	[95% Conf.	Interval]
x1	1.03188	.0200175	51.55	0.000	.992646	1.071113
x2	2.079477	.0398469	52.19	0.000	2.001378	2.157575
a1	1.679568	.1606557	10.45	0.000	1.364688	1.994447
b1	2.396211	.162133	14.78	0.000	2.078436	2.713985
_cons	6.132005	.192499	31.85	0.000	5.754714	6.509296

```
. glm y x1 x2 a2 b2, noheader
Iteration 0:   log likelihood = -867.97304
```

y	Coef.	OIM Std. Err.	z	P>|z|	[95% Conf.	Interval]
x1	1.012303	.0211807	47.79	0.000	.9707899	1.053817
x2	2.048303	.0420299	48.73	0.000	1.965926	2.13068
a2	1.383699	.1681084	8.23	0.000	1.054213	1.713185
b2	2.130001	.1691149	12.59	0.000	1.798542	2.46146
_cons	6.619024	.1915367	34.56	0.000	6.243619	6.994429

However, the results do not agree with what we know about the data. This is a well-known phenomenon where regression coefficients are attenuated toward the null; estimated coefficients for error-prone proxy covariates tend toward zero.

One approach to this particular problem involves using the fact that we have replicate (error prone) measures. Since we obtained replicates of the error-prone measures, we have some way of estimating the error in those proxy variables. The simulation–extrapolation method estimates that error and then runs a series of experiments. In each experiment, the approach generates even more measurement error, adds it to the error-prone covariates, and then fits the model. After running these experiments, the approach gains knowledge about how the fitted coefficients are related to the measurement error. This is the simulation part of the algorithm. The extrapolation part is in forecasting what the coefficients would be for no measurement error, thus extrapolating the relationship built up through simulation. The results of the approach along with a plot to illustrate the extrapolation help to illustrate the technique.

(*Continued on next page*)

```
. simex (y=x1 x2) (w3: a1 a2) (w4: b1 b2), brep(299) seed(12345)
Simulation extrapolation                         No. of obs          =        500
                                                 Bootstraps reps     =        299
Residual df   =        495                       Wald F(4,495)       =    1559.86
                                                 Prob > F            =     0.0000
Variance Function: V(u) = 1                         [Gaussian]
Link Function     : g(u) = u                        [Identity]

------------------------------------------------------------------------------
             |              Bootstrap
           y |      Coef.   Std. Err.      t    P>|t|     [95% Conf. Interval]
-------------+----------------------------------------------------------------
          x1 |   1.001971    .0196916    50.88   0.000     .9632812     1.04066
          x2 |   2.045304    .0391092    52.30   0.000     1.968463    2.122145
          w3 |   2.740266    .2480518    11.05   0.000     2.252901     3.22763
          w4 |   3.799953    .2673007    14.22   0.000      3.27477    4.325137
       _cons |    5.15145    .2221262    23.19   0.000     4.715024    5.587877
------------------------------------------------------------------------------
```

The graph generated by the `simexplot` command is given in figure 18.1. Bootstrap standard errors are calculated in this example; see the documentation of the software for other options. Note how the estimated coefficients are much closer to the values used to generate the original data even though the estimation was performed without considering the `x3` and `x4` covariates.

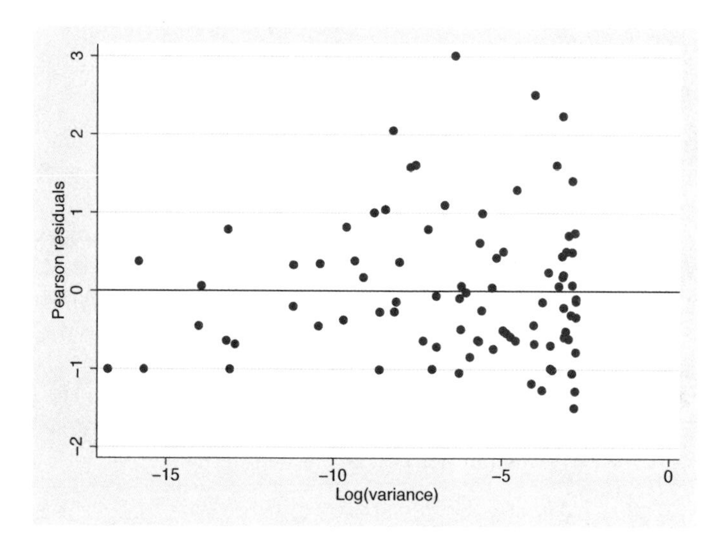

Figure 18.1: Simulation–extrapolation results

Finally, another great source for developments in longitudinal and panel data analysis is the Stata documentation. Specifically, the *Longitudinal/Panel Data Reference Manual* (StataCorp 2005) includes statistical derivations, technical descriptions, documentation, and example usage for many different models.

Part VII

Stata Software

19 Programs for Stata

Contents

The knowledge presented in the text would not be useful without the practical application of available supporting software. Readers are not required to use the following software, but other software will not address all the issues described here.

This chapter describes commands for the Stata software program. The entire chapter is devoted to Stata software and makes many references to Stata documentation (manuals, help files, the web site, and NetCourses). We follow the notational and typographic conventions of the Stata reference manuals; [U] refers to the *Stata User's Guide* and [R] refers to the *Stata Base Reference Manual*.

In addition to this documentation on the `glm` command in Stata, we also used other user-written programs. Interested readers may locate, download, and install such programs by using Stata's `net search` facilities.

19.1 The glm command

19.1.1 Syntax

glm *depvar* [*indepvars*] [*if*] [*in*] [*weight*] [, *options*]

options	description
Model	
family(*familyname*)	distribution of *depvar*; default is family(gaussian)
link(*linkname*)	link function; default is canonical link for family() specified
Model 2	
noconstant	suppress constant term
exposure(*varname*)	include ln(*varname*) in model with coefficient constrained to 1
offset(*varname*)	include *varname* in model with coefficient constrained to 1
mu(*varname*)	use *varname* as the initial estimate for the mean of *depvar*
init(*varname*)	synonym for mu(*varname*)
SE/Robust	
vce(*vcetype*)	*vcetype* may be oim, robust, opg, bootstrap, jackknife, eim, jackknife1, hac, or unbiased
t(*varname*)	variable name corresponding to t
robust	synonym for vce(robust)
cluster(*varname*)	adjust standard errors for intragroup correlation
vfactor(#)	multiply variance matrix by scalar #
disp(#)	quasilikelihood multiplier
scale(x2|dev|#)	set the scale parameter
Reporting	
level(#)	set confidence level; default is level(95)
eform	report exponentiated coefficients
Max options	
irls	use iterated, reweighted least-squares optimization of the deviance
maximize_options	control the maximization process; seldom used
fisher(#)	use the Fisher scoring Hessian or EIM
search	search for good starting values
†noheader	suppress header table from above coefficient table
†notable	suppress coefficient table
†nodisplay	suppress the output; iteration log is still displayed

familyname	description
gaussian	Gaussian (normal)
igaussian	inverse Gaussian
binomial [*varname*$_N$ | #$_N$]	Bernoulli/binomial
poisson	Poisson
nbinomial [#$_k$]	negative binomial
gamma	gamma

linkname	description
identity	identity
log	log
logit	logit
probit	probit
cloglog	complementary log-log
power #	power
opower #	odds power
nbinomial	negative binomial
loglog	log-log
logc	log-complement

†noheader and nodisplay do not appear in the dialog box.

depvar and *indepvars* may contain time-series operators; see [U] **11.4.3 Time-series varlists**.

bootstrap, by, jackknife, rolling, statsby, stepwise, and xi are allowed; see [U] **11.1.10 Prefix commands**.

fweights, aweights, iweights, and pweights are allowed; see [U] **11.1.6 weight**.

See [U] **20 Estimation and postestimation commands** for additional capabilities of estimation commands.

19.1.2 Description

glm fits generalized linear models. It can fit models using either IRLS (maximum quasi-likelihood) or Newton–Raphson (maximum likelihood) optimization, which is the default. Previous versions of glm used only IRLS.

See [R] **logistic** and [R] **regress** for lists of related estimation commands.

19.1.3 Options

Model

family(*familyname*) specifies the distribution of *depvar*; family(gaussian) is the default.

`link(`*linkname*`)` specifies the link function; the default is the canonical link for the `family()` specified.

Model 2

`noconstant`, `exposure(`*varname*`)`, `offset(`*varname*`)`; see [R] **estimation options**.

`mu(`*varname*`)` specifies *varname* as the initial estimate for the mean of *depvar*. This method can be useful with models that experience convergence difficulties, such as `family(binomial)` models with power or odds-power links. `init(`*varname*`)` is a synonym.

SE/Robust

`vce(`*vcetype*`)`; see [R] ***vce_option***. In addition to the standard *vcetypes*, `glm` allows the following alternatives:

`vce(eim)` specifies that the expected information matrix estimate of variance be used.

`vce(jackknife1)` specifies that the one-step jackknife estimate of variance be used.

`vce(hac` *kernel* [#]`)` specifies that a heteroskedastic and autocorrelation consistent (HAC) variance estimate be used. HAC refers to the general form for combining weighted matrices to form the variance estimate. There are three kernels built into `glm`. *kernel* is a user-written program or one of

`nwest | gallant | anderson`

If # is not specified, $N-2$ is assumed.

`vce(unbiased)` specifies that the unbiased sandwich estimate of variance be used. This option implies the `robust` option.

`t(`*varname*`)` specifies the variable name corresponding to `t`; see [TS] **tsset**. `glm` does not need to know `t()` in all cases, though it does if `vce(hac ...)` is specified. Then you can either specify the time variable with `t()` or you can `tsset` your data before calling `glm`. When the time variable is required, `glm` assumes that the observations are spaced equally over time.

`robust`, `cluster(`*varname*`)`; see [R] **estimation options**.

`vfactor(#)` specifies a scalar by which to multiply the resulting variance matrix. This option allows you to match output with other packages, which may apply degrees-of-freedom or other small-sample corrections to estimates of variance.

`disp(#)` multiplies the variance of *depvar* by # and divides the deviance by #. The resulting distributions are members of the quasilikelihood family.

`scale(x2 | dev | #)` overrides the default scale parameter. This option is allowed only with Hessian (information matrix) variance estimates.

By default, `scale(1)` is assumed for the discrete distributions (binomial, Poisson, and negative binomial), and `scale(x2)` is assumed for the continuous distributions (Gaussian, gamma, and inverse Gaussian).

`scale(x2)` specifies that the scale parameter be set to the Pearson chi-squared (or generalized chi-squared) statistic divided by the residual degrees of freedom, which McCullagh and Nelder (1989) recommend as a good general choice for continuous distributions.

`scale(dev)` sets the scale parameter to the deviance divided by the residual degrees of freedom. This provides an alternative to `scale(x2)` for continuous distributions and overdispersed or underdispersed discrete distributions.

`scale(#)` sets the scale parameter to #. For example, using option `scale(1)` in `family(gamma)` models results in exponential-errors regression. Use of `link(log)` rather than the default `link(power -1)` for `family(gamma)` essentially reproduces Stata's `streg, dist(exp) nohr` command (see [ST] **streg**) if all the observations are uncensored.

Table 19.1: Resulting standard errors

Options	Estimated variance matrix	Reference
nothing	EIM Hessian	Equation 3.32
`irls`	OIM Hessian	Equation 3.51[a]
`opg`	OPG	Section 3.6.2
`vce(robust)`	Sandwich	Section 3.6.3
`irls vce(robust)`	Sandwich	Section 3.6.3[b]
`vce(robust) unbiased`	Unbiased sandwich	Section 3.6.5
`irls vce(robust) unbiased`	Unbiased sandwich	Section 3.6.3[b]
`vce(robust) cluster()`	Modified sandwich	Section 3.6.4
`irls vce(robust) cluster()`	Modified sandwich	Section 3.6.4[b]
`vce(robust) unbiased cluster()`	Mod. unbiased sandwich	Section 3.6.6
`irls vce(robust) unbiased cluster()`	Mod. unbiased sandwich	Section 3.6.6[b]
`nwest()`	Weighted sandwich	Section 3.6.7
`jknife`	Usual jackknife	Section 3.6.8
`jknife1`	One-step jackknife	Section 3.6.8.2
`jknife cluster()`	Variable jackknife	Section 3.6.8.4
`bstrap`	Usual bootstrap	Section 3.6.9.1
`bstrap cluster()`	Grouped bootstrap	Section 3.6.9.2

[a] If the canonical link is specified, then the EIM is the same as the OIM. Output from the program will still label the variance matrix as EIM.

[b] Since the program assumes that the sandwich estimate of variance is calculated using the EIM instead of the OIM (even when the two are equal), the output will be labeled "Semi-Robust" instead of "Robust".

Reporting

level(#); see [R] **estimation options**.

eform displays the exponentiated coefficients and corresponding standard errors and confidence intervals. For family(binomial) link(logit) (i.e., logistic regression), exponentiation results in odds ratios; for family(poisson) link(log) (i.e., Poisson regression), exponentiated coefficients are rate ratios.

Max options

irls requests iterated, reweighted least-squares (IRLS) optimization of the deviance instead of Newton–Raphson optimization of the log likelihood. If the irls option is not specified, the optimization is carried out using Stata's ml commands, in which case all options of ml maximize are also available.

maximize_options: difficult, technique(*algorithm_spec*), iterate(#), [no]log, trace, gradient, showstep, hessian, shownrtolerance, tolerance(#), ltolerance(#), gtolerance(#), nrtolerance(#), nonrtolerance, from(*init_specs*); see [R] **maximize**. These options are seldom used.

fisher(#) specifies the number of Newton–Raphson steps that should use the Fisher scoring Hessian or expected information matrix (EIM) before switching to the observed information matrix (OIM). This option is useful only for Newton–Raphson optimization (and not when using irls).

search specifies that the command should search for good starting values. This option is useful only for Newton–Raphson optimization (and not when using irls).

The following options are available with glm but are not shown in the dialog box:

noheader suppresses the header information from the output. The coefficient table is still displayed.

notable suppresses the table of coefficients from the output. The header information is still displayed.

nodisplay suppresses the output. The iteration log is still displayed.

19.2 The predict command after glm

19.2.1 Syntax

`predict` [*type*] *newvar* [*if*] [*in*] [, *statistic options*]

statistic	description
Main	
mu	calculate $g^{-1}(\mathbf{x}\widehat{\boldsymbol{\beta}})$; the default
xb	calculate the linear prediction $\eta = \mathbf{x}\widehat{\boldsymbol{\beta}}$
eta	synonym of `xb`
stdp	calculate standard error of the linear prediction
anscombe	calculate Anscombe (1953) residuals
cooksd	calculate Cook's distance
deviance	calculate deviance residuals
hat	calculate diagonals of the "hat" matrix
likelihood	calculate a weighted average of standardized deviance and standardized Pearson residuals
pearson	calculate Pearson residuals
response	calculate differences between the observed and fitted outcomes
score	first derivative of the log likelihood with respect to $\mathbf{x}_j\boldsymbol{\beta}$
working	calculate working residuals

options	description
Options	
nooffset	modify calculations to ignore offset variable
adjusted	adjust deviance residual to speed up convergence
standardized	multiply residual by the factor $(1-h)^{-1/2}$
studentized	multiply residual by one over the square root of the estimated scale parameter
modified	modify denominator of residual to be a reasonable estimate of the variance of *depvar*

These statistics are available both in and out of sample; type `predict ... if e(sample) ...` if wanted only for the estimation sample.

`mu`, `xb`, `stdp`, and `score` are the only statistics allowed with `svy` estimation results.

19.2.2 Options

Main

`mu`, the default, specifies that `predict` calculate $g^{-1}(\mathbf{x}\widehat{\boldsymbol{\beta}})$, the inverse link of the linear prediction.

`xb` calculates the linear prediction $\eta = \mathbf{x}\widehat{\boldsymbol{\beta}}$.

`eta` is a synonym for `xb`.

`stdp` calculates the standard error of the linear prediction.

`anscombe` calculates the Anscombe (1953) residuals to produce residuals that closely follow a normal distribution.

`cooksd` calculates Cook's distance, which measures the aggregate change in the estimated coefficients when each observation is left out of the estimation.

`deviance` calculates the deviance residuals. Deviance residuals are recommended by McCullagh and Nelder (1989) and by others as having the best properties for examining the goodness of fit of a GLM. They are approximately normally distributed if the model is correct. They may be plotted against the fitted values or against a covariate to inspect the model's fit. Also see the `pearson` option below.

`hat` calculates the diagonals of the "hat" matrix as an analog to simple linear regression.

`likelihood` calculates a weighted average of standardized deviance and standardized Pearson (described below) residuals.

`pearson` calculates the Pearson residuals. Pearson residuals often have markedly skewed distributions for nonnormal family distributions. Also see the `deviance` option above.

`response` calculates the differences between the observed and fitted outcomes.

`score` calculates the equation-level score, $\partial \ln L / \partial(\mathbf{x}_j \boldsymbol{\beta})$.

`working` calculates the working residuals, which are response residuals weighted according to the derivative of the link function.

Options

`nooffset` is relevant only if you specified `offset(`*varname*`)` for `glm`. It modifies the calculations made by `predict` so that they ignore the offset variable; the linear prediction is treated as $\mathbf{x}_j\mathbf{b}$ rather than as $\mathbf{x}_j\mathbf{b} + \text{offset}_j$.

`adjusted` adjusts the deviance residual to speed up the convergence to the limiting normal distribution. The adjustment deals with adding to the deviance residual a higher-order term that depends on the variance function family. This option is allowed only when `deviance` is specified.

`standardized` requests that the residual be multiplied by the factor $(1-h)^{-1/2}$, where h is the diagonal of the hat matrix. This operation is done to account for the correlation between *depvar* and its predicted value.

`studentized` requests that the residual be multiplied by one over the square root of the estimated scale parameter.

`modified` requests that the denominator of the residual be modified to be a reasonable estimate of the variance of *depvar*. The base residual is multiplied by the factor $(k/w)^{-1/2}$, where k is either one or the user-specified dispersion parameter and w is the specified weight (or one if left unspecified).

Table 19.2: Statistics for predict

Statistic	Description	Reference
`mu`	inverse link; the default	$g^{-1}(\mathbf{X}\widehat{\boldsymbol{\beta}})$, see table A.2
`xb`	linear prediction	$\widehat{\eta} = \mathbf{X}\widehat{\beta}$
`eta`	linear prediction (synonym for `xb`)	$\widehat{\eta} = \mathbf{X}\widehat{\beta}$
`stdp`	standard error	standard error of $\widehat{\eta}$
`anscombe`	Anscombe residual	See sec. 4.5.5 and table A.10
`cooksd`	Cook's distance	See section 4.2.1
`deviance`	deviance residual	See sec. 4.5.6 and table A.11
`hat`	hat diagonal	See (4.5)
`likelihood`	likelihood residual	See sec. 4.5.8
`pearson`	Pearson residual	See sec. 4.5.3
`response`	response residual	See sec. 4.5.1
`score`	score residual	See sec. 4.5.9
`working`	working residual	See sec. 4.5.2

19.3 User-written programs

This section is for advanced users who need to extend the functionality of the included software. You should be comfortable with programming in Stata to at least the level taught in Stata NetCourse 151, Introduction to Stata Programming; see http://www.stata.com/netcourse/nc151.html for information.

The functionality of the `glm` program may be extended in key areas. Users may write their own programs for the variance function, for the link function, and for the Newey–West kernel weights.

The following sections present the format of the user-written programs. Once written, these programs can be used in the `glm` program by specifying the program name in the appropriate option.

19.3.1 Global macros available for user-written programs

User-written programs will need access to information about the data and to the model specification to compute the necessary information. In the following table, we present a list of global macros along with the information that they carry. User-written programs may use these macros to access needed information.

19.3.2 User-written variance functions

Users may write their own programs for variance functions. To take advantage of all the functionality of the `glm` command and for the predictions available postestimation, the program for variance functions should be able to calculate various quantities as well as set macros for output.

The following gives the outline of the program:

Listing 19.1: Skeleton code for a user-written variance program

```
1  program define program-name
2      args todo eta mu return
3
4      if `todo' == -1 {
5          Macros are set for the output
6
7          Where arguments are defined as
8          eta    = indicator variable for whether each obs is in-sample
9          mu     = <ignored>
10         return = <ignored>
11
12         global SGLM_vt "title assigned here"
13         global SGLM_vf "subtitle showing function assigned here"
14
15         error checking code for range of y
16         limit checks based on `eta' sample definition
17
18         global SGLM_mu "program to enforce boundary conditions on μ"
19
20         exit
21     }
22     if `todo' == 0 {
23         η is defined
24
25         Where arguments are defined as
26         eta    = variable name to define
27         mu     = μ
28         return = <ignored>
29
30         generate double variable named `eta'
31
32         exit
33     }
34     if `todo' == 1 {
35         V(μ) is defined
36
37         Where arguments are defined as
38         eta    = η
39         mu     = μ
```

```
40            return = variable name to define
41
42            generate double variable named 'return'
43
44            exit
45        }
46        if 'todo' == 2 {
47            ∂V(μ)/∂μ is defined
48
49            Where arguments are defined as
50            eta    = η
51            mu     = μ
52            return = variable name to define
53
54            generate double variable named 'return'
55
56            exit
57        }
58        if 'todo' == 3 {
59            deviance is defined
60
61            Where arguments are defined as
62            eta    = η
63            mu     = μ
64            return = variable name to define
65
66            generate double variable named 'return'
67            exit
68        }
69        if 'todo' == 4 {
70            Anscombe residual is defined
71
72            Where arguments are defined as
73            eta    = η
74            mu     = μ
75            return = variable name to define
76
77            generate double variable named 'return'
78            or issue an error if no support for Anscombe residuals
79
80            exit
81        }
82        if 'todo' == 5 {
83            Ln-likelihood ℒ is defined
84
85            Where arguments are defined as
86            eta    = η
87            mu     = μ
88            return = variable name to define
```

```
89
90            generate double variable named 'return'
91            or issue an error if there is no true likelihood
92
93            exit
94        }
95        if 'todo' == 6 {
96            ρ3(θ)/6 for adjusted deviance residuals is defined
97
98            Where arguments are defined as
99            eta    = η
100           mu     = μ
101           return = variable name to define
102
103           generate double variable named 'return'
104           or issue an error if no support for adjusted deviance residuals
105
106           exit
107       }
108       noisily display as error "Unknown glm variance function"
109       error 198
110   end
```

19.3.3 User-written programs for link functions

Users may write their own programs for link functions. To take advantage of all the functionality of the `glm` command and for the predictions available postestimation, the program for link functions should be able to calculate various quantities and set macros for output.

The following gives the outline of the program:

Listing 19.2: Skeleton code for a user-written link program

```
1   program define program-name
2       args todo eta mu return
3
4       if 'todo' == -1 {
5           Macros are set for the output
6
7           Where arguments are defined as
8           eta    = <ignored>
9           mu     = <ignored>
10          return = <ignored>
11
12          global SGLM_lt "title assigned here"
13          global SGLM_lf "subtitle showing function assigned here"
14
15          exit
```

```
16        }
17        if `todo' == 0 {
18             η is defined from g(μ)
19
20             Where arguments are defined as
21             eta    = variable name to define
22             mu     = μ
23             return = <ignored>
24
25             generate double variable named `eta'
26
27             exit
28        }
29        if `todo' == 1 {
30             μ is defined from g^-1(η)
31
32             Where arguments are defined as
33             eta    = η
34             mu     = variable name to define
35             return = <ignored>
36
37             generate double variable named `mu'
38
39             exit
40        }
41        if `todo' == 2 {
42             ∂μ/∂η is defined
43
44             Where arguments are defined as
45             eta    = η
46             mu     = μ
47             return = variable name to define
48
49             generate double variable named `return'
50
51             exit
52        }
53        if `todo' == 3 {
54             ∂^2μ/∂η^2 is defined
55
56             Where arguments are defined as
57             eta    = η
58             mu     = μ
59             return = variable name to define
60
61             generate double variable named `return'
62
63             exit
64        }
```

```
65        noisily display as error "Unknown glm link function"
66        exit 198
67    end
```

19.3.4 User-written programs for Newey–West weights

Users may write their own kernel evaluators for the Newey–West variance estimate. The program to be developed is an rclass program that will be passed four arguments. The program will return information via one scalar and two macros.

The following gives the outline of the program:

Listing 19.3: Skeleton code for user-written Newey–West kernel weights program

```
1    program define program-name , rclass
2        args G j
3            Where arguments are defined as
4            G    = maximum lag
5            j    = current lag
6
7        your code goes here
8
9        return scalar wt    = computed weight value assigned here
10       return local setype   "Newey-West"
11       return local sewtype "name of kernel assigned here"
12   end
```

After defining the program, you can use the nwest() option to specify that program for calculating the weights.

As an example, recall from section 3.6.7 the definition of the Tukey–Hanning kernel. Using the definitions of the scale factor C and maximum lag G given in table A.8, we may develop the following do-file for Stata:

Listing 19.4: Example code for user-written Tukey–Hanning weights kernel

```
1    capture program drop thanning
2    program define thanning, rclass
3        args lag j
4
5        tempname z
6        scalar `z' = `j'/(`lag'+1.0)
7
8        return scalar wt = (1.0 + cos(_pi*`z'))/2
9        return local setype  "Newey-West"
10       return local sewtype "Tukey-Hanning"
11   end
12
13   glm price weight displ foreign mpg, nwest(thanning 1)
```

19.4 Remarks

19.4.1 Equivalent commands

Some `family()` and `link()` combinations with the `glm` command result in a model that is already fitted by Stata:

Table 19.3: Equivalent Stata commands

`family()`	`link()`	Options	Equivalent Stata command
`gaussian`	`identity`	*nothing* | `irls` | `irls vce(oim)`	`regress`
`gaussian`	`identity`	`t(`*var*`) vce(nwest nwest #)` `vfactor(`$\#_v$`)`	`newey, t(`*var*`) lag(#)`
`binomial`	`cloglog`	*nothing* | `irls vce(oim)`	`cloglog`
`binomial`	`probit`	*nothing* | `irls vce(oim)`	`probit`
`binomial`	`logit`	*nothing* | `irls` | `irls vce(oim)`	`logit`
`poisson`	`log`	*nothing* | `irls` | `irls vce(oim)`	`poisson`
`nbinomial`	`log`	*nothing* | `irls vce(oim)`	`nbreg`
`gamma`	`log`	`scale(1)`	`streg, dist(exp) nohr`

In general, if a Stata command already exists for a family and link combination, then the results will match for the coefficients (will be close to numeric differences in the optimization) and for the standard errors (will be close to numeric differences in the optimization) if `irls` is not specified or if both `irls` and `oim` are specified. And, if the link is the canonical link for the specified family, the standard errors will match if just `irls` is specified since the EIM is equal to the OIM in this case. See section 3.1 for details.

19.4.2 Special comments on family(Gaussian) models

Although `glm` can be used to fit linear regression (`family(Gaussian) link(identity)` models) and, in fact, does so by default, it is better to use the `regress` command because it is quicker and because many postestimation commands are available to explore the adequacy of the fit.

19.4.3 Special comments on family(binomial) models

The binomial distribution can be specified as (1) `family(binomial)`, (2) `family(binomial #)`, or (3) `family(binomial` *varname*`)`.

In case 2, # is the value of the binomial denominator N, the number of trials. Specifying `family(binomial 1)` is the same as specifying `family(binomial)`; both mean that y has the Bernoulli distribution with values 0 and 1 only.

In case 3, *varname* is a variable containing the binomial denominator, thus allowing the number of trials to vary across observations.

For `family(binomial) link(logit)` models, we recommend also using the command `logistic`. Both produce the same answers, and each command provides useful postestimation commands (not found in both commands).

Log-log regression may be performed by subtracting the response from one and modeling as a complementary log-log model. Proportional log-log data may be transformed by transposing the proportion of successes and failures.

19.4.4 Special comments on family(nbinomial) models

The negative binomial distribution can be specified as (1) `family(nbinomial)`, or the ancillary parameter may be specified in (2) `family(nbinomial #)`.

`family(nbinomial)` is equivalent to `family(nbinomial 1)`. #, often called k, enters the variance and deviance functions; typical values range between .01 and 2.

`family(nbinomial) link(log)` models, also known as negative binomial regression, are used for data with an overdispersed Poisson distribution. Although `glm` can be used to fit such models, Stata's maximum-likelihood `nbreg` command is probably preferable. Under the `glm` approach, we must search for value of k that results in the deviance-based dispersion being 1. `nbreg`, on the other hand, finds the maximum likelihood estimate of k and reports its confidence interval.

The default ancillary parameter value of 1 is the same as geometric regression. A negative binomial distribution with a scale of one is a geometric distribution. Conversely, a negative binomial distribution parameterized as we have shown in the text with a scale of zero is a Poisson distribution.

19.4.5 Special comment on family(gamma) link(log) models

`glm` with `family(gamma)` and `link(log)` is identical to exponential regression. However, censoring is not available with this method. Censored exponential regression may be modeled using `glm` with `family(poisson)`. The log of the original response is entered into a Poisson model as an offset, whereas the new response is the censor variable. The result of such modeling is identical to the log relative hazard parameterization of the `streg` command in Stata.

Maximum-likelihood exponential regression estimation via `glm` with the `scale(1)` option will report a log likelihood that differs from the log likelihood reported by the `streg` command. The difference is documented in the manual and amounts to the `streg` command recentering the log likelihood to allow a comparison to Weibull regression log likelihoods.

A Tables

Table A.1: Variance functions

Family (distribution)	Variance $V(\mu)$	Range restrictions	$\partial V(\mu)/\partial\mu$
Gaussian	1	$\begin{cases} \mu \in \Re \\ y \in \Re \end{cases}$	0
Bernoulli	$\mu(1-\mu)$	$\begin{cases} 0 < \mu < 1 \\ 0 \le y \le 1 \end{cases}$	$1-2\mu$
Binomial(k)	$\mu(1-\mu/k)$	$\begin{cases} 0 < \mu < k \\ 0 \le y \le k \end{cases}$	$1-2\mu/k$
Poisson	μ	$\begin{cases} \mu > 0 \\ y \ge 0 \end{cases}$	1
Gamma	μ^2	$\begin{cases} \mu > 0 \\ y > 0 \end{cases}$	2μ
Inverse Gaussian	μ^3	$\begin{cases} \mu > 0 \\ y > 0 \end{cases}$	$3\mu^2$
Negative binomial(α)	$\mu+\alpha\mu^2$	$\begin{cases} \mu > 0 \\ y \ge 0 \end{cases}$	$1+2\alpha\mu$
Power(k)	μ^k	$\begin{cases} \mu > 0 \\ k \ne 0,1,2 \end{cases}$	$k\mu^{k-1}$
Quasi	$V(\mu)$		$\dfrac{\partial V(\mu)}{\partial\mu}$

Table A.2: Link and inverse link functions ($\eta = \mathbf{X}\boldsymbol{\beta} + \text{offset}$)

Link name		Link function $\eta = g(\mu)$	Inverse link $\mu = g^{-1}(\eta)$	Range of $\widehat{\mu}$
Identity		μ	η	$\widehat{\mu} \in \Re$
Logit		$\ln\left\{\mu/(1-\mu)\right\}$	$e^\eta/(1+e^\eta)$	$\widehat{\mu} \in (0,1)$
Log		$\ln(\mu)$	$\exp(\eta)$	$\widehat{\mu} > 0$
Negative binomial(α)		$\ln\{\mu/(\mu+1/\alpha)\}$	$e^\eta/\{\alpha(1-e^\eta)\}$	$\widehat{\mu} > 0$
Log-complement		$\ln(1-\mu)$	$1-\exp(\eta)$	$\widehat{\mu} < 1$
Log-log		$-\ln\{-\ln(\mu)\}$	$\exp\{-\exp(-\eta)\}$	$\widehat{\mu} \in (0,1)$
Complementary log-log		$\ln\{-\ln(1-\mu)\}$	$1-\exp\{-\exp(\eta)\}$	$\widehat{\mu} \in (0,1)$
Probit		$\Phi^{-1}(\mu)$	$\Phi(\eta)$	$\widehat{\mu} \in (0,1)$
Reciprocal		$1/\mu$	$1/\eta$	$\widehat{\mu} \in \Re$
Power($\alpha = -2$)		$1/\mu^2$	$1/\sqrt{\eta}$	$\widehat{\mu} > 0$
Power(α)	$\begin{cases} \alpha \neq 0 \\ \alpha = 0 \end{cases}$	$\begin{cases} \mu^\alpha \\ \ln(\mu) \end{cases}$	$\begin{cases} \eta^{1/\alpha} \\ \exp(\eta) \end{cases}$	$\widehat{\mu} \in \Re$
Odds power(α)	$\begin{cases} \alpha \neq 0 \\ \alpha = 0 \end{cases}$	$\begin{cases} \dfrac{\mu/(1-\mu)^\alpha - 1}{\alpha} \\ \ln\left(\dfrac{\mu}{1-\mu}\right) \end{cases}$	$\begin{cases} \dfrac{(1+\alpha\eta)^{1/\alpha}}{1+(1+\alpha\eta)^{1/\alpha}} \\ \dfrac{e^\eta}{1+e^\eta} \end{cases}$	$\widehat{\mu} \in (0,1)$

Table A.3: First derivatives of link functions ($\eta = \mathbf{X}\boldsymbol{\beta} + \text{offset}$)

Link name		Link $\eta = g(\mu)$	First derivatives $\triangle = \partial\eta/\partial\mu$
Identity		μ	1
Logit		$\ln\{\mu/(1-\mu)\}$	$1/\{\mu(1-\mu)\}$
Log		$\ln(\mu)$	$1/\mu$
Negative binomial(α)		$\ln\{\alpha\mu/(1+\alpha\mu)\}$	$1/(\mu+\alpha\mu^2)$
Log-complement		$\ln(1-\mu)$	$-1/(1-\mu)$
Log-log		$-\ln\{-\ln(\mu)\}$	$-1/\{\mu\ln(\mu)\}$
Complementary log-log		$\ln\{-\ln(1-\mu)\}$	$\{(\mu-1)\ln(1-\mu)\}^{-1}$
Probit		$\Phi^{-1}(\mu)$	$1/\phi\{\Phi^{-1}(\mu)\}$
Reciprocal		$1/\mu$	$-1/\mu^2$
Power($\alpha = -2$)		$1/\mu^2$	$-2/\mu^3$
Power(α)	$\begin{cases} \alpha \neq 0 \\ \alpha = 0 \end{cases}$	$\begin{cases} \mu^\alpha \\ \ln(\mu) \end{cases}$	$\begin{cases} \alpha\mu^{\alpha-1} \\ 1/\mu \end{cases}$
Odds power(α)	$\begin{cases} \alpha \neq 0 \\ \alpha = 0 \end{cases}$	$\begin{cases} \dfrac{\{\mu/(1-\mu)\}^\alpha - 1}{\alpha} \\ \ln\left(\dfrac{\mu}{1-\mu}\right) \end{cases}$	$\begin{cases} \dfrac{\mu^{\alpha-1}}{(1-\mu)^{\alpha+1}} \\ \dfrac{1}{\mu(1-\mu)} \end{cases}$

Table A.4: First derivatives of inverse link functions ($\eta = \mathbf{X}\boldsymbol{\beta} + \text{offset}$)

Link name		Inverse link $\mu = g^{-1}(\eta)$	First derivatives $\nabla = \partial\mu/\partial\eta$
Identity		η	1
Logit		$e^\eta/(1+e^\eta)$	$\mu(1-\mu)$
Log		$\exp(\eta)$	μ
Negative binomial(α)		$e^\eta/\{\alpha(1-e^\eta)\}$	$\mu+\alpha\mu^2$
Log-complement		$1-\exp(\eta)$	$\mu-1$
Log-log		$\exp\{-\exp(-\eta)\}$	$-\mu\ln(\mu)$
Complementary log-log		$1-\exp\{-\exp(\eta)\}$	$(\mu-1)\ln(1-\mu)$
Probit		$\Phi(\eta)$	$\phi(\eta)$
Reciprocal		$1/\eta$	$-\mu^2$
Power($\alpha=-2$)		$1/\sqrt{\eta}$	$-\mu^3/2$
Power(α)	$\alpha \neq 0$	$\eta^{1/\alpha}$	$\frac{1}{\alpha}\mu^{1-\alpha}$
	$\alpha = 0$	$\exp(\eta)$	μ
Odds power(α)	$\alpha \neq 0$	$\dfrac{(1+\alpha\eta)^{1/\alpha}}{1+(1+\alpha\eta)^{1/\alpha}}$	$\dfrac{\mu(1-\mu)}{1+\alpha\eta}$
	$\alpha = 0$	$\dfrac{e^\eta}{1+e^\eta}$	$\mu(1-\mu)$

Table A.5: Second derivatives of link functions where $\eta = \mathbf{X}\boldsymbol{\beta} + \text{offset}$ and $\triangle = \partial\eta/\partial\mu$

Link name	Link $\eta = g(\mu)$	Second derivatives $\partial^2\eta/\partial\mu^2$
Identity	μ	0
Logit	$\ln\{\mu/(1-\mu)\}$	$\mu\triangle^2$
Log	$\ln(\mu)$	$-\triangle^2$
Negative binomial(α)	$\ln\{\alpha\mu/(1+\alpha\mu)\}$	$-\triangle^2(1+2\alpha\mu)$
Log-complement	$\ln(1-\mu)$	$-\triangle^2$
Log-log	$-\ln\{-\ln(\mu)\}$	$\{1+\ln(\mu)\}\triangle^2$
Complementary log-log	$\ln\{-\ln(1-\mu)\}$	$-\{1+\ln(1-\mu)\}\triangle^2$
Probit	$\Phi^{-1}(\mu)$	$\eta\triangle^2$
Reciprocal	$1/\mu$	$-2\triangle/\mu$
Power($\alpha = -2$)	$1/\mu^2$	$-3\triangle/\mu$
Power(α) $\begin{cases} \alpha \neq 0 \\ \alpha = 0 \end{cases}$	$\begin{cases} \mu^\alpha \\ \ln(\mu) \end{cases}$	$\begin{cases} (\alpha-1)\triangle/\alpha \\ -\triangle^2 \end{cases}$
Odds power(α) $\begin{cases} \alpha \neq 0 \\ \alpha = 0 \end{cases}$	$\begin{cases} \dfrac{\{\mu/(1-\mu)\}^\alpha - 1}{\alpha} \\ \ln\left(\dfrac{\mu}{1-\mu}\right) \end{cases}$	$\begin{cases} \triangle\left(\dfrac{1-1/\alpha}{1-\mu} + \alpha + 1\right) \\ \mu\triangle^2 \end{cases}$

Table A.6: Second derivatives of inverse link functions where $\eta = \mathbf{X}\boldsymbol{\beta} + \text{offset}$ and $\nabla = \partial\mu/\partial\eta$

Link name		Inverse link $\mu = g^{-1}(\eta)$	Second derivatives $\partial^2\mu/\partial\eta^2$
Identity		η	0
Logit		$e^\eta/(1+e^\eta)$	$\nabla(1-2\mu)$
Log		$\exp(\eta)$	∇
Negative binomial(α)		$e^\eta/\{\alpha(1-e^\eta)\}$	$\nabla(1+2\alpha\mu)$
Log-complement		$1-\exp(\eta)$	∇
Log-log		$\exp\{-\exp(-\eta)\}$	$\nabla\{1+\ln(\mu)\}$
Complementary log-log		$1-\exp\{-\exp(\eta)\}$	$\nabla\{1+\ln(1-\mu)\}$
Probit		$\Phi(\eta)$	$-\nabla\eta$
Reciprocal		$1/\eta$	$-2\nabla\mu$
Power($\alpha=-2$)		$1/\sqrt{\eta}$	$3\nabla^2/\mu$
Power(α)	$\alpha \neq 0$	$\eta^{1/\alpha}$	$\nabla(\frac{1}{\alpha}-1)/\mu^\alpha$
	$\alpha = 0$	$\exp(\eta)$	∇
Odds power(α)	$\alpha \neq 0$	$\dfrac{(1+\alpha\eta)^{1/\alpha}}{1+(1+\alpha\eta)^{1/\alpha}}$	$\nabla\left\{1-2\mu-\dfrac{\alpha}{1+\alpha\eta}\right\}$
	$\alpha = 0$	$\dfrac{e^\eta}{1+e^\eta}$	$\nabla(1-2\mu)$

Table A.7: Log likelihoods

Family	$\mathcal{L}$
Gaussian	$-\frac{1}{2}\sum\left\{\frac{(y_i-\mu_i)^2}{\phi}+\ln(2\pi\phi)\right\}$
Bernoulli	$\phi\sum_{i:y_i=0}\ln\left(1-\frac{\mu_i}{2}\right)+\phi\sum_{i:y_i\neq 0}\ln\left(\frac{\mu_i}{2}\right)$
Binomial(k)	$\phi\sum\left\{\ln\Gamma(k_i+1)-\ln\Gamma(y_i+1)-\ln\Gamma(k_i-y_i+1)\right.$ $\left.+y_i\ln\left(\frac{\mu_i}{k_i}\right)+(\mu_i-y_i)\ln\left(1-\frac{\mu_i}{y_i}\right)\right\}$
Poisson	$\phi\sum\{-\mu_i+y_i\ln(\mu_i)-\ln\Gamma(y_i+1)\}$
Gamma	$-\frac{1}{\phi}\sum\left\{\frac{y_i}{\mu_i}-\ln\left(\frac{\phi}{\mu_i}\right)-\frac{\phi-1}{\phi}\ln(y_i)+\frac{1}{\phi}\ln\Gamma(\phi)\right\}$
Inverse Gaussian	$-\frac{1}{2}\sum\left\{\frac{(y_i-\mu_i)^2}{y_i\mu_i^2\phi}+\ln\left(\phi y_i^3\right)+\ln(2\pi)\right\}$
Negative binomial(α)	$\phi\sum\left\{\ln\Gamma\left(\frac{1}{\alpha}+y_i\right)-\ln\Gamma(y_i+1)-\ln\Gamma\left(\frac{1}{\alpha}\right)\right.$ $\left.-\frac{1}{\alpha}\ln(1+\alpha\mu_i)+y_i\ln\left(\frac{\alpha\mu_i}{\alpha\mu_i+1}\right)\right\}$

Table A.8: Weight functions (kernels) for weighted sandwich variance estimates

Estimator	Maximum lag G	Sandwich weights $\omega(z)$
Newey–West	q	$\begin{cases} 1-\lvert z\rvert & \text{if } \lvert z\rvert \le 1 \\ 0 & \text{otherwise} \end{cases}$
Gallant	q	$\begin{cases} 1-6z^2+6\lvert z\rvert^3 & \text{if } 0 \le \lvert z\rvert \le 1/2 \\ 2(1-\lvert z\rvert)^3 & \text{if } 1/2 \le \lvert z\rvert \le 1 \end{cases}$
Anderson	q	$\dfrac{3}{(6\pi z/5)^2}\left\{\dfrac{\sin(6\pi z/5)}{6\pi z/5}-\cos(6\pi z/5)\right\}$

Table A.9: Pearson residuals

Family (variance function)	Pearson residual (r_P)
Gaussian	$y_i - \widehat{\mu}_i$
Bernoulli	$\dfrac{y_i - \widehat{\mu}_i}{\sqrt{\widehat{\mu}_i(1-\widehat{\mu}_i)}}$
Binomial(k)	$\dfrac{y_i - \widehat{\mu}_i}{\sqrt{\widehat{\mu}_i(1-\widehat{\mu}_i/k_i)}}$
Poisson	$\dfrac{y_i - \widehat{\mu}_i}{\sqrt{\widehat{\mu}_i}}$
Gamma	$\dfrac{y_i - \widehat{\mu}_i}{\sqrt{\widehat{\mu}_i^2}}$
Inverse Gaussian	$\dfrac{y_i - \widehat{\mu}_i}{\sqrt{\widehat{\mu}_i^3}}$
Negative binomial(α)	$\dfrac{y_i - \mu_i}{\sqrt{\widehat{\mu}_i + \alpha\widehat{\mu}_i^2}}$
Power(k)	$\dfrac{y_i - \widehat{\mu}_i}{\sqrt{\widehat{\mu}_i^k}}$

Table A.10: Anscombe residuals

Family (variance function)	Anscombe residual (r_A)
Gaussian	$y_i - \widehat{\mu}_i$
Bernoulli	$\dfrac{1.5\left[y_i^{2/3}\,{}_2F_1(y_i) - \mu_i^{2/3}\,{}_2F_1(\mu_i)\right]}{(\widehat{\mu}_i - \widehat{\mu}_i^2)^{1/6}}$
Binomial(k)	$\dfrac{1.5\left[y_i^{2/3}\,{}_2F_1(y_i/k_i) - \mu_i^{2/3}\,{}_2F_1(\mu_i/k_i)\right]}{(\widehat{\mu}_i - \widehat{\mu}_i^2/k_i)^{1/6}}$
Poisson	$\dfrac{\frac{3}{2}(y_i^{2/3} - \widehat{\mu}_i^{2/3})}{\widehat{\mu}_i^{1/6}}$
Gamma	$\dfrac{3(y_i^{1/3} - \widehat{\mu}_i^{1/3})}{\widehat{\mu}_i^{1/3}}$
Inverse Gaussian	$\dfrac{\ln y_i - \ln \widehat{\mu}_i}{\sqrt{\widehat{\mu}_i}}$
Negative binomial(α)	$\dfrac{{}_2F_1(-\alpha y_i) - {}_2F_1(-\alpha\widehat{\mu}_i) + 1.5\left[y_i^{2/3} - \widehat{\mu}_i^{2/3}\right]}{(\widehat{\mu}_i + \alpha\widehat{\mu}_i^2)^{1/6}}$
Power(k)	$\dfrac{3\widehat{\mu}_i^{-k/6}\left(\widehat{\mu}_i^{1-k/3} - y_i^{1-k/3}\right)}{k-3}$

Table A.11: Squared deviance residuals and deviance adjustment factors $\rho_3(\theta)$

Family (variance function)	Squared deviance residual $\widehat{d}_i^2$	$\rho_3(\theta)$
Gaussian	$(y_i - \widehat{\mu}_i)^2$	0
Bernoulli	$\begin{cases} -2\ln(1-\widehat{\mu}_i) & \text{if } y_i = 0 \\ -2\ln(\widehat{\mu}_i) & \text{if } y_i = 1 \end{cases}$	†
Binomial(k)	$\begin{cases} 2k_i \ln\left(\frac{k_i}{k_i-\widehat{\mu}_i}\right) & \text{if } y_i = 0 \\ 2y_i \ln\left(\frac{y_i}{\widehat{\mu}_i}\right) + 2(k_i - y_i)\ln\left(\frac{k_i-y_i}{k_i-\widehat{\mu}_i}\right) & \text{if } 0 < y_i < k_i \\ 2k_i \ln\left(\frac{k_i}{\widehat{\mu}_i}\right) & \text{if } y_i = k_i \end{cases}$	$\frac{1-2\widehat{\mu}_i/k_i}{\sqrt{k_i\widehat{\mu}_i(1-\widehat{\mu}_i)}}$
Poisson	$\begin{cases} 2\widehat{\mu}_i & \text{if } y_i = 0 \\ 2\left\{y_i \ln\left(\frac{y_i}{\widehat{\mu}_i}\right) - (y_i - \widehat{\mu}_i)\right\} & \text{otherwise} \end{cases}$	$\frac{1}{\sqrt{\widehat{\mu}_i}}$
Gamma	$-2\left\{\ln\left(\frac{y_i}{\widehat{\mu}_i}\right) - \frac{y_i - \widehat{\mu}_i}{\widehat{\mu}_i}\right\}$	$\frac{2}{\sqrt{\widehat{\phi}}}$
Inverse Gaussian	$\frac{(y_i - \widehat{\mu}_i)^2}{\widehat{\mu}_i^2 y_i}$	$3\widehat{\mu}_i^{7/2}/\widehat{\phi}^2$
Neg. binomial(α)	$\begin{cases} 2\ln(1+\alpha\widehat{\mu}_i)/\alpha & \text{if } y_i = 0 \\ 2y_i \ln\left(\frac{y_i}{\widehat{\mu}_i}\right) - \frac{2}{\alpha}(1+\alpha y_i)\ln\left(\frac{1+\alpha y_i}{1+\alpha\widehat{\mu}_i}\right) & \text{otherwise} \end{cases}$	‡
Power(k)	$\frac{2y_i}{(1-k)(y_i^{1-k} - \widehat{\mu}_i^{1-k})} - \frac{2}{(2-k)(y_i^{2-k} - \widehat{\mu}_i^{2-k})}$	‡

† Not recommended for Bernoulli. In general, not recommended for binomial families with $k < 10$.
‡ Not available.

Table A.12: Cameron–Windmeijer Kullback–Leibler divergence

Family	K–L divergence
Gaussian	$\sum \dfrac{(y_i - \mu_i)^2}{\sigma^2}$
Bernoulli	$-2\sum\{y_i \ln \mu_i + (1 - y_i)\ln(1 - \mu_i)\}$
Binomial(k)	$2\sum \left\{ y_i \ln \dfrac{y_i}{\mu_i} + (k_i - y_i)\ln \dfrac{k_i - y_i}{k_i - \mu_i} \right\}$
Poisson	$2\sum \left\{ y \ln \left(\dfrac{y_i}{\mu_i}\right) - (y_i - \mu_i) \right\}$
Gamma	$-2\phi \sum \left\{ \ln \left(\dfrac{y_i}{\mu_i}\right) + \dfrac{y_i - \mu_i}{\mu_i} \right\}$
Inverse Gaussian	$\sum \dfrac{(y_i - \mu_i)^2}{\mu_i^2 y_i}$

Table A.13: Cameron–Windmeijer R^2

Family	$R^2_{\text{Cameron\&Windmeijer}}$
Gaussian	$1 - \dfrac{\sum(y_i - \widehat{\mu}_i)^2}{\sum(y_i - \overline{y})^2}$
Bernoulli	$1 - \dfrac{\sum \widehat{\mu}_i \ln(\widehat{\mu}_i) + (1 - \widehat{\mu}_i)\ln(1 - \widehat{\mu}_i)}{n\left\{\overline{y}\ln(\overline{y}) + (1 - \overline{y})\ln(1 - \overline{y})\right\}}$
Binomial(k)	$1 - \dfrac{\sum \widehat{\mu}_i \ln(\widehat{\mu}_i) + (k_i - \widehat{\mu}_i)\ln(k_i - \widehat{\mu}_i)}{n\left\{\overline{y}\ln(\overline{y}) + (k_i - \overline{y})\ln(k_i - \overline{y})\right\}}$
Poisson	$1 - \dfrac{\sum y_i \ln(y_i/\widehat{\mu}_i) - (y_i - \widehat{\mu}_i)}{\sum y_i \ln(y_i/\overline{y})}$
Gamma	$1 - \dfrac{\sum y_i \ln(y_i/\widehat{\mu}_i) + (y_i - \widehat{\mu}_i)/\widehat{\mu}_i}{\sum \ln(y_i/\overline{y})}$
Inverse Gaussian	$1 - \dfrac{\sum(y_i - \widehat{\mu}_i)^2/(\widehat{\mu}_i^2 y_i)}{\sum(y_i - \overline{y})^2/(\overline{y}^2 y)}$

Table A.14: Interpretation of power links

Gaussian	
`link(power 1)`	canonical identity (standard OLS regression)
`link(power 0)`	log-normal
Binomial	
`link(power 0)`	log-binomial; used with incidence rates
Poisson	
`link(power 0)`	canonical log link
`link(power 1)`	identity link
Gamma	
`link(power -1)`	canonical inverse link
`link(power 0)`	log link
`link(power 1)`	identity link
Inverse Gaussian	
`link(power -2)`	canonical inverse quadratic
`link(power 0)`	log link
`link(power 1)`	identity link
Negative binomial	
`link(power 0)`	log link (common parameterization of negative binomial)
Continuous distributions	
`link(power 2)`	square link
`link(power .5)`	square root link

References

Akaike, H. 1973. Information theory and an extension of the maximum likelihood principle. In *Second International Symposium on Information Theory*, ed. B. N. Petrov and F. Csaki, 267–281. Budapest, Hungary: Akademiai Kiado.

Albright, R. L., S. R. Lerman, and C. F. Manski. 1977. Report on the development of an estimation program for the multinomial probit model.

Anderson, J. A. 1984. Regression and ordered categorical variables. *Journal of the Royal Statistical Society, Series B* 46: 1–30.

Andrews, D. W. K. 1991. Heteroskedasticity and autocorrelation consistent covariance matrix estimation. *Econometrica* 59: 817–858.

Anscombe, F. J. 1972. Contribution to the discussion of H. Hotelling's paper. *Journal of the Royal Statistical Society, Series B* 15: 229–230.

Atkinson, A. C. 1988. A note on the generalized information criterion for choice of model. *Biometrika* 67: 413–418.

Bai, Z., P. Krishnaiah, N. Sambamoorthi, and L. Zhao. 1992. Model selection for log-linear models. *Sankhyā* B54: 200–219.

Barndorff-Nielsen, O. 1976. Factorization of likelihood functions for full exponential families. *Journal of the Royal Statistical Society, Series B* 38: 37–44.

Belsley, D. A., E. Kuh, and R. E. Welsch. 1980. *Regression Diagnostics: Identifying Influential Data and Sources of Collinearity*. New York: Wiley.

Ben-Akiva, M., and S. R. Lerman. 1985. *Discrete Choice Analysis: Theory and Application to Travel Demand*. Cambridge, MA: MIT Press.

Berkson, J. 1944. Application of the logistic function to bio-assay. *Journal of the American Statistical Association* 39: 357–365.

Berndt, E. K., B. H. Hall, R. E. Hall, and J. A. Hausman. 1974. Estimation and inference in nonlinear structural models. *Annals of Economic and Social Measurement* 3/4: 653–665.

Binder, D. A. 1983. On the variances of asymptotically normal estimators from complex surveys. *International Statistical Review* 51: 279–292.

Bliss, C. I. 1934. The method of probits. *Science* 79: 38–39, 409–410.

Boes, S., and R. Winkelmann. 2006. Ordered response models. *Allgemeines Statistisches Archiv* 90: 167–181.

von Bortkewitsch, L. 1898. *Das Gesetz der Kleinen Zahlen*. Leipzig: Teubner.

Brant, R. 1990. Assessing proportionality in the proportional odds model for ordinal logistic regression. *Biometrics* 46: 1171–1178.

Breslow, N. E. 1990. Tests of hypotheses in overdispersed Poisson regression and other quasi-likelihood models. *Journal of the American Statistical Association* 85: 565–571.

———. 1996. Generalized linear models: Checking assumptions and strengthening conclusions. *Statistica Applicata* 8: 23–41.

Brown, D. 1992. A graphical analysis of deviance. *Applied Statistics* 41: 55–62.

Cameron, A., and P. Trivedi. 1998. *Regression Analysis of Count Data*. New York: Cambridge University Press.

Cameron, A. C., and F. A. G. Windmeijer. 1997. An R-squared measure of goodness of fit for some common nonlinear regression models. *Journal of Econometrics* 77: 329–342.

Carroll, R. J., S. Wang, D. G. Simpson, A. J. Stromberg, and D. Ruppert. 1998. The sandwich (robust covariance matrix) estimator. Technical report. http://www.stat.tamu.edu/ftp/pub/rjcarroll/sandwich.pdf.

Cleves, M., W. W. Gould, and R. G. Gutierrez. 2004. *An Introduction to Survival Analysis Using Stata*. Rev. ed. College Station, TX: Stata Press.

Collett, D. 1991. *Modelling Binary Data*. New York: Chapman & Hall.

Consul, P. C., and F. Famoye. 1992. Generalized Poisson regression model. *Communications in Statistics—Theory and Methods* 21: 89–109.

Cordeiro, G. M., and P. McCullagh. 1991. Bias correction in generalized linear models. *Journal of the Royal Statistical Society, Series B* 53: 629–643.

Cox, D. R. 1983. Some remarks on overdispersion. *Biometrika* 70: 269–274.

Cox, D. R., and E. J. Snell. 1968. A general definition of residuals. *Journal of the Royal Statistical Society, Series B* 30: 248–275.

Cragg, J. G., and R. Uhler. 1970. The demand for automobiles. *Canadian Journal of Economics* 3: 386–406.

Daganzo, C. 1979. *Multinomial Probit: The Theory and Its Application to Demand Forecasting*. New York: Academic Press.

Davidian, M., and R. J. Carroll. 1987. Variance function estimation. *Journal of the American Statistical Association* 82: 1079–1091.

Dean, C., and J. F. Lawless. 1989. Tests for detecting overdispersion in Poisson regression models. *Journal of the American Statistical Association* 84: 467–472.

Doll, R., and A. B. Hill. 1966. Mortality of British doctors in relation to smoking; observations on coronary thrombosis. In *Epidemiological Approaches to the Study of Cancer and Other Chronic Diseases*, ed. W. Haenszel, vol. 19, 204–268. National Cancer Institute Monograph.

Dyke, G. V., and H. D. Patterson. 1952. Analysis of factorial arrangements when the data are proportions. *Biometrics* 8: 1–12.

Efron, B. 1978. Regression and ANOVA with zero-one data: Measures of residual variation. *Journal of the American Statistical Association* 73: 113–121.

———. 1981. Nonparametric estimates of standard error: The jackknife, the bootstrap and other methods. *Biometrika* 68: 589–599.

Famoye, F. 1995. Generalized binomial regression model. *Biometrical Journal* 37: 581–594.

Famoye, F., and E. H. Kaufman, Jr. 2002. Generalized negative binomial regression model. *Journal of Applied Statistical Sciences* 11: 289–296.

Fisher, R. A. 1922. On the mathematical foundations of theoretical statistics. *Philosophical Transactions of the Royal Society* 222: 309–368.

———. 1934. Two new properties of mathematical likelihood. *Proceedings of the Royal Society* A144: 285–307.

Francis, B., M. Green, and C. Payne, ed. 1993. *The GLIM System*. New York: Oxford University Press.

Fu, V. K. 1998. Estimating generalized ordered logit models. *Stata Technical Bulletin* 44: 27–30. Reprinted in *Stata Technical Bulletin Reprints*, vol. 8, pp. 160–164. College Station, TX: Stata Press.

Gail, M. H., W. Y. Tan, and S. Piantadosi. 1988. Tests for no treatment effect in randomized clinical trials. *Biometrika* 75: 57–64.

Gallant, A. R. 1987. *Nonlinear Statistical Models*. New York: Wiley.

Ganio, L. M., and D. W. Schafer. 1992. Diagnostics for overdispersion. *Journal of the American Statistical Association* 87: 795–804.

Gould, W., J. Pitblado, and W. Sribney. 2006. *Maximum Likelihood Estimation with Stata*. 3rd ed. College Station, TX: Stata Press.

Green, P. J. 1984. Iteratively reweighted least squares for maximum likelihood estimation, and some robust and resistant alternatives. *Journal of the Royal Statistical Society, Series B* 46: 149–192.

Greene, W. 2002. *LIMDEP, Version 8.0: Econometric Modeling Guide*. Plainview, NY: Econometric Software.

———. 2003. *Econometric Analysis*. 5th ed. Upper Saddle River, NJ: Prentice Hall.

Hansen, M., and B. Yu. 2001. Model selection and minimum description length principle. *Journal of the American Statistical Association* 96: 746–774.

Hardin, J. W. 2003. The sandwich estimate of variance. In *Advances in Econometrics: Maximum Likelihood of Misspecified Models: Twenty Years Later*, ed. T. Fomby and C. Hill, 45–73. New York: Elsevier.

Hardin, J. W., and R. J. Carroll. 2003. Measurement error, GLMs, and notational conventions. *Stata Journal* 3: 329–341.

Hardin, J. W., and M. A. Cleves. 1999. Generalized linear models: Extensions to the binomial family. *Stata Technical Bulletin* 50: 21–25. Reprinted in *Stata Technical Bulletin Reprints*, vol. 9, pp. 140–160. College Station, TX: Stata Press.

Hardin, J. W., and J. M. Hilbe. 2002. *Generalized Estimating Equations*. London: Chapman & Hall/CRC.

Hastie, T., and R. Tibshirani. 1986. Generalized additive models. *Statistical Science* 1: 297–318.

Hausman, J. 1978. Specification tests in econometrics. *Econometrica* 46: 1251–1271.

Hilbe, J. M. 1993a. Generalized linear models. *Stata Technical Bulletin* 11: 20–28. Reprinted in *Stata Technical Bulletin Reprints*, vol. 2, pp. 149–159. College Station, TX: Stata Press.

———. 1993b. Log Negative Binomial Regression as a Generalized Linear Model. *Technical Report 26* Graduate College, Arizona State University.

———. 1994. Generalized linear models. *American Statistician* 48: 255–265.

———. 2000. Two-parameter log-gamma and log-inverse Gaussian models. *Stata Technical Bulletin* 53: 31–32. Reprinted in *Stata Technical Bulletin Reprints*, vol. 9, pp. 273–275. College Station, TX: Stata Press.

———. 2007. *Negative Binomial Regression*. Cambridge: Cambridge University Press.

Hilbe, J. M., and W. Greene. 2007. Count response regression models. In *Epidemiology and Medical Statistics, Elsevier Handbook of Statistics Series*, ed. C. Rao, J. Miller, and D. Rao. London: Elsevier.

Hilbe, J. M., and J. W. Hardin. 2007. Generalized estimating equations for longitudinal panel analysis. In *Handbook of Longitudinal Research: Design, Measurement, and Analysis*, ed. S. Menard. New York: Elsevier.

Hilbe, J. M., and W. Linde-Zwirble. 1995. Random number generators. *Stata Technical Bulletin* 28: 20–21. Reprinted in *Stata Technical Bulletin Reprints*, vol. 5, pp. 118–121. College Station, TX: Stata Press.

Hilbe, J. M., and B. A. Turlach. 1995. Generalized linear models. In *XploRe: An Interactive Statistical Computing Environment*, ed. W. Härdle, S. Klinke, and S. Turlach, 195–222. New York: Springer.

Hines, R. J. O., and E. M. Carter. 1993. Improved added variable and partial residual plots for detection of influential observations in generalized linar models. *Applied Statistics* 42: 3–20.

Hinkley, D. V. 1977. Jackknifing in unbalanced situations. *Technometrics* 19: 285–292.

Hosmer, D. W., Jr., and S. Lemeshow. 2000. *Applied Logistic Regression*. 2nd ed. New York: Wiley.

Huber, P. J. 1967. The behavior of maximum likelihood estimates under nonstandard conditions. In *Proceedings of the Fifth Berkeley Symposium on Mathematical Statistics and Probability*, vol. 1, 221–233. Berkeley, CA: University of California Press.

Jain, G., and P. Consul. 1971. A generalized negative binomial distribution. *SIAM Journal of Applied Mathematics* 21: 501–513.

Joe, H., and R. Zhu. 2005. Generalized Poisson distribution: The property of mixture of Poisson and comparison with negative binomial distribution. *Biometrical Journal* 47: 219–229.

Kendall, M., and A. Stuart. 1979. *The Advanced Theory of Statistics: Volume 2, Inference and Relationship*. 4th ed. London: Griffin.

Kish, L., and M. R. Frankel. 1974. Inference from complex samples. *Journal of the Royal Statistical Society, Series A* 36: 1–37.

Lambert, D. 1992. Zero-inflated Poisson regression, with an application to defects in manufacturing. *Technometrics* 34: 1–14.

Lambert, D., and K. Roeder. 1995. Overdispersion diagnostics for generalized linear models. *Journal of the American Statistical Association* 90: 1225–1236.

Lawless, J. F. 1987. Negative binomial and mixed Poisson regression. *The Canadian Journal of Statistics* 15: 209–225.

Lerman, S. R., and C. F. Manski. 1981. On the use of simulated frequencies to approximate choice probabilities. In *Structural Analysis of Discrete Data with Econometric Applications*, vol. 1. Cambridge, MA: MIT Press.

Liang, K.-Y., and S. L. Zeger. 1986. Longitudinal data analysis using generalized linear models. *Biometrika* 73: 13–22.

Lindsey, J., and B. Jones. 1998. Choosing among generalized linear models applied to medical data. *Statistics in Medicine* 17: 59–68.

Lindsey, J. K. 1997. *Applying Generalized Linear Models*. Berlin: Springer.

Long, J. S. 1997. *Regression Models for Categorical and Limited Dependent Variables*. Thousand Oaks, CA: Sage.

Long, J. S., and L. H. Ervin. 1998. Correcting for heteroskedasticity with heteroskedasticity consistent standard errors in the linear regression model: Small sample considerations. http://www.indiana.edu/~jsl650/files/hccm/98TAS.pdf.

Long, J. S., and J. Freese. 2000. Scalar measures of fit for regression models. *Stata Technical Bulletin* 56: 34–40. Reprinted in *Stata Technical Bulletin Reprints*, vol. 10, pp. 197–205.

———. 2006. *Regression Models for Categorical Dependent Variables Using Stata*. 2nd ed. College Station, TX: Stata Press.

Lumley, T., and P. Heagerty. 1999. Weighted empirical adaptive variance estimators for correlated data regression. *Journal of the Royal Statistical Society, Series B* 61: 459–477.

MacKinnon, J. G., and H. White. 1985. Some heteroskedasticity consistent covariance matrix estimators with improved finite sample properties. *Journal of Econometrics* 29: 305–325.

Maddala, G. S. 1983. *Limited-Dependent and Qualitative Variables in Econometrics*. Cambridge: Cambridge University Press.

———. 1992. *Introduction to Econometrics*. 2nd ed. New York: MacMillan.

Marquardt, D. W. 1963. An algorithm for least squares estimation of nonlinear parameters. *Journal of the Society for Industrial and Applied Mathematics* 11: 431–441.

McCullagh, P., and J. A. Nelder. 1989. *Generalized Linear Models*. 2nd ed. London: Chapman & Hall.

McFadden, D. 1974. Conditional logit analysis of qualitative choice behavior. In *Frontiers of Econometrics*, ed. P. Zarembka, 105–142. New York: Academic Press.

McKelvey, R. D., and W. Zavoina. 1975. A statistical model for the analysis of ordinal level dependent variables. *Journal of Mathematical Sociology* 4: 103–120.

Milicer, H., and F. Szczotka. 1966. Age at menarche in Warsaw girls in 1965. *Human Biology* 38: 199–203.

Miller, R. G. 1974. The jackknife—a review. *Biometrika* 61: 1–15.

Mullahy, J. 1986. Specification and testing of some modified count data models. *Journal of Econometrics* 33: 341–365.

Nelder, J. A., and D. Pregibon. 1987. An extended quasi-likelihood function. *Biometrika* 74: 221–232.

Nelder, J. A., and R. W. M. Wedderburn. 1972. Generalized linear models. *Journal of the Royal Statistical Society, Series A* 135: 370–384.

Neuhaus, J. M. 1992. Statistical methods for longitudinal and clustered designs with binary responses. *Statistical Methods in Medical Research* 1: 249–273.

Neuhaus, J. M., and N. P. Jewell. 1993. A geometric approach to assess bias due to omitted covariates in generalized linear models. *Biometrika* 80: 807–815.

Neuhaus, J. M., J. D. Kalbfleisch, and W. W. Hauck. 1991. A comparison of cluster-specific and population-averaged approaches for analyzing correlated binary data. *International Statistical Review* 59: 25–35.

Newey, W. K., and K. D. West. 1987. A simple, positive semi-definite, heteroskedasticity and autocorrelation consistent covariance matrix. *Econometrica* 55: 703–708.

———. 1994. Automatic lag selection in covariance matrix estimation. *Review of Economic Studies* 61: 631–653.

Newson, R. 1999. rglm—Robust variance estimates for generalized linear models. *Stata Technical Bulletin* 50: 27–33. Reprinted in *Stata Technical Bulletin Reprints*, vol. 9, pp. 19–23. College Station, TX: Stata Press.

Nyquist, H. 1991. Restricted estimation of generalized linar models. *Applied Statistics* 40: 133–141.

Oakes, D. 1999. Direct calculation of the information matrix via the EM algorithm. *Journal of the Royal Statistical Society, Series B* 61: 479–482.

Parzen, E. 1957. On consistent estimates of the spectrum of a stationary time series. *Annals of Mathematical Statistics* 28: 329–348.

Pierce, D. A., and D. W. Schafer. 1986. Residuals in generalized linear models. *Journal of the American Statistical Association* 81: 977–986.

Poisson, S. D. 1837. *Recherches sur la probabilité des jugements en matière criminelle et en materière civile, précédés des règles générales du calcul des probabilités.* Paris: Bachelier.

Pregibon, D. 1980. Goodness of link tests for generalized linear models. *Applied Statistics* 29: 15–24.

———. 1981. Logistic regression diagnostics. *Annals of Statistics* 9: 705–724.

Quenouille, M. H. 1949. Approximate tests of correlation in time-series. *Journal of the Royal Statistical Society, Series B* 11: 68–84.

Rabe-Hesketh, S., A. Pickles, and C. Taylor. 2000. Generalized linear latent and mixed models. *Stata Technical Bulletin* 53: 47–57. Reprinted in *Stata Technical Bulletin Reprints*, vol. 9, pp. 293–307. College Station, TX: Stata Press.

Raftery, A. 1996. Bayesian model selection in social research. In *Sociological Methodology*, ed. P. V. Marsden, vol. 25, 111–163. Oxford: Basil Blackwell.

Rasch, G. 1960. *Probabilistic Models for some Intelligence and Attainment Tests.* Copenhangen: Danmarks Paedogogiske Institut.

Rogers, W. 1992. Probability weighting. *Stata Technical Bulletin* 8: 15–17. Reprinted in *Stata Technical Bulletin Reprints*, vol. 2, pp. 126–129. College Station, TX: Stata Press.

Royall, R. M. 1986. Model robust confidence intervals using maximum likelihood estimators. *International Statistical Review* 54: 221–226.

Smith, P. J., and D. F. Heitjan. 1993. Testing and adjusting for departures from nominal dispersion in generalized linear models. *Applied Statistics* 42: 31–41.

StataCorp. 2005. *Stata Statistical Software Release 9, Longitudinal/Panel Data Reference Manual.* College Station, TX: Stata Press.

Veall, M., and K. Zimmermann. 1992. Pseudo-R^2 in the ordinal probit model. *Journal of Mathematical Sociology* 16: 333–342.

Vuong, Q. H. 1989. Likelihood ratio tests for model selection and non-nested hypotheses. *Econometrica* 57: 307–333.

Wacholder, S. 1986. Binomial regression in GLIM: Estimating risk ratios and risk differences. *American Journal of Epidemiology* 123: 174–184.

Wahrendorf, J., H. Becher, and C. C. Brown. 1987. Bootstrap comparison of non-nested generalized linear models: Applications in survival analysis and epidemiology. *Applied Statistics* 36: 72–81.

Wedderburn, R. W. M. 1974. Quasi-likelihood functions, generalized linear models, and the Gauss–Newton method. *Biometrika* 61: 439–447.

Wheatley, G. A., and G. H. Freeman. 1982. A method of using the proportion of undamaged carrots or parsnips to estimate the relative population densities of carrot fly (*Psila rosae*) larvae, and its practical applications. *Annals of Applied Biology* 100: 229–244.

White, H. 1980. A heteroskedasticity-consistent covariance matrix estimator and a direct test for heteroskedasticity. *Econometrica* 48: 817–838.

Williams, D. A. 1982. Extra-binomial variation in logistic linear models. *Applied Statistics* 31: 144–148.

———. 1987. Generalized linear model diagnostics using the deviance and single case deletions. *Applied Statistics* 36: 181–191.

Williams, R. W. 2006. `gologit2`: Generalized ordered logit/partial proportional odds models for ordinal dependent variables. *Stata Journal* 6: 58–82.

Wu, C. F. J. 1986. Jackknife, bootstrap and other resampling methods in regression analysis. *Annals of Statistics* 14: 1261–1295.

Zeger, S. L., and M. R. Karim. 1991. Generalized linear models with random effects; a Gibbs sampling approach. *Journal of the American Statistical Association* 86: 79–86.

Zeger, S. L., K.-Y. Liang, and P. S. Albert. 1988. Models for longitudinal data: A generalized estimating equation approach. *Biometrics* 44: 1049–1060.

Author index

T

U

V

W

Y

Z

Subject index

L

M

N

O

S

T

U

V

W

X

Z